William Cavazza

François Roure

Wim Spakman

Gerard M. Stampfli

Peter A. Ziegler

(Editors)

The TRANSMED Atlas
The Mediterranean Region from Crust to Mantle

William Cavazza
François Roure
Wim Spakman
Gerard M. Stampfli
Peter A. Ziegler
(Editors)

The TRANSMED Atlas
The Mediterranean Region
from Crust to Mantle

Geological and Geophysical Framework
of the Mediterranean and the Surrounding Areas

*A publication of the Mediterranean Consortium
for the 32nd International Geological Congress*

With 44 Figures, 37 in Color

 Springer

Editors

Professor Dr. William Cavazza

Department of Earth and Geoenvironmental Sciences, University of Bologna
Piazza Porta San Donato 1, 40126 Bologna, Italy
E-mail: cavazza@geomin.unibo.it, Internet: http://www.geomin.unibo.it/

Dr. François Roure

Institut Français du Petrole, 1–4 Avenue de Bois Préau, 92 Rueil-Malmaison Cedex, France
E-mail: francois.roure@ifp.fr, Internet: http://www.ifp.fr

Professor Dr. Wim Spakman

Faculty of Earth Sciences, University of Utrecht, P.O. Box 80.021, 3508 TA Utrecht, The Netherlands
E-mail: wims@geo.uu.nl, Internet: http://www.geo.uu.nl/

Professor Dr. Gerard M. Stampfli

Department of Earth Sciences, University of Lausanne, BFSH2 - 1015 Lausanne, Switzerland
E-mail: gerard.stampfli@igp.unil.ch, Internet: http://www-sst.unil.ch/

Professor Dr. Peter A. Ziegler

Department of Geosciences, University of Basel, Bernoullistrasse 32, 4056 Basel, Switzerland
Internet: http://www.unibas.ch/geo

ISBN 978-3-642-62355-4   ISBN 978-3-642-18919-7 (eBook)
DOI 10.1007/978-3-642-18919-7

Reference to this publication should be made as follows:
(i) entire publication: Cavazza W, Roure F, Spakman W, Stampfli GM, Ziegler PA (eds)(2004) The TRANSMED Atlas – The Mediterranean Region from Crust to Mantle. Springer, Berlin Heidelberg
(ii) individual chapters or transects: Schmid SM, Fuegenschuh B, Kissling E, Schuster R (2004) TRANSMED Transect IV. In: Cavazza W, Roure F, Spakman W, Stampfli GM, Ziegler PA (eds) (2004) The TRANSMED Atlas – The Mediterranean Region from Crust to Mantle. Springer, Berlin Heidelberg

Library of Congress Control Number: 2004106760

Additional material to this book can be downloaded from http://extras.springer.com

springeronline.com
© Springer-Verlag Berlin Heidelberg 2004
Originally published by Springer-Verlag Berlin Heidelberg New York in 2004
Softcover reprint of the hardcover 1st edition 2004

Cover design: Erich Kirchner, Heidelberg
Typesetting: Büro Stasch (*stasch@stasch.com*) · Klaus Häringer, Bayreuth
Production: Luisa Tonarelli
Printing and binding: Stürtz AG, Würzburg

Printed on acid-free paper   32/3141/LT – 5 4 3 2 1 0

Foreword

In the Spring of 2000 the idea of a consortium of Mediterranean countries supporting the Italian bid to host the 32nd International Geological Congress took off during a geological fieldtrip on the slopes of Mount Vesuvius hosted by Prof. Bruno D'Argenio (University of Naples) with the sponsorship of SMED (the UNESCO-CNR Office for Scientific and Technological Cooperation with Mediterranean Countries). On that occasion, the head of the Italian delegation to the coming 31st IGC Prof. Gian Battista Vai championed the notion that – had the bid been accepted – such cooperation should have not only translated into the participation of the Mediterranean countries in the organization of the future congress, but also should have been a springboard for launching a scientific project focused on the Mediterranean region and whose results had to be presented at the congress.

During the 31st IGC in Rio de Janeiro, after the designation of Florence by the IUGS Council as the venue for the 32nd IGC, the Mediterranean Consortium was set up. In its full configuration, the Consortium was an association of thirty-one Mediterranean and nearby countries. Along with Italy, they are: Albania, Algeria, Austria, Bosnia-Herzegovina, Bulgaria, Croatia, Cyprus, Egypt, France, Greece, Hungary, Iran, Iraq, Israel, Jordan, Lebanon, Libya, Macedonia, Malta, Morocco, Palestine, Romania, Saudi Arabia, Serbia and Montenegro, Slovakia, Slovenia, Spain, Switzerland, Syria, Tunisia, and Turkey.

Each member country nominated a National Representative who served as a liaison between his/her national geological community and the IGC Organizing Committee. The National Representatives disseminated information on the congress and stimulated the submission of proposals for scientific sessions, short courses, workshops and fieldtrips from their national Earth sciences communities. Three Mediterranean Consortium representatives sat on the Advisory Board of the 32nd IGC, representing the Mediterranean countries of Europe, North Africa and the Middle East, thus providing additional input for the organization of the congress.

This publication is the main outcome of the TRANSMED Project, a scientific research program sponsored by the Organizing Committee of the 32nd IGC within the framework of the activities of the Mediterranean Consortium. The Project kicked off at the end of 2001 and in about two years generated a number of transects depicting the lithospheric and mantle structure across selected, representative regions of the Mediterranean domain and adjoining areas. This was accomplished integrating surface geology, seismic profiles and mantle tomography, both on land and at sea. The goal is to provide the international geoscientist with an updated, supranational overview of the geological and geophysical structure of the complex Mediterranean domain. It is my hope that this scientific and editorial initiative will be useful both to the Earth scientists unfamiliar with the Mediterranean and to those willing to put the results of their own research within a wider framework.

Attilio Boriani

President, 32nd International Geological Congress

Preface

"The whole is simpler than the sum of its parts."
Josiah Willard Gibbs (1839–1903)

As already noted by Schwan (1997), the geological literature on the Mediterranean region is forbidding for the outsiders, being published in a galaxy of outlets, including regional journals, geological survey reports and academic theses written in at least twenty languages. Despite several publications summarizing specific or broader aspects of Mediterranean geology (e.g. Biju-Duval and Montadert 1977; Dixon and Robertson 1984; Stanley and Wezel 1985; Morris and Tarling 1996; Durand et al. 1999), it is therefore hardly surprising that until now there was no coherent synthesis adequately covering this wide region. This volume-cum-CD aims at filling the gap.

Geological research on the Mediterranean region is currently experiencing quite a dynamic period, characterized by the transition from disciplinary to multidisciplinary research, as well as from national to international investigations. In order to synthesize and integrate the vast disciplinary and national datasets which are available it is necessary to implement maximum interaction among geoscientists of different extractions. The creation of project-oriented task forces in universities and other research institutions, as well as the development of large international cooperation programs, is instrumental in pursuing such a multidisciplinary and supranational approach.

This publication is the main result of the TRANSMED Project, an international scientific cooperation program which brought together sixty-two structural geologists, geophysicists, marine geologists, petrologists, sedimentologists, stratigraphers, paleogeographers, and petroleum geologists coming from eighteen countries and working for the petroleum industry, academia, and other institutions, both public and private.

The TRANSMED Atlas (printed volume plus CD-ROM) strives to provide an updated overview of the geological and geophysical fabric of the Mediterranean region. The printed volume contains three chapters: an introductory chapter on the main geological and geophysical features, a chapter on the lithospheric structure as imaged by mantle tomography, and a chapter on the paleogeographic-paleotectonic evolution of the study area. The CD-ROM includes eight lithospheric transects across significant domains of the Mediterranean region and the surrounding areas. Each transect was drawn at 1:1 000 000 scale (with no vertical exaggeration) in two versions: chronolithostratigraphic (rock units are divided solely according to their age) and tectonic (rock units are divided according to their tectonic affiliation). Chronostratigraphic subdivisions follow the International Stratigraphic Chart by UNESCO-IUGS (2000); tectonic affiliations follow with some modifications the scheme developed for the North American Continent-Ocean Transects Program (see Speed 1991).

The transects provide a comparative view of the complex Phanerozoic structure of the Mediterranean region and the surrounding areas using a standardized format, and portray the nature and sequence of events in the tectonic evolution with a tectonic coding scheme. Each transect is accompanied by an explanatory text with figures as well as by a series of clickable insets (seismic lines, well logs, lithochronostratigraphic charts, detailed maps, etc.) providing data in support of the interpretation shown in the transects. All transects were drawn following the same legends although some leeway was given to the various working groups to accommodate the varying amounts of data and detail available for the different regions.

The TRANSMED Atlas is geared towards (*i*) all Earth scientists working in the Mediterranean domain, particularly those willing to put the results of their own research within a wider framework, (*ii*) those unfamiliar with the Mediterranean, and (*iii*) university teachers looking for a source of synthetic information for a course in Mediterranean regional geology. We hope this compilation will serve as a synthesis of the current state of knowledge on the geology of the Mediterranean domain and will stimulate further research in this geologically fascinating region.

W. Cavazza, F. Roure, W. Spakman, G. M. Stampfli, P. A. Ziegler

Bologna, Paris, Utrecht, Lausanne, and Basle, May 2004

Acknowledgements

Dimitrios Papanikolaou first suggested that the network of the Mediterranean Consortium should tackle not only aspects pertaining to the organization of the 32nd International Geological Congress but also should work on a scientific project on the entire Mediterranean area. This seminal idea was then elaborated among the soon-to-become editors of this publication who agreed that an efficient and visually stimulating way to portray the geological-geophysical fabric of the area was by drawing lithospheric cross-sections across representative areas. This proposal was accepted by the Organizing Commmittee of the 32nd International Geological Congress which supported the TRANSMED Project in many ways. We are most grateful for this as it is uncommon that the organizing committee of a congress actually sponsors a scientific research project. Additional support was provided by the Italian National Research Council (CNR) through its Office for Scientific and Technological Cooperation with Mediterranean Countries. Most participants in the Project received additional support by other funding agencies and institutions to which thanks are also due.

During the two years of the project, the participants met in Amsterdam, Athens, Barcelona, Basle, Bologna, Florence, Istanbul, Lausanne, Nice, Paris, Rabat and Sofia. The meetings in Athens and Sofia were sponsored by the General Secretariat for Civil Protection of the Ministry of Interior, Public Administration and Decentralization of the Hellenic Republic and by the Bulgarian Academy of Sciences, respectively.

The editors would like to thank the participants in the TRANSMED Project not only for their contributions to this publication, but also for their enthusiasm for the objectives of the Project over two years. Particular thanks go to the Working Group Coordinators, who managed the often difficult task of synthesizing geological and geophysical data across political boundaries. We would also like to thank the external reviewers for their efforts, as well as the numerous geoscientists who at various stages provided expertise and advice. Several public and private organizations have provided original data for this atlas.

We thank the following organizations and publishers which granted permission to reproduce copyrighted illustrations: American Geophysical Union, Blackwell Publishing Ltd., Bulgarian Academy of Sciences, Elsevier Ltd., Geological Society of America, Geological Society of London, Muséum National d'Histoire Naturelle (Paris), Österreichische Geologische Gesellschaft.

Contents

Part One – Printed Volume

3   **The TRANSMED Transects in Space and Time: Constraints on the
     Paleotectonic Evolution of the Mediterranean Domain** 53
     Gérard M. Stampfli · Gilles Borel

Contents

Part Two – CD-ROM

---

[1] For orientation see Fig. 0.1 (page XVII) showing the study area and the geological transects.

III   Transect III:
Massif Central – Provence – Gulf of Lion – Provençal Basin – Sardinia –
Tyrrhenian Basin – Southern Apennines – Apulia – Adriatic Sea –
Albanides – Balkans – Moesian Platform

Eugenio Carminati · Carlo Doglioni · Andrea Argnani · Gabriela Carrara
Christo Dabovski · Nikola Dumurdzanov · Maurizio Gaetani · Georgi Georgiev
Alain Mauffret · Shaquir Nazai · Renzo Sartori † · Veronica Scionti
Davide Scrocca · Michel Séranne · Luigi Torelli · Ivan Zagorchev

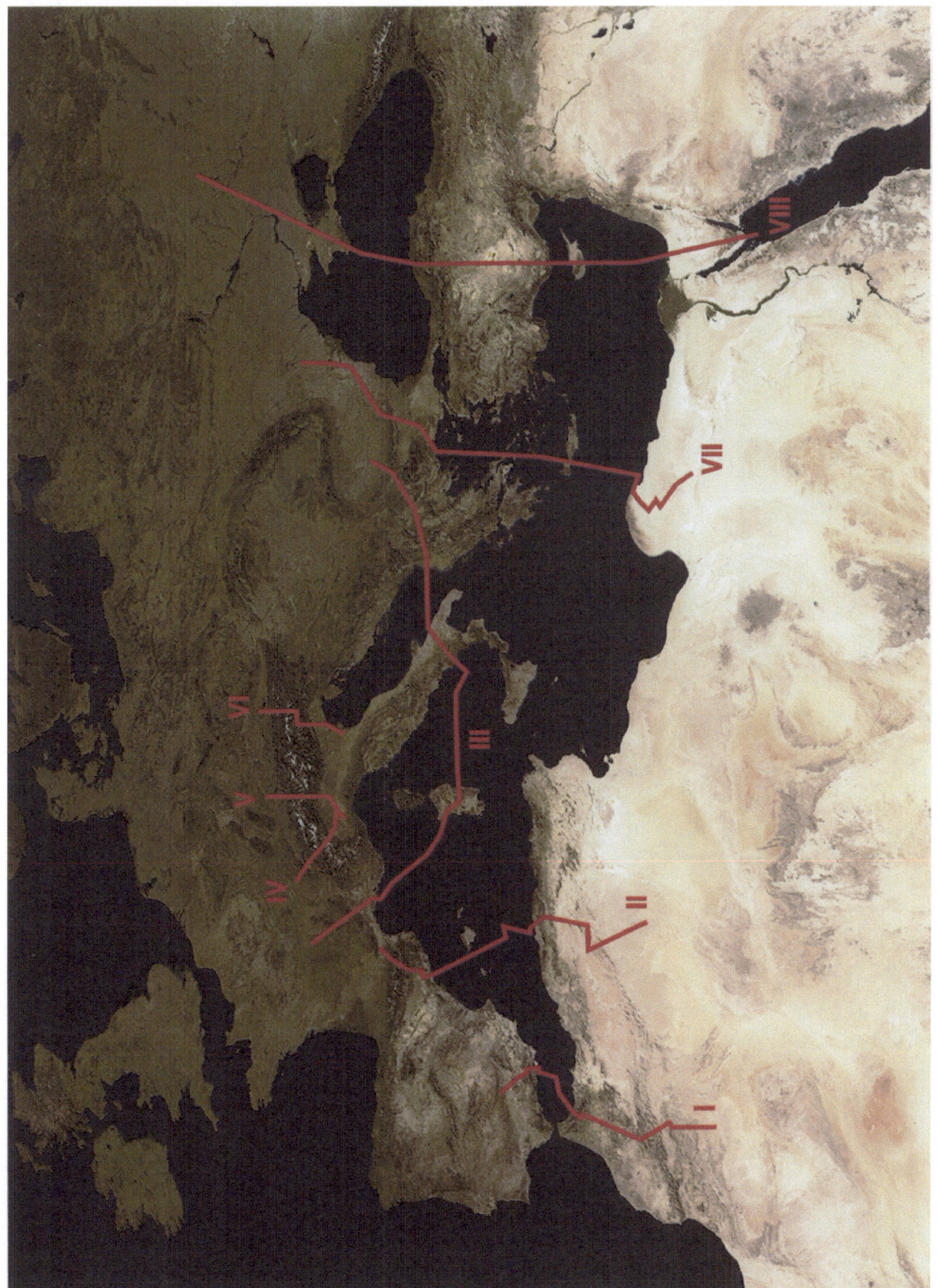

Fig. 0.1. Map showing the study area and the geological transects described in the CD portion of the publication

Contributing Authors

Fernando Alvarez-Lobato

Department of Geology
University of Salamanca
37008 Salamanca, Spain
fernando@usal.es

Maria-Luisa Arboleya

Department of Geology
Autonomous University of Barcelona
08193 Bellaterra, Spain
marialuisa.arboleya@uab.es

Andrea Argnani

Marine Geology Section
Institute for Marine Sciences (ISMAR), CNR
Via Gobetti 101
40129 Bologna, Italy
argnani@igm.bo.cnr.it

Puy Ayarza

Department of Geology
University of Salamanca
37008 Salamanca, Spain
puy@usal.es

Narimane Benaouali

Geology-Geochemistry Division
Institut Français du Pétrole
1–4, avenue de Bois-Préau
92852 Rueil-Malmaison Cedex, France
narimane.benaouali@ifp.fr

Gilles Borel

Geological State Museum, UNIL-BFSH2
1015 Lausanne, Switzerland
Gilles.Borel@igp.unil.ch

Rabah Bracène

Sonatrach Exploration
Avenue du 1er Novembre
Boumerdès, Algeria

Eugenio Carminati
(Working Group Coordinator)

Department of Earth Sciences
University of Rome "La Sapienza"
Piazzale Aldo Moro 5
00185 Rome, Italy
eugenio.carminati@uniroma1.it

Gabriela Carrara

Department of Earth Sciences
University of Parma
Parco Area delle Scienze 157A
43100 Parma, Italy
gabriela.carrara@unipr.it

William Cavazza

Department of Earth
and Geoenvironmental Sciences
University of Bologna
Piazza Porta San Donato 1
40126 Bologna, Italy
cavazza@geomin.unibo.it

Ahmed Chalouan

Department of Geology
University Mohamed V
Av. Ibn Batouta, BP 1014
Rabat 10000, Morocco
chalouan@fsr.ac.ma

Menchu Comas

Department of Geodynamics
Andalusian Institute of Earth Sciences
Univ. of Granada-CSIC
18071 Granada, Spain
mcomas@goliat.ugr.es

Ana Crespo-Blanc

Andalusian Institute of Earth Sciences
Consejo Superior de Investigationes
Cientificas (CSIC)
18071 Granada, Spain
acrespo@goliat.ugr.es

Christo Dabovski

Geological Institute
Bulgarian Academy of Sciences
Acad. G. Bonchev str. bldg. 24
1113 Sofia, Bulgaria
cndab@geology.bas.bg

Carlo Doglioni
(Working Group Coordinator)

Department of Earth Sciences
University of Rome "La Sapienza"
Piazzale Aldo Moro 5
00185 Rome, Italy
carlo.doglioni@uniroma1.it

Nikola Dumurdzanov

Faculty of Mining and Geology
University of Skopie
Skopie, Macedonia
nika@unet.com.mk

Hassan El-Bargathi

TotalFinaElf E&P Libya
Dhat El Imad Complex
P.O. Box 91171
Tripoli, Libya
hassan.el-bargathi@totlib.tlmail.com

Ahmed El-Hawat

Department of Earth Sciences
Garyounis University
P.O. Box 543
Benghazi, Libya
ashawat@LTTNET.NET

Manel Fernàndez

Institute of Earth Sciences "Jaume Almera"
Consejo Superior de Investigationes Cientificas (CSIC)
Barcelona, Spain
mfernandez@ija.csic.es

Dominique Frizon de Lamotte
(Working Group Coordinator)

Department of Earth and Environmental Sciences
(CNRS UMR7072)
University of Cergy-Pontoise
8, avenue du Parc
95031 Cergy-Pontoise Cedex, France
Dominique.Frizon-de-Lamotte@geol.u-cergy.fr

Bernhard Fügenschuh

Department of Geosciences
University of Basle
Bernoullistrasse 32
4056 Basle, Switzerland
bernhard.fuegenschuh@unibas.ch

Maurizio Gaetani

Department of Earth Sciences
University of Milan
Via Mangiagalli 34
20133 Milan, Italy
maurizio.gaetani@unimi.it

Georgi Georgiev

Department of Geology and Paleontology
University of Sofia
15 Tzar Osvoboditel Blvd.
1504 Sofia, Bulgaria
gigeor@gea.uni-sofia.bg

Dumitriu Ioane

Faculty of Geology and Geophysics
University of Bucharest
Bd. N. Balcescu no. 1
Bucharest, Romania
ioane@gg.unibuc.ro

Manuel Julivert

Department of Geology
Autonomous University of Barcelona
08193 Bellaterra, Spain
manuel.julivert@uab.es

Oksana Khriachtchevskaia

Technology Center, Ukrgeofisika
Sophia Perovska 10
Kiev, 03057 Ukraine

Eduard Kissling

Institute of Geophysics, ETH Hoenggerberg
8093 Zürich, Switzerland
kissling@tomo.ig.erdw.ethz.ch

Haralambos D. Kranis

Faculty of Geology, University of Athens
Panepistimioupoli, Zografou,
Athens 15784, Greece
hkranis@geol.uoa.gr

Yossi Mart

Recanati Institute for Marine Studies, University of Haifa
Haifa 31905, Israel
y.mart@research.haifa.ac.il

Alain Mauffret

Laboratory of Tectonics, University Pierre et Marie Curie
4 Place Jussieu
75256 Paris Cedex, France
mauffret@lgs.jussieu.fr

André Michard

Laboratory of Geology, Ecole Normale Supérieure
24 rue Lhomond
75231 Paris Cedex 05, France
andremichard@wanadoo.fr

Josep Anton Muñoz

Department of Geodynamics and Geophysics
University of Barcelona
08028 Barcelona, Spain
josep@geo.ub.es

Shaquir Nazai

Oil and Gas Institute
Fieri, Albania

Gheorghe Oaie

Geological Institute of Romania
Caransebes St. 1
8344 Bucharest, Romania

Adel Obeidi

GeoQuest
Schlumberger Overseas S.A.
Dhat El Imad Complex
P.O. Box 91931, Tripoli, Libya
aobeidi@tripoli.oilfield.slb.com

Aral Okay

Eurasia Institute of Earth Sciences
Istanbul Technical University, Ayazaga
80626 Istanbul, Turkey
okay@itu.edu.tr

Dimitrios Papanikolaou
(Working Group Coordinator)

Faculty of Geology, University of Athens
Panepistimioupoli, Zografou
Athens 15784, Greece
dpapan@geol.uoa.gr

Dimitriu Radu

Geological Institute of Romania
Caransebes St. 1
8344 Bucharest, Romania

Charles Robert-Charrue

Institute of Geology, University of Neuchâtel
CP 2, 2007 Neuchâtel, Switzerland
Charles.Robert-Charrue@unine.ch

Alastair H. F. Robertson

School of Geosciences, Grant Institute
University of Edinburgh
West Mains Road
Edinburgh EH9 3JW, United Kingdom
Alastair.Robertson@ed.ac.uk

Eduard Roca
(Working Group Coordinator)

Department of Geodynamics and Geophysics
University of Barcelona
08028 Barcelona, Spain
eduard@geo.ub.es

François Roure

Geology-Geochemistry Division
Institut Français du Pétrole
1-4 Avenue de Bois Préau
92 Rueil-Malmaison Cedex, France
francois.roure@ifp.fr

Bertrand Saint-Bezar

Department of Earth Sciences (CNRS FRE 2566)
Univ. Paris-Sud, Bat. 504
91 405 Orsay cedex, France

Aline Saintot

Department of Tectonics/Structural Geology
Free Universiteit
De Boelelaan 1085
1081HV Amsterdam, The Netherlands
saia@geo.vu.nl

Renzo Sartori †

Department of Earth and Geoenvironmental Sciences
University of Bologna
V. Zamboni 67
40126 Bologna, Italy

Stefan Schmid
(Working Group Coordinator)

Department of Geosciences, University of Basle
Bernoullistrasse 32
CH 4056 Basle, Switzerland
stefan.schmid@unibas.ch

Ralf Schuster

Geological Survey of Austria
Rasumofskygasse 23
1031 Vienna, Austria
Ralf.Schuster@cc.geolba.ac.at

Veronica Scionti

Department of Earth Sciences
University of Rome "La Sapienza"
Piazzale Aldo Moro 5
00185 Rome, Italy

Davide Scrocca

Institute for Environmental Geology
and Geoengineering (IGAG), CNR
Rome, Italy
d.scrocca@igag.cnr.it

Antoneta Seghedi

Geological Institute of Romania
Caransebes St. 1
8344 Bucharest, Romania
antoneta@ns.igr.ro

Michel Séranne

UMR 5573 Geophysics-Tectonics-Sedimentology
case courier 060, University Montpellier II
34095 Montpellier cedex 05, France
seranne@dstu.univ-montp2.fr

Wim Spakman

Vening Meinesz Research School of Geodynamics
Department of Earth Sciences, Utrecht University
Budapestlaan 4
3584 CD Utrecht, The Netherlands
wims@geo.uu.nl

Gerard Stampfli

Department of Earth Sciences, University of Lausanne
BFSH2 - 1015 Lausanne, Switzerland
gerard.stampfli@igp.unil.ch

Randell Stephenson
(Working Group Coordinator)

Dept. Tectonics/Structural Geology, Free University
De Boelelaan 1085
1081HV Amsterdam, The Netherlands
ster@geo.vu.nl

Sergiy Stovba

Technology Center, Ukrgeofisika
Sophia Perovska 10
Kiev, 03057 Ukraine
stovba@geofiz.kiev.ua

Antonio Teixell

Department of Geology
Autonomous University of Barcelona
08193 Bellaterra, Spain
antonio.teixell@uab.es

Luigi Torelli

Department of Earth Sciences
University of Parma
Parco Area delle Scienze 157A
43100 Parma, Italy
luigi.torelli@unipr.it

Jaume Vergés

Inst. of Earth Sciences "Jaume Almera"
Consejo Superior de Investigationes Cientificas (CSIC)
08028 Barcelona, Spain
jverges@ija.csic.es

Rinus Wortel

Vening Meinesz Research School of Geodynamics
Department of Earth Sciences
Utrecht University,
Budapestlaan 4
3584 CD Utrecht, The Netherlands
wortel@geo.uu.nl

Ivan Zagorchev

Geological Institute
Bulgarian Academy of Sciences
Acad. G. Bonchev str. bldg. 24
1113 Sofia, Bulgaria
zagor@router.geology.bas.bg

Hermann Zeyen

Department of Earth Sciences
(CNRS FRE 2566)
Univ. Paris-Sud
Bat. 504
91 405 Orsay cedex, France
zeyen@geol.u-psud.fr

Peter A. Ziegler

Department of Geosciences
University of Basle
Bernoullistrasse 32
CH 4056 Basle, Switzerland

Mahmoud Zizi

ONAREP
avenue Al Fadila
BP 8030-Agdal, Rabat, Morocco
zizi@onhym.com

External Reviewers

Chapter 1

The Mediterranean Area and the Surrounding Regions: Active Processes, Remnants of Former Tethyan Oceans and Related Thrust Belts

William Cavazza · François Roure · Peter A. Ziegler

Abstract

The Mediterranean domain provides a present-day geodynamic analog for the final stages of a continent-continent collisional orogeny. Over this area, oceanic lithospheric domains originally present between the Eurasian and African-Arabian plates have been subducted and partially obducted, except for the Ionian basin and the southeastern Mediterranean. A number of interconnected, yet discrete, Mediterranean orogens have been traditionally considered collectively as the result of an "Alpine" orogeny, when instead they are the result of diverse tectonic events spanning some 250 Myr, from the late Triassic to the Quaternary. To further complicate the picture, throughout the prolonged history of convergence between the two plates, new oceanic domains have been formed as back-arc basins either (*i*) behind active subduction zones during Permian-Mesozoic time, or (*ii*) associated with slab roll-back during Neogene time, when during advanced stages of lithospheric coupling the rate of active subduction was reduced. The closure of these heterogenous oceanic domains produced a system of discrete orogenic belts which vary in terms of timing of deformation, tectonic setting and internal architecture, and cannot be interpreted as the end product of a single Alpine orogenic cycle.

Similarly, the traditional paleogeographic notion of a single – albeit complex – Tethyan ocean extending from the Caribbean to the Far East and whose closure produced the Alpine-Himalayan orogenic belt must be discarded altogether. Instead, the present-day geological configuration of the Mediterranean region is the result of the opening and subsequent consumption of two major oceanic basins – the Paleotethys (mostly Paleozoic) and the Neotethys (late Paleozoic-Mesozoic) – and of additional smaller oceanic basins, such as the Atlantic Alpine Tethys, within an overall regime of prolonged interaction between the Eurasian and the African-Arabian plates. Paradoxically, the Alps, that have long been considered as the classic example of Tethys-derived orogen, are instead the product of the consumption of an eastward extension of the central Atlantic ocean, the middle Jurassic Alpine Tethys, and of the North Atlantic ocean, the middle Cretaceous Valais ocean.

1.1 Introduction

From the pioneering studies of Marsili – who singlehandedly founded the field of oceanography with the publication in 1725 of the *Histoire physique de la mer*, a scientific best-seller of the time (Sartori 2003) – to the technologically most advanced cruises of the R/V *JOIDES Resolution*, the Mediterranean Sea has represented a crucible of scientific discoveries. Similarly, the peri-Mediterranean terranes on land have been accurately surveyed over the centuries by amateur, academic and industrial geologists alike, and some locations have represented geological training grounds for several generations of Earth scientists.

Although relatively small on a global scale – its area being roughly equivalent to half of that of the People's Republic of China – the Mediterranean region has an exceedingly complex geological structure. For example, tectonic activity here spans from the Panafrican orogeny (Precambrian) of the Gondwanan, northern Africa craton to the destructive present-day seismicity along the North Anatolian Fault. Many important ideas and influential geological models were developed based on research undertaken in the Mediterranean region. For example, the Alps are the most studied orogen in the world, their structure has been elucidated in great detail for the most part and has served as an orogenic model applied to other collisional orogens. Ophiolites and olistostromes were defined and studied for the first time in this region. The Mediterranean Sea has possibly the highest density of DSDP/ODP sites in the world, and extensive research on its Messinian deposits and on their on-land counterparts provided a spectacular example for the generation of widespread basinal evaporites. Other portions of this region are less well understood and are now receiving much international attention.

Apart from its historical and cultural importance as a crossroad among various religions, trade routes and civilizations, the Mediterranean constitues also a geological transition between the Middle East and the Atlantic, as well as between Europe and Africa. For example, the Mediterranean represents a proxy of the long-lasting interactions between Eurasia and Gondwana, with successive episodes of continental break-up and oceanic development, subduction, continental collision and orogeny. The Neotethyan oceanic domain initiated during the Permian in the eastern part of the Mediterranean, separating progressively the Gondwanan continental margin to the south from the Eurasian margin to the north. Opening in a scissor-like fashion, the Neotethys widened toward the east where it was connected with the world ocean, whereas it remained closed to the west until the onset of spreading in the central Atlantic during the Jurassic. As a result, oceanic crust in the former Tethyan domains and in currently allochthonous ophiolitic units still preserved in Alpine thrust belts display a wide range of ages, depending on whether they are derived from the Neotethys (in areas already oceanic before the Jurassic) or from younger Piedmont-Ligurian-Carpathian oceanic domains once connected with the Atlantic (see Stampfli and Borel, this volume). The present-day Mediterranean Sea is a composite array of oceanic and continental domains of various affinities, including remnants of the Neotethys in the easternmost Mediterranean and the Ionian Sea, neoformed oceanic basins in the western Mediterranean, and shallow epicontinental seas like the Adriatic.

This contribution is by no means intended as a thorough description of the geological structure of the Mediterranean region. This and the following two chapters by Spakman and Wortel and by Stampfli and Borel aim at (i) providing the reader unfamiliar with the Mediterranean domain with an updated, yet opinionated, overview of such complex area, particularly in terms of description of those geological-geophysical features which the authors deem representative, and (ii) setting the stage for the TRANSMED CD-ROM which contains detailed information on the majority of the most significant areas of the Mediterranean. Fulfilling these tasks clearly involved (over)simplification of a complex matter and in some cases rather drastic choices had to be made among different explanations and/or models proposed by various authors. Similarly, only the main references are cited and the interested reader should refer to the CD for further details on the vast research dedicated to the area. In the search for clarity and conciseness we plead guilty of deliberate simplifications and omissions.

1.2 Mediterranean Fold-and-thrust Belts

The Mediterranean domain is dominated geologically by a system of connected fold-and-thrust belts and associated foreland and back-arc basins (Fig. 1.1). These belts cannot be interpreted as the end product of a single "Alpine" orogenic cycle as they vary in terms of timing, tectonic setting and internal architecture (see, for example, Dixon and Robertson 1984; Ziegler and Roure 1996). Instead, the major suture zones of this area are the result of

complex tectonic events which closed different oceanic basins of variable size and age (see Stampfli and Borel, this volume). In addition, some Mediterranean foldbelts developed by inversion of intracontinental rift zones (e.g. Atlas, Iberian Chain, Provence-Languedoc, Crimea). The Pyrenees – somehow transitional between these two end members – evolved out of a continental transform rift zone.

A large wealth of data – including deep seismic soundings, seismic tomographies, paleomagnetic and gravity data, and palinspastic reconstructions – constrains the lithospheric structure of the various elements of the Mediterranean Alpine orogenic system and indicates that the late Mesozoic and Paleogene convergence between Africa-Arabia and Europe has totalled hundreds of kilometers. Such convergence was accomodated by the subduction of oceanic and partly continental lithosphere (de Jong et al. 1993), as indicated also by the existence of lithospheric slabs beneath the major fossil and modern subduction zones (e.g. Spakman et al. 1993; Wortel and Spakman 2000; Spakman and Wortel, this publ.). The Mediterranean orogenic system features several belts of tectonized and obducted ophiolitic rocks which are located along often narrow suture zones within the allochthon and represent remnants of former extensional basins. Some elements of the Mediterranean orogenic system, such as the Pyrenees and the Greater Caucasus, may comprise local ultramafic rock bodies but are devoid of true ophiolitic sutures. Following is a very concise description – largely abstracted from the text accompanying the TRANSMED transects – of the main fold-and-thrust belts of the Mediterranean orogenic system.

The **Pyrenees** run between the Bay of Biscay and the Gulf of Lion for a length of about 450 km and a width of 50–100 km. In spite of some differences in terms of chronology and structural style, the Pyrenees are physically linked to the Languedoc-Provence orogen of southern France and – ultimately – to the western Alps. Overall, the Pyrenees are characterized by an asymmetric and bivergent, V-shaped upper crustal wedge which developed along the collision zone between the Iberian and the Eurasian plates, where limited continental subduction of the Iberian lower crust and lithospheric mantle underneath the European plate occurred. The northern wedge is formed by a northward directed stack of thrusts; the southern wedge is wider and shows greater displacement and cumulative shortening. The analysis of kinematic indicators point to a convergence nearly orthogonal to the Iberia-Eurasia plate boundary through most of the tectonic evolution of this area (see TRANSMED Transect II). There is no consensus on the exact nature and geometry of the deeper crustal levels beneath the Pyrenees: previous studies (e.g. ECORS Pyrenees Team 1988; Torné et al. 1989) pointed to a narrow and thinner lithospheric root than the one shown by Roca et al. (this publ.). At any rate, the Pyrenees are characterized by a limited

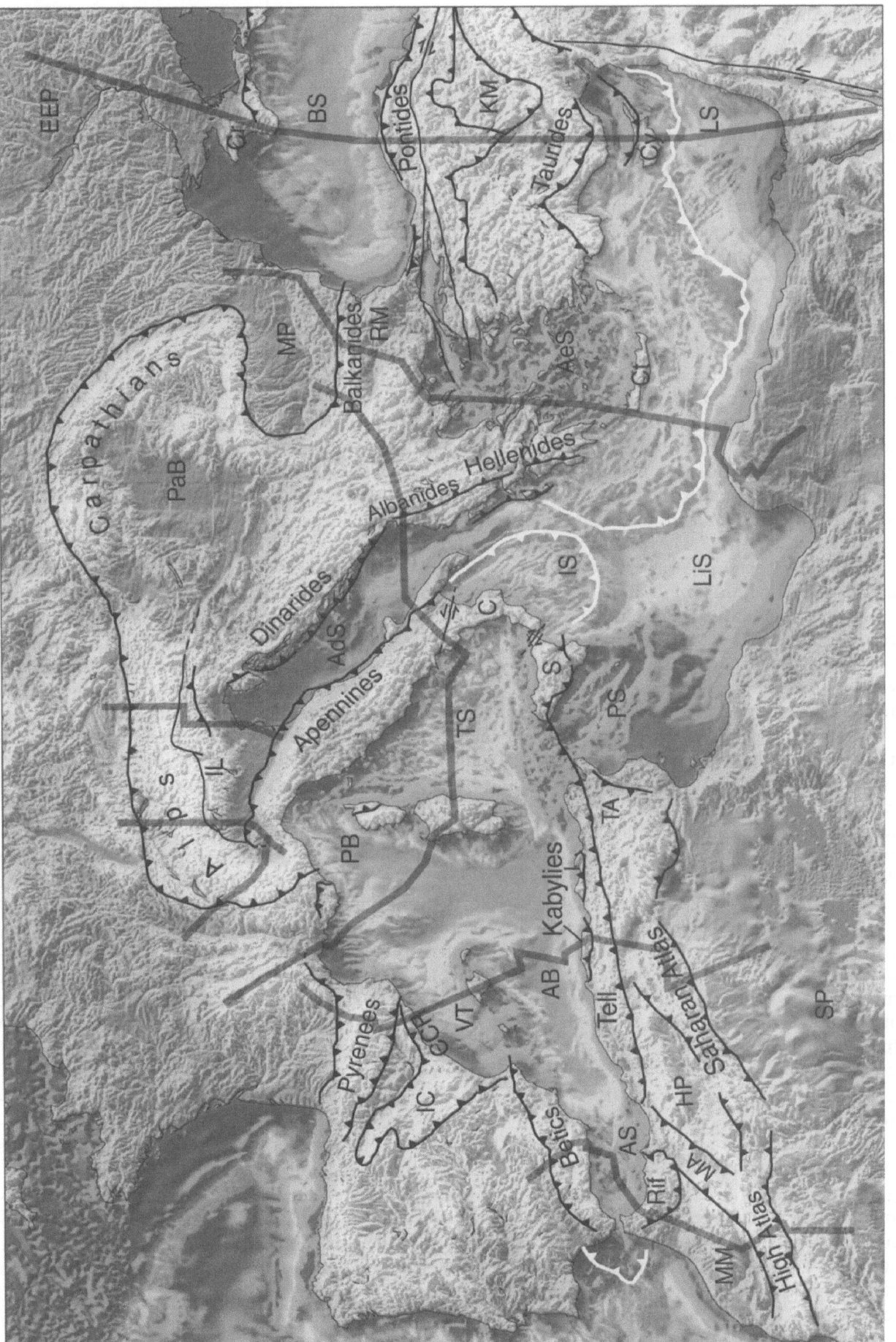

Fig. 1.1. Digital terrain model of the Mediterranean region with major, simplified geological structures. *White thrust symbols* indicate the submarine deformation front along the Ionian and eastern Mediterranean accretionary wedges. Shown are the traces of the eight TRANSMED transects included and discussed in the CD-ROM. *AB*, Algerian basin; *AS*, Alboran Sea; *AdS*, Adriatic Sea; *AeS*, Aegean Sea; *BS*, Black Sea; *C*, Calabria-Peloritani terrane; *CCR*, Catalan Coast Range; *Cr*, Crimea; *Ct*, Crete; *Cy*, Cyprus; *EEP*, East European platform; *HP*, High Plateaux; *KM*, Kirsehir Massif; *IC*, Iberian Chain; *IL*, Insubric line; *IS*, Ionian Sea; *LS*, Levant Sea; *LiS*, Libyan Sea; *MA*, Middle Atlas; *MM*, Moroccan Meseta; *MP*, Moesian platform; *PB*, Provençal basin; *PaB*, Pannonian basin; *PS*, Pelagian Shelf; *RM*, Rhodope Massif; *S*, Sicilian Maghrebides; *SP*, Saharan platform; *TA*, Tunisian Atlas; *TS*, Tyrrhenian Sea; *VT*, Valencia trough (from Cavazza and Wezel 2003, modified)

crustal root, in agreement with relatively small lithospheric contraction during the late Senonian-Paleogene Pyrenean orogeny. In contrast to the Iberian Moho, the European Moho shallows toward the Axial Zone of the Pyrenees, pointing to a pre-orogenic thinning of the European crust in the vicinity of former Albian rift basins. As pointed out by Ziegler and Roure (1996), some Alpine-age Mediterranean chains (western and eastern Carpathians, parts of the Apennines) are also characterized by relatively shallow crustal roots and by a Moho which shallows progressively toward their internal zones. However, unlike the Pyrenees where crustal thinning in the European foreland predates the orogeny, such geometry of the Moho probably results from the extensional collapse of the internal parts of these orogens, involving structural inversion of thrust faults and lower-crust exhumation on the footwalls of metamorphic core complexes.

The Pyrenees originated from the tectonic inversion of Triassic-Cretaceous rift systems that had developed during the fragmentation of southern Variscan Europe and western Tethys in conjunction with the break-up of Pangea, the opening of the central Atlantic and the Bay of Biscay, and the resulting rotation of Iberia (see Roca et al. and Stampfli and Borel, this publ., for a review). Convergence occurred from late Santonian to middle Miocene times as the African-Iberian plates moved generally northward against Europe. The Pyrenees are flanked by two main foreland basins, the Aquitaine basin to the north and the Ebro basin to the south (see TRANSMED Transect II).

From west to east, the **Alps** extend from the Gulf of Genoa to the Vienna basin, where their connection with the Carpathians can be traced only in the subsurface. Based on tectonic vergence, the Alps can be subdivided across tectonic strike into (*i*) a Europe-vergent belt of Cretaceous-Neogene age and (*ii*) the Southern Alps, a subordinate and shallower south-verging fold-and-thrust belt of Neogene age (see Dal Piaz et al. 2003, and contributions in Moores and Fairbridge 1997, for an introduction to the Alps). Overall, the Alpine orogen is highly asymmetric, being volumetrically dominated by north-vergent tectonic structures, whereas the Southalpine structures are surficial (see Pfiffner et al. 1996; Roure et al. 1996; TRANSALP Working Group 2002; Schmid et al., this publ.).

The Alps record the closure of different oceanic basins during the Late Cretaceous and Cenozoic convergence of the African (or Apulian) and European plates. Despite being the birthplace of cylindrism, the Alps display significant along-strike changes in their overall architecture, thus supporting the notion that the oceanic and continental paleogeographic domains from which the Alpine tectonic units derive were arranged in a rather non-cylindrical fashion (see Stampfli and Borel, this publ., and references therein). This is reflected by the areal arrangement of the Alpine terranes, in the deep structure of the Alps, and in the different ages of metamorphism (see Schmid et al., this publ., for a review). For example, the eastern Alps are largely made up of tectonic units derived from Apulia, the Austroalpine nappes, while the western Alps are nearly exclusively composed of more external, and tectonically lower units of the European margin, the Briançonnais terrane (a microcontinent rifted off Europe and separated from it by the narrow Valais ocean, which probably did not extend into the eastern Alps) and the intervening oceanic units. The main metamorphic events are also irregularly distributed in time and space (Frey et al. 1999): Tertiary in the western Alps, Cretaceous in the Austroalpine units of the eastern Alps. The western Alps include outcrops of blueschists and coesite-bearing, eclogite-facies rocks formed at pressures of up to 30 kbar at depths which may have reached 100 km (Compagnoni 2003 and references therein). The lithosphere is thicker (ca. 200 km) in the western Alps, while it is in the order of 140 km along the central and eastern Alps. This supports the notion that collisional coupling was stronger to the west.

Within this context, the Alps can be considered the product of two discrete orogenies: a Cretaceous one related to the closure of an embayment of the Meliata ocean into Apulia, followed by a Tertiary one due to the closure of the Alpine Tethys between Apulia and Europe (Haas et al. 1995; Stampfli et al. 2001a, b). The former affected only what are now the Eastern Alps: this implies that the Austroalpine wedge contains two sutures, namely the Meliata and the Penninic sutures, both of which are associated with HP/LT rocks (Thöni 1999).

The Alps continue eastward into the **Carpathians**, a broad (ca. 1500 km long) arcuate orogen which extends from Slovakia to Romania through Poland and Ukraine. To the south, the Carpathians merge with the east-west trending, north-verging Balkanides through a complex north-trending wrench system. Three major tectonic assemblages are recognized (see, for example, Royden and Horvath 1988): the Inner Carpathians, made of Hercynian basement and Permian-lower Cretaceous rocks; the tectonic mélange of the Pieniny Klippen belt; and the Outer Carpathians, a stack of rootless nappes made of early Cretaceous to early Miocene turbidites. All these units are thrust towards the foreland and partly override shallow-marine/continental deposits of the foredeep. Two distinct major compressional events are recognized (e.g. Ellouz and Roca 1994): thrusting of the Inner Carpathians took place at the end of the Early Cretaceous, while the Outer Carpathians underwent thrusting in the late Oligocene-Miocene. The present-day arcuate shape of this complex mountain belt is mostly the product of Neogene eastward slab retreat (e.g. Linzer 1996) and displacements along shear zones. The recent seismic activity in the Romanian sector of the Carpathians – the most severe seismic hazard in Europe today –

is inferred to be the final expression of such slab roll-back, involving delamination of the lower mantle-lithosphere from the continental foreland (Sperner et al. 2004).

The **Balkanides** are an east-west-trending, north-verging thrust belt located between the Moesian platform to the north and the Rhodope Massif to the south. Underneath the Black Sea, the Balkanides continue with a NW-SE trend. From north to the south, three domains can be recognized: the Fore-Balkan, Stara Planina (Balkans s.s.), and Srednogorie (Georgiev et al. 2001; see TRANSMED Transects III and VII). The Fore-Balkan is the transitional zone between the Balkan thrust belt front to the south and the Moesian platform, i.e. the foreland, to the north. The transitional character is indicated by the lower degree of deformation of the foredeep deposits affected only by the late stages of the orogeny and by the lack of Jurassic-Lower Cretaceous flysch. The Balkans (Stara Planina) are a complex system of north-verging nappes thrust over the South Carpathian units and the southern margin of the Moesian platform. Key characteristics of the Balkans are: (*i*) widespread Mesozoic to Early Tertiary flysch; (*ii*) general lack of products of Alpine magmatic activity; (*iii*) intense mid-Eocene compressional deformations in the central and eastern segments; (*iv*) relatively thick continental crust (38–34 km), gradually thinning toward the Moesian platform. The Balkans include thrust sheets of mid-Cretaceous, Late Cretaceous and Paleogene age. The Srednogorie zone is a segment of a Late Cretaceous magmatic belt that extends from former Yugoslavia, through Romania and Bulgaria into Turkey and the Lesser Caucasus. This belt was interpreted by Boccaletti et al. (1974) as the remnant of a volcanic island arc related to northward subduction of Tethyan oceanic crust beneath the Eurasian continent. The Srednogorie zone in the area crossed by TRANSMED Transects III and VII is characterized by thick Upper Cretaceous volcano-sedimentary successions and numerous intrusive bodies of island-arc signature, and by main Late Cretaceous compressional deformation followed by mid-Eocene north-verging thrusting over the Balkans s.s.

The stable Adriatic (Apulian) platform is flanked to the east by the **Dinarides** which continue to the southeast into the **Albanides** and **Hellenides**. The Dinarides-Albanides-Hellenides are a fairly continuous orogenic belt connected with the southern Alps to the north. It derives from the collision in the Tertiary between the Adriatic promontory and the Serbo-Macedonian-Rhodope block(s). Ophiolites are widespread and crop out along two parallel belts; these ophiolites were obducted in the late Jurassic and then involved in the Paleogene Alpine collision (Pamic et al. 2002). The Dinarides-Albanides are bordered to the west by a foredeep lying in the eastern Adriatic basin, filled by Eocene-Quaternary turbiditic sediments (see Frasheri et al. 1996 for the Albanian sector). South of the Scutari-Pec transversal

tectonic structure, the Albanides are characterized by thin-skinned thrust sheets which are detached from their basement at the level of Triassic evaporites. The Hellenides are bordered to the southeast by the Antalya convex zone, which separates them from the Taurides in the east. The Dinarides-Hellenides are the birthplace of the now abandoned concept of geosyncline, elaborated by Aubouin and co-workers in the 1960s.

The **Apennines** of Italy are one of the youngest mountain chains in the world as they formed during Neogene-Quaternary time. In addition, during the last 3 Myr the Apennines have experienced significant shortening (80–200 km), uplift (up to 2.5 km) at fast rates (up to about 1 mm a^{-1} in the Apuane Alps, Calabria and the Peloritani Mts. of NE Sicily), and considerable subsidence (up to 5 km) at a relatively fast rate in the related peri-Adriatic foredeeps (e.g. 2.5 mm a^{-1} in the Po plain) (Vai and Martini 2001).

Along their length the Apennines can be separated into two somewhat different arcuate segments: the northern segment is convex – and vergent – toward the northeast, the southern segment is convex – and vergent – toward the southeast. Structurally, the northern segment is a fairly regular orogenic wedge with in-sequence thrusts and significant piggy-back basins; conversely, the southern segment is more complex and has duplexes and widespread out-of-sequence thrusts. The southern segment was tectonically overridden by the Calabria-Peloritani terrane (Bonardi et al. 2001), a fragment of the Alpine Chain drifted off the Corsica-Sardinia block during the opening of the Tyrrhenian Sea (see following section). Only a few outcrops along the Tuscany coast of the Tyrrhenian Sea could suggest the existence of a similar terrane in the northen segment of the Apennines.

The Apennines feature a series of detached sedimentary nappes involving Triassic-Paleogene shallow water and pelagic, mostly carbonate series and late Oligocene-Miocene turbidites, deposited in an eastward migrating foreland basin. A nappe made of ophiolitic mélange (Liguride unit) is locally preserved along the Tyrrhenian coast. The Apennines have low structural and morphological relief, involve crustally shallow (mainly sedimentary Mesozoic-Tertiary) rocks, and have been characterized by the coexistence during Neogene-Quaternary time of compression in the frontal (northeastern) portion of the orogenic wedge and widespread extension in its rear portion. The Apennines were generated by limited subduction of the Adriatic sub-plate toward the west. See Vai and Martini (2001) and Elter et al. (2003) for further details.

The rock units of the **Betic Cordillera** of Spain and the **Rif** of northern Morocco form together an arc-shaped mountain belt encompassing the Alboran basin and referred to as the the Gibraltar arc (see Frizon de Lamotte et al., this publ.). This orogenic system developed during late Mesozoic to Cenozoic convergence and strike-slip movements between NW Africa and Iberia, and – notwithstanding a rather complex and locally cumbersome tectonostratigraphic nomenclature – it can be broadly subdivided into External Zones, Flysch nappes and Internal Zones. The Guadalquivir and Pre-Rif foreland basins fringe the orogenic system north and south of the Gibraltar arc.

The *External Zones* represent the deformed southern Iberian and northwestern African paleomargins of the Alpine Tethys (see Stampfli and Borel, this publ.) and consist of autochthonous, parautochthonous and/or allochthonous non-metamorphic Mesozoic and Tertiary sedimentary covers which were detached during Miocene time from a Variscan basement, the Iberian and Moroccan Meseta, respectively. The *Flysch nappes* are derived from an oceanic or semi-oceanic basin overlain by Jurassic to Miocene deep-water sediments. These units, now deformed in an accretionary prism presently located in the western Betics and along the northern part of Africa from the Strait of Gibraltar to Tunisia, demonstrate the existence of a long oceanic basin fringing the North African paleomargin. The *Internal Zones* consist mainly of three nappe complexes of variable metamorphic grade, two of which can be traced all along the Gibraltar arc. The Internal Zones record an Alpine-age tectonometamorphic evolution, which includes an early high-pressure low-temperature event. These relics of a former orogenic wedge are now *megaboudins* of the Alpine metamorphic belt, stretched and separated during later extensional episodes (see below); they are the remnants of older subduction processes and are present not only in the Betic-Rif orocline, but also in the Kabylies of Algeria (Caby et al. 2001) and in the Calabria-Peloritani terrane of Italy (Bonardi et al. 2001). Starting from the early Miocene, the Internal Zones were thrust onto the Flysch nappes, followed by the development of a thin-skinned fold-and-thrust belt in the External Zones. Contemporaneous crustal extension affected the center of the Gibraltar orocline, leading to the development of the Alboran Sea (see following section), which is floored by metamorphic rocks of the Internal Zones (Comas et al. 1999).

The **Tell** of Algeria and the Rif are parts of the Maghrebides, a coherent mountain belt longer than 2 500 km running along the coasts of NW Africa and the northern coast of the island of Sicily, which belongs geologically to the African continent (see Elter et al. 2003, for an outline of the Sicilian Maghrebides). The Tell is mostly composed of rootless south-verging thrust sheets mainly emplaced in Miocene time. The internal (northern) portion of the Tell is characterized by the Kabylies, small blocks of European lithosphere composed of a Paleozoic basement complex nonconformably overlain by Triassic-Eocene, mostly carbonate sedimentary rocks.

In the Maghrebian domain, despite the presence of contractional deformation at upper crustal levels, the lithosphere and crust do not show evidence of significant thickening. Gravity and heat-flow models (see TRANSMED Transect II) point to fairly constant crustal and lithospheric thicknesses of 35 km and 175–185 km that decrease rapidly near the coastline to reach low values offshore in Algerian basin. Such configuration could be explained by the presence of a detachment at the base of the upper crust and by the subduction of the underlying levels beneath the Iberian plate. However, mantle tomographic (Spakman et al. 1993), seismological (Buforn et al. 1995), gravity and heat-flow (Roca et al., this publ.) data argue against the presence of a subducting slab along the northern border of Africa. This does not imply that such subduction zone did not exist in the past. In fact, seismic tomography provides evidence for a deep detached slab below the Tell (Carminati et al. 1998b; Piromallo and Morelli 2003), in agreement with the petrological features of Neogene magmatism (e.g. Coulon et al. 2002). From this viewpoint, such detached slab is the remnant of the subduction responsible for the formation of the entire Maghrebides Chain.

Two major mountain belts characterize the geological structure of Turkey: the Pontides and the Taurides (for a review, see Stephenson et al., this publ.). The **Pontides** are a west-east-trending mountain belt traceable for more than 1 200 km from the Strandja zone at the Turkey-Bulgaria border to the Lesser Caucasus; they are separated from the Kirsehir Massif to the south by the Izmir-Ankara-Erzincan ophiolite belt. The Pontides display important lithologic and structural variations along strike. The bulk of the Pontides is made of a complex continental fragment (Sakarya Zone) characterized by widespread outcrops of deformed and partly metamorphosed Triassic subduction-accretion complexes overlain by early Jurassic-Eocene sedimentary rocks. The structure of the Pontides is complicated by the presence of a smaller intra-Pontide ophiolite belt marking the suture between an exotic terrane (the so-called Istanbul Zone) and the Sakarya Zone, which in turn is separated from the Taurides to the south by the Izmir-Ankara suture. The Istanbul Zone has been interpreted as a portion of the Moesian platform which, prior to the Late Cretaceous opening of the west Black Sea was situated south of the Odessa shelf and collided with the Anatolian margin in the Paleogene (Okay et al. 1994; Göncüoglu et al. 2000).

The **Taurides** are made of both allochthonous and, subordinately, autochtonous rocks. The widespread allochthonous rocks form both metamorphic and non-metamorphic nappes, mostly south-vergent, emplaced through multiphase thrusting between the Campanian and the ?Serravallian. The stratigraphy of the Taurides consists of rocks ranging in age from Cambrian to Miocene, with a characteristic abundance of thick carbonate successions. The Izmir-Ankara suture represents the former plate boundary between the Pontides to the north and the Anatolide-Tauride block to the south, and it is marked by a zone of southward imbrication of Upper Cretaceous and Triassic accretionary complexes (see TRANSMED Transect VIII). The collision of these two continental domains started in the early Tertiary, while the continuing convergence and contractional tectonism persisted until the late Miocene (Okay et al. 2001). The appearance of the North Anatolian transform fault in the late Miocene marks the end of the north-south collision between Eurasia and Africa-Arabia.

Most syntheses of the geology of the Mediterranean region have focused on the orogenic belts and have largely disregarded the large marginal intraplate rift/wrench basins located along the adjacent cratons of Africa-Arabia and Europe, ranging in age from Paleozoic to Cenozoic. Peritethyan extensional basins are instead key elements for understanding the complex evolution of this area as their sedimentary and structural records document in detail the transfer of extensional and compressional stress from plate boundaries into intraplate domains (see contributions in Roure 1994, Stampfli et al. 2001b, and Ziegler et al. 2001b). The development of the peritethyan rift/wrench basins and passive margins can be variably related to the opening of the Tethyan system of oceanic basins and the Atlantic and Indian oceans (Stampfli and Borel, this publ.). Some of these basins are still preserved whereas others were structurally inverted during the development of the Alpine-Mediterranean system of orogenic belts or were ultimately incorporated into it. Examples of inversion include the Iberian Chain and Catalonian Coast Range (Fig. 1.1) which formed during the Paleogene phases of the Pyrenean orogeny through inversion of a long-lived Mesozoic rift system which developed in discrete pulses during the break-up of Pangea, the opening of the Alpine Tethys and the north Atlantic ocean (Salas et al. 2001). The Mesozoic rift basins of the High Atlas of Morocco and Algeria underwent a first mild phase of inversion during the Senonian followed by more intense deformation during the late Eocene. This main inversion phase has been interpreted as resulting either from far-field stress transferred across the oceanic crust of the Maghrebian Tethys in response to its accelerated northward subduction beneath Iberia (Frizon de Lamotte et al. 2000) or from the arrival of an unspecified obstacle in the trench (Ziegler et al. 2002). After detachment of the Kabylian terrane from Iberia and its Langhian collision with North Africa, increased coupling between the Kabylian orogenic wedge and its foreland gave rise to further inversion of the Atlas troughs during Serravallian-Tortonian times and, after slab detachment, during the Pliocene-Quaternary.

1.3 Mediterranean Marine Basins

The modern marine basins of the Mediterranean domain are heterogeneous both in terms of age and geological structure. They are floored by

1. thick continental lithosphere (Adriatic Sea),
2. continental lithosphere thinned to a variable extent (Alboran Sea, Valencia trough, Aegean Sea) up to denuded mantle (central Tyrrhenian Sea),
3. relics of the Permo-Triassic Neotethyan oceanic domains (Ionian and Libyan Seas, E Mediterranean), and
4. oceanic crust of back-arc basins of Late Cretaceous-Paleogene age (Black Sea) or Neogene age (Algero-Provençal basin).

In detail, several of these basins have a more complex structure: for example, only the central, areally subordinate portion of the Black Sea is made of oceanic crust – which, in turn, can be subdivided in two smaller oceanic domains of different ages – whereas all the rest of it is made of stretched continental crust.

1. The **Adriatic Sea** is floored by 30–35 km thick continental crust whose upper portion is mostly made of a thick succession of Permian-Paleogene platform and basinal carbonates. The Adriatic Sea is fringed to the west and east by the flexural foredeep basins of the Apennines and Dinarides, respectively, where several kilometers of synorogenic sediments were deposited during the Oligocene-Quaternary. The Mesozoic Adriatic domain has been considered a continental promontory of the African plate (e.g., Channel et al. 1979; Muttoni et al. 2001); this domain – also referred to as Adria – includes not only what is now the Adriatic Sea but also portions of the Southern Alps, Istria, Gargano and Apulia. The southern Adriatic Sea (crossed by TRANSMED Transect III) is characterized by the updoming of the Adriatic lithosphere both on land (in the Puglia peninsula) and offshore Italy (Carminati et al., this publ.). A broad anticline of 100–150 km wavelength exposes part of the foreland in the Puglia peninsula; such updoming probably constitutes a forebulge related to the subduction of thick Adriatic lithosphere under the southern Apennines since the Pliocene (Doglioni et al. 1994).
2. The **Alboran Sea** is floored by thinned continental crust (down to a minimum of 15 km) and is bounded to the north, west and south by the Betic-Rif orocline. Where sampled by drilling or dredging, the basement of the Alboran Sea consists of metamorphic rocks similar to those of the internal domain of the Rif-Betics. The acoustic basement along several deep seismic profiles is locally formed by volcanic rocks (ba-

saltic andesites to rhyolites) from calc-alkaline series 10 Ma old. (Frizon de Lamotte et al., this publ.). During the Miocene, considerable extension in the Alboran domain and in the adjacent internal domain of the Betic-Rif occurred coevally with thrusting in the more external zones of these mountain belts. Syn-rift sediments are early Burdigalian to late Serravallian in age (Comas et al. 1999). Such late-orogenic extension can be interpreted as the result of westward roll-back of the subducted African lithospheric slab whereby thickened continental crust extends rapidly as the subduction zone retreats (Lonergan and White 1997; Gutscher et al. 2002).

The **Valencia trough** is located between the Iberian mainland and the Balearic Islands. Together with the Liguro-Provençal basin to the NE, it constitutes the oldest western Mediterranean basin, although the Valencia trough displays younger syn-rift deposits, thus indicating a progressive southwestward rift propagation from southern France (Camargue, Gulf of Lion) (Roca 2001). Overall, both basins are characterized by water depths of up to 2 200 m and by an Oligocene-to-Recent sedimentary fill ranging in thickness between 2 and 6 km. The Valencia trough is floored by continental crust which was consolidated during the Variscan orogeny and was extended during the Mesozoic rifting phases preceding the middle Jurassic opening of the oceanic Atlantic-Tethys basin and the mid-Cretaceous opening of the Bay of Biscay-Pyrenean basin (see Stampfli and Borel, this publ.). The Mesozoic rift basins in the area of the Valencia trough were inverted and uplifted during latest Cretaceous-Paleogene time, thus inducing the development of a major unconformity. Finally, the Variscan basement and its Mesozoic sedimentary cover underwent extension starting from the late Chattian (Roca et al., this publ.). Main extensional deformation in the Valencia trough took place during late Oligocene-Aquitanian times although most faults were also active during the entire duration of the early Miocene, as indicated by large lateral thickness variations; a few coastal faults were active throughout the Miocene.

The **Aegean Sea** is located in the upper plate of the Hellenic subduction zone. The arcuate structure of the southern Aegean Sea features the geological and geophysical characteristics typical of an island arc. A Benioff plane defined by seismicity dips from the Hellenic trench towards the NNE as deep as 180 km, and a calcalkaline volcanic arc outlines its curvature. South of the volcanic arc the southern portion of the Aegean Sea is a fore-arc basin (see TRANSMED Transect VII). Crustal-scale extension in this region has been accommodated by shallow-dipping detachment faults, has started at least in the early Miocene, and continues today in areas like the Corinth-Patras rift

and the southern Rhodope Massif in western Turkey. Miocene extension was accompanied by exhumation of metamorphic rocks in core complexes and by the intrusion of granitoid and monzonitic magmas at upper crustal levels. The northern Aegean Sea is characterized by a complex fault pattern resulting from east-west-trending strike-slip movements related to the westward propagation of the North Anatolian fault and from north-south-trending extension. According to Jolivet (2001), the engine for Aegean extension is gravitational collapse of a thick crust, allowed by extensional boundary conditions provided by slab retreat. From this viewpoint, the rather recent tectonic "extrusion" of Anatolia added only a rigid component to the long-lasting crustal collapse in the Aegean region.

The **Tyrrhenian Sea** is the youngest Mediterranean basin: the oldest sedimentary deposits filling the rift-related grabens along its western and eastern margins are ?Serravallian-Tortonian, thus marking the age of the onset of extension in this region (e.g. Kastens et al. 1990; Mattei et al. 2002). The development of the Tyrrhenian basin has been interpreted as resulting from back-arc extension above the NW-dipping Ionian subduction zone, possibly with a significant component of passive subduction (see Carminati et al., this publ., for further details). The presence of young basaltic bodies in the deepest portions of the Tyrrhenian basin has been somehow overemphasized in the past, leading to a vision of the central Tyrrhenian basin as underlain by true oceanic crust. The results of the reprocessing of preexisting geophysical data integrated with more recent acquisitions indicate that basalts in the central Tyrrhenian are volumetrically limited and that the deepest portions of the basin are mostly made of denuded serpentinized mantle overlain by a veneer of sediments (TRANSMED Transect III). At the scale of the entire Tyrrhenian Sea, the vast majority of the basin is floored by stretched continental crust forming rotated blocks bounded by listric, crustal-scale faults flattening close to the Moho. This structural configuration is particularly clear in the western part of the basin whereas along the Italian peninsula normal faults tend to have a higher dip angle and the stretching factor is lower. North of the 41st parallel the Tyrrhenian Sea shows only a limited degree of crustal stretching.

3. The exact nature of the lithosphere underlying the **Ionian-Libyan Sea** and the **eastern Mediterranean** has been the topic of much debate, being interpreted either as a relic of oceanic crust (Biju-Duval et al. 1977; Vai 1994) or as thinned continental crust (Giese et al. 1982). The timing of the opening of these connected basins has also been a matter of discussion, with age attributions ranging from the late Paleozoic (Vai 1994) to the Cretaceous (Dercourt et al. 1985, 1993, 2000).

For the eastern Mediterranean most authors indicate a Late Triassic or Early Jurassic opening (Garfunkel and Derin 1984; Sengor et al. 1984; Robertson et al. 1996). The presence of a continental crust in the Ionian basin was postulated mainly on the evidence of the overall thickness of the crust (ca. 20 km; deduced from the dispersion of seismic surface waves) and the low heat flow. The issue of the nature of the crust in the deep portions of the Eastern Mediterranean basins was solved by De Voogd et al. (1992) with a two-ship refraction and oblique deep seismic survey showing a relatively thin crust (8–11 km) overlain by a thick pile of sediments (up to 10 km) (see also Finetti in press). Recent palinspastic reconstructions based on all available geological-geophysical evidence (e.g. Stampfli and Borel, this volume) point to the presence of old (Permian?) oceanic crust underneath a thick pile of Mesozoic and Cenozoic sediments which hampers direct sampling and dating. The Ionian-Libyan Sea and the eastern Mediterranean are currently being subducted beneath the Calabria-Peloritani terrane of southernmost Italy (see Bonardi et al. 2001, for a review) and the Crete-Cyprus arcs, respectively (Fig. 1.1).

4. The more than 2000 m deep **Black Sea** is partly floored by oceanic crust and probably represents the remnant of a composite Cretaceous-Eocene back-arc basin which developed on the upper plate during north-dipping subduction of the Neotethys (see Stampfli and Borel, this volume). Seismic studies and field evidence in the regions surrounding the Black Sea indicate that its geological make-up is the result of the post-rift coalescence of two different extensional basins (Zonenshain and LePichon 1986; Finetti et al. 1988; Robinson 1997). Rifting of the western Black Sea began in the middle Early Cretaceous (Okay et al. 1994) with the separation of a lithospheric fragment (the Istanbul Zone of Okay and Tüysüz 1999) from the Odessa shelf, i.e. the offshore continuation of the Moesian platform of Romania and Bulgaria. Opening of the western Black Sea came to an end during the early Eocene when the southward drifting of the Istanbul Zone led to collision with Anatolia to form the western Pontides. The age of rifting of the eastern Black Sea is not as well constrained because the relevant stratigraphy is poorly exposed. Nonetheless, several lines of evidence (see Spadini et al. 1996, for a review) support the hypothesis that the eastern Black Sea developed between the Paleocene and the middle Eocene. The crustal structure of the Black Sea is well constrained by a number of geophysical studies (e.g. Belousov et al. 1988). Beneath the central western Black Sea the Moho rises to a depth of about 20 km, including as much as 15 km of post-rift sedimentary fill; the Moho depth increases to 40–45 km both to the north

(Russian platform) and to the south (Pontides). Beneath the eastern Black Sea the Moho rises to about 25 km and the thickness of the post-rift succession is about 13 km. Clear-cut magnetic anomalies are absent in both sub-basins, possibly the effect of the huge thickness of post-rift sediments.

Rifting in the **Provençal basin** area occurred at least from the Oligocene (34 Ma) to the middle Aquitanian (21 Ma), according to the age of the sediments drilled in the Gulf of Lion, just beneath the break-up unconformity (Gorini et al. 1993). Drifting and the creation of the central, oceanic portion of the basin took place in the Burdigalian, as indicated by paleomagnetic data (Vigliotti and Langenheim 1995) and by the transition from syn-rift to post-rift subsidence of its margins (Vially and Trémolières 1996; Roca 2001). It is a commonly held notion that the Provençal basin is a partly oceanised back-arc basin produced by the southeastward roll-back of the Apennines-Maghrebides subduction (see Carminati et al., this publ., for further details). Paleomagnetic studies (Alvarez et al. 1974; Vigliotti and Langenheim 1995; Speranza et al. 2002) indicate a counterclockwise rotation of the Corsica-Sardinia block between 19 Ma and 16 Ma (Burdigalian), synchronous with the formation of oceanic crust in the Provençal basin. In a pre-rotation fit, Corsica is to be considered adjacent to Provence and the original position of Sardinia was in the centre of the present-day Provençal basin, far from the continental slope of the Gulf of Lion.

The **Algerian basin** is located between the Balearic block and the North African coast and morphologically represents the continuation of the Provençal basin toward the southwest. In the absence of deep drilling in the Algerian basin, little is known on the age and characteristics of its sedimentary fill whose stratigraphy has been inferred by correlation with better known areas nearby (Mauffret et al. 1973; Sans and Sàbat 1993). However, the prominent difference in the thickness of Miocene sediments between the Provençal basin (4 km) and the Algerian basin (1.8 km) suggests that the latter is younger (Roca et al., this publ.). Despite the absence of direct evidence, the thin crust (4–6 km) of the Algerian basin is probably oceanic (Hinz 1972) and can be compared to the oceanic crust (about 5 km thick) of the Provençal basin to the northeast (Pascal et al. 1993). The age of the Algerian basin must be pre-Messinian, as the oldest dated deposits filling the basin are Messinian. The unloaded depth of the basement gives an apparent age of 20 Ma whereas the 125 mW m^{-2} heat flow of the East Alboran basin corresponds to an apparent age of 16 Ma. Tomographic studies (Carminati et al. 1998a,b) and the magmatic history (Maury et al. 2000) suggest an age comprised between 15 Ma and 10 Ma.

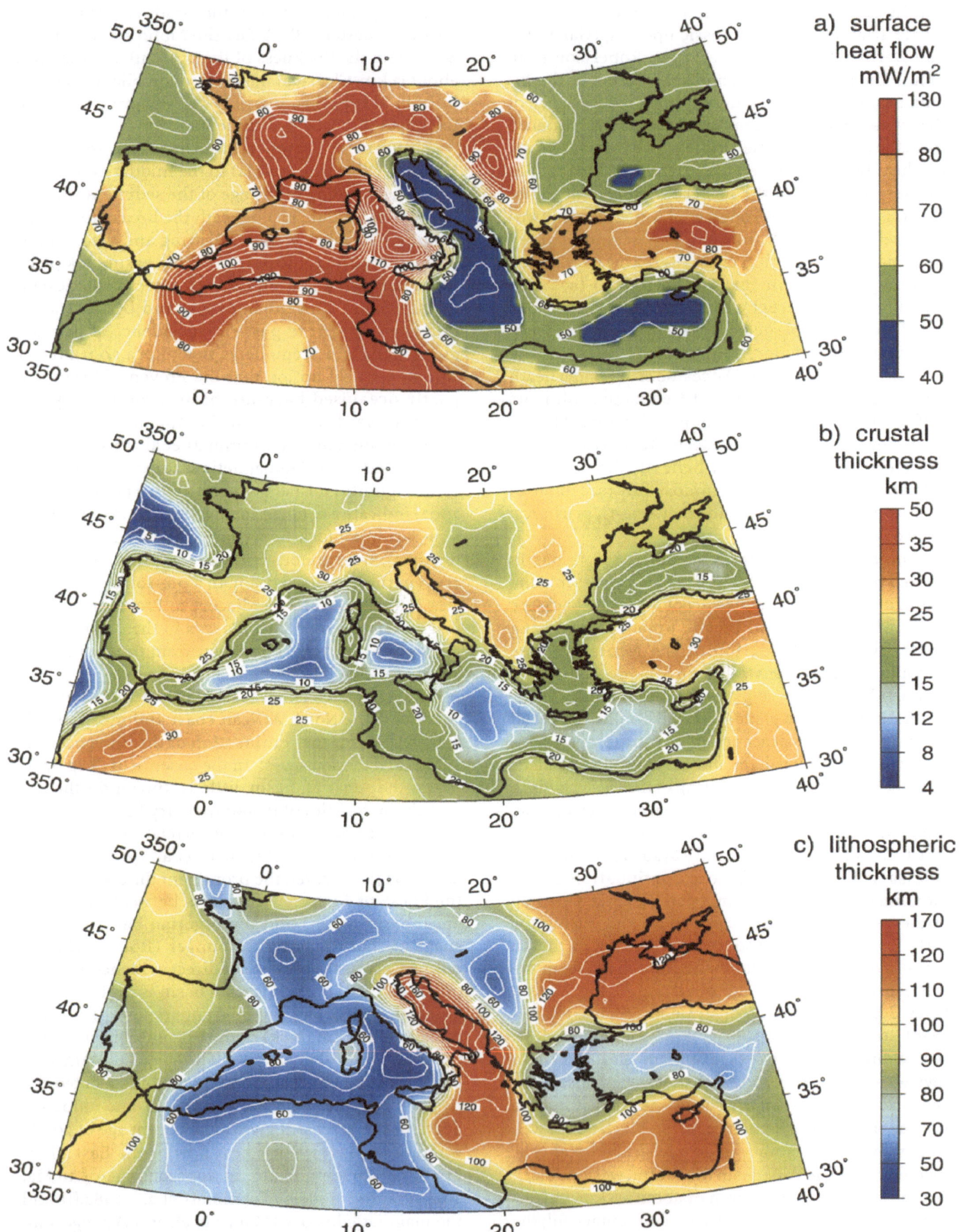

Fig. 1.2. a Surface heat-flow data (mW m⁻²; contour interval 5 mW m⁻²); **b** calculated crustal thickness (contour interval 2.5 km); **c** calculated lithospheric thickness, contour interval 10 km. After Jiménez-Munt et al. (2003)

1.4 Geological-geophysical Baseline

1.4.1 Heat Flow

Figure 1.2a is a schematic surface heat-flow map of the Mediterranean area and the surrounding regions. There is a good correlation between lithospheric thickness (Fig. 1.2c) and heat-flow values. The entire southern portion of western and central Europe is subject to normal or high heat flow, with higher values in south-central France (Massif Central), northern Switzerland and the Pannonian basin. The western Mediterranean basins show the highest heat-flow values: 60–100 mW m^{-2} in the Gulf of Lion and Provençal basin, around 100 mW m^{-2} in the southern part of the Algerian basin, and values in excess of 120 mW m^{-2} in the deepest portion of the Tyrrhenian Sea, thus suggesting that heat flow is inversely proportional to the age of rifting/drifting of the basins (Carminati et al., this publ.). Other regions of relatively high heat flow are Anatolia and the Aegean Sea.

Low heat flows (< 50 mW m^{-2}) are present in the Adriatic and Ionian Seas and in the Eastern Mediterranean, all areas characterized by old lithosphere (either continental – Adriatic – or oceanic – Ionian and eastern Mediterranean) and by thick sedimentary successions. Another area of low heat flow is the Black Sea, where up to 15 km of sediments have blanketed the floor of this composite Cretaceous-Eocene back-arc basin (see Stephenson et al., this publ.).

1.4.2 Crustal and Lithospheric Structure

Figure 1.2b and c depict the overall structure of the crust and lithosphere, respectively, over the Mediterranean region. The depth of the Moho varies approximately between 5 and 50 km. Minimum crustal thickness of about 5–15 km is found in oceanic domains such as – from west to east – the Algero-Provençal basin, the Tyrrhenian Sea, the Ionian Sea, and the E Mediterranean basins (Jiménez-Munt et al. 2003). The crust is thick beneath the orogenic belts, like the Atlas, the Alps, and the Dinarides; a broad region of crustal thickening is present in eastern Anatolia. Significant crustal thinning is present in the Pannonian basin.

More detail on the crustal structure of the western Mediterranean (and NW European) region is shown in Fig. 1.3, a compilation of the most recent sources (Dèzes and Ziegler 2002). At this scale it is possible to appreciate details otherwise lost. These include (*i*) the significant crustal root of the Pyrenees, (*ii*) the extremely shallow Moho in areas of the Tyrrhenian Sea, where the mantle has been denuded (see Transect III, this publ.), and (*iii*) crustal thickening along the frontal part of the northern Apennines produced by thrust stacking and syntectonic sedimentation.

Lithospheric thickness (Fig. 1.2c) reaches minimum values in the Algero-Provençal and Tyrrhenian basins, both sites of Neogene extension. Conversely, the much older (Permian-Triassic?) Ionian and Levant oceanic basins have a thicker lithosphere.

1.4.3 Gravity

Figure 1.4 shows the gravity field over most of the Mediterranean region, with a few exceptions in eastern Europe, Asia Minor and the Middle East (Wybraniec et al. 2004). The figure shows Bouguer anomalies on land and free-air anomalies offshore. The general patterns of the Mediterranean gravity field can be explained to a large extent by the variation in Moho depth as well as by crustal-level igneous intrusions and the presence of large sedimentary basins. For instance, all major mountain ranges, including the Alps, Apennines, Carpathians, Pyrenees, and Betic Cordillera, are associated with strong negative Bouguer anomalies (shown in blue-violet color), a natural consequence of crustal thickening beneath the mountains. Sedimentary accumulations, such as the Po basin south of the Alps and north of the northern Apennines, are clearly associated with large negative anomalies. The prominent lows in most of the Iberian peninsula may indicate low-density mantle, as significant crustal thickening beneath the elevated plateau has not been documented. The active subduction zone in the region from the Hellenic arc to Cyprus is also associated with large negative anomalies.

A belt of strong positive Bouguer gravity anomalies (shown in red) along the Tyrrhenian coast of Italy may be the result of crustal thinning in the inner portion of the Apenninic orogenic wedge (see Elter et al. 2003 for a review of the Apennines). The marked high centered in the northern Aegean Sea can be attributed to the extensional tectonic regime characterizing the area during the Neogene (see TRANSMED Transect VII).

Figure 1.5 shows the Bouguer gravity anomalies for the Mediterranean Sea and most of the adjacent regions on land. Overall, the gravity fabric of the Mediterranean Sea is characterized in almost all deep physiographic basins by strong to very strong positive Bouguer anomalies. In the deep basins west of the Corsica-Sardinia block $\Delta g''$ values range between 100 and 180 mGal. The Algero-Provençal basin and most of the Ligurian Sea have $\Delta g''$ values in excess of 140 mGal, typical of the oceanic crust. Along the Iberian coast the gravity values follow the bathymetry, also outlining with a certain detail the Balearic continental block. In the Alboran Sea Bouguer anomaly values range from about –20 mGal at the Gibraltar arc to +40 mGal where it merges with the Algerian basin; these values agree with the stretched continental crust flooring the basin (Frizon de Lamotte et al., this publ.). The western Mediterranean basins are

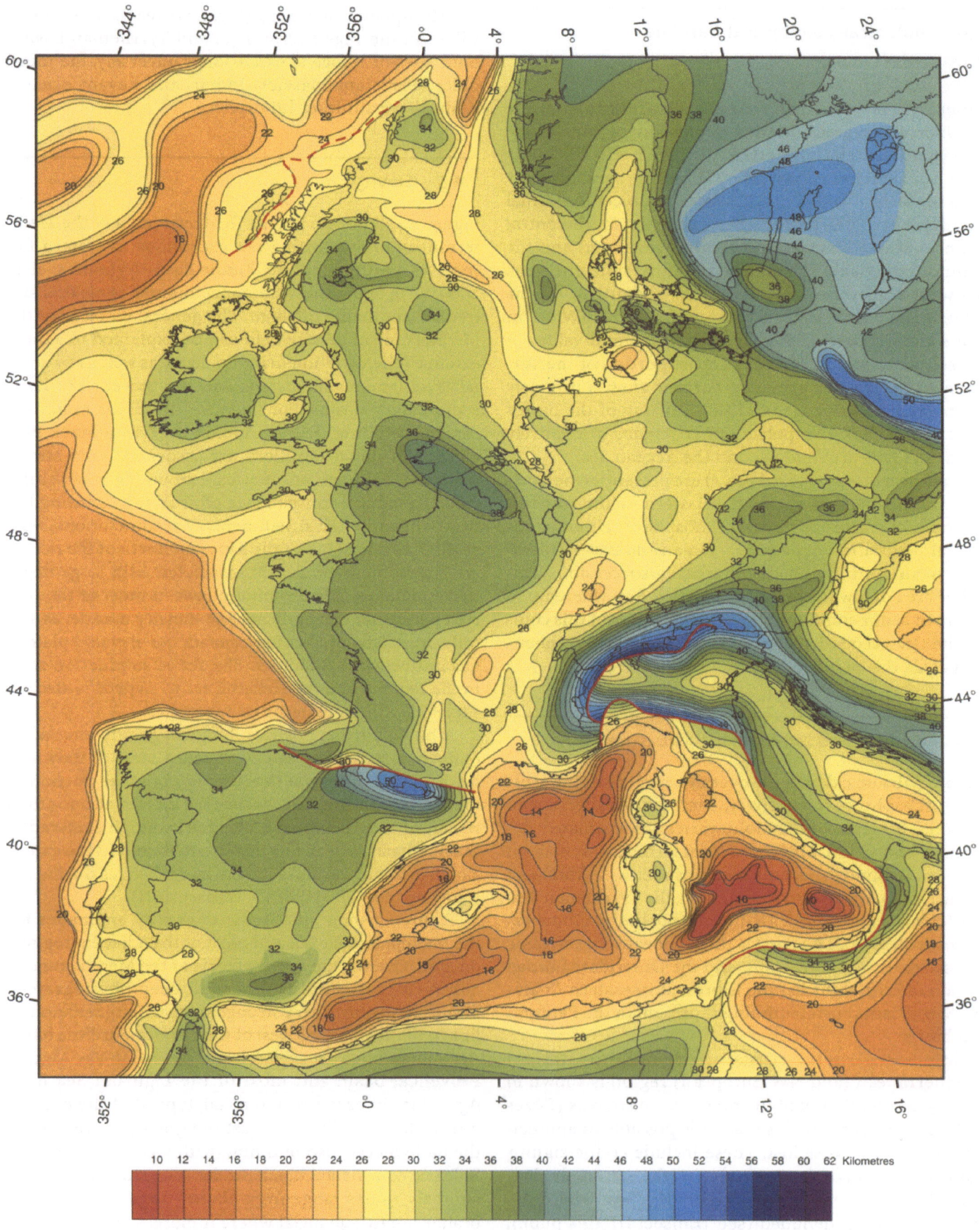

Projection: Lambert Azimuthal Equal Area; Centre: 04°.00"/48°.00"; Region : W/E/N/S = 350°/28°/62°/34°; Ellipsoide wgs-84

Fig. 1.3. Map of the depth of the Mohorovicic discontinuity in western Europe. From Dèzes and Ziegler (2002)

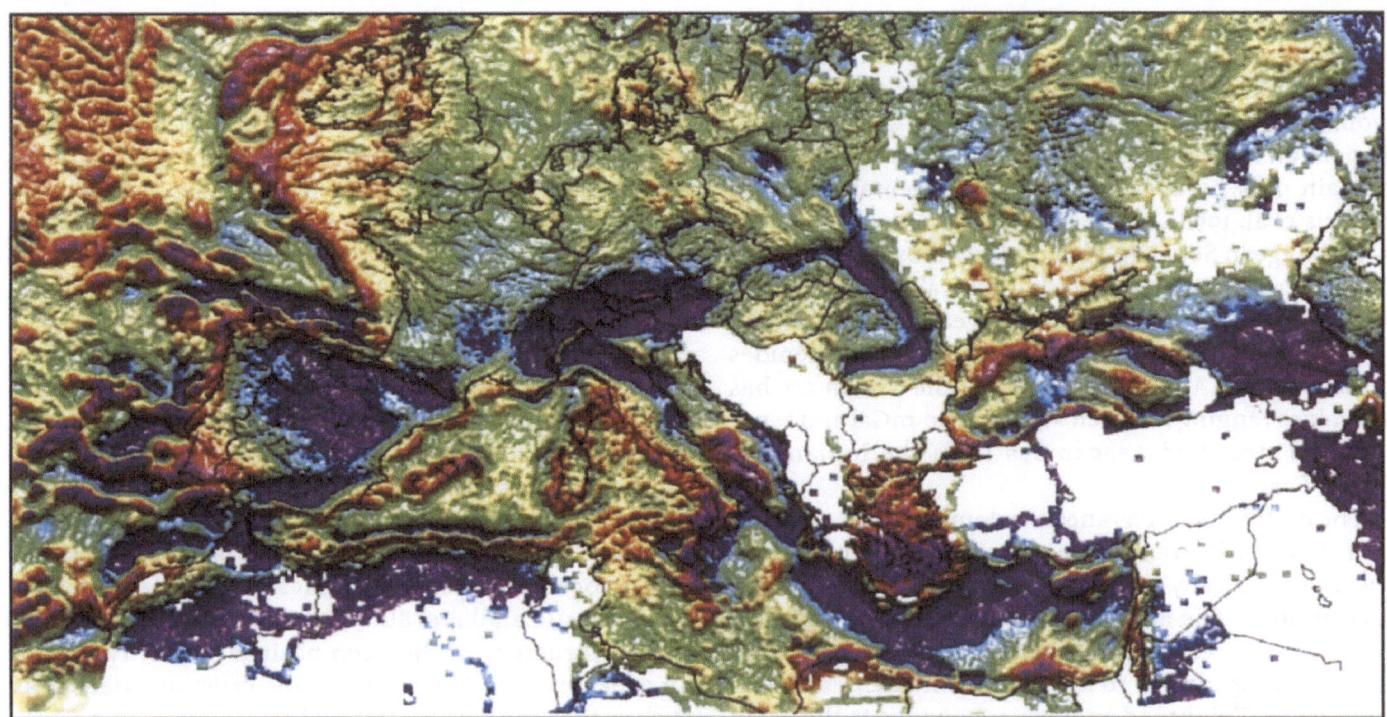

Fig. 1.4. Gravity map showing Bouguer anomalies on land and free-air anomalies offshore (after Wybraniec et al. 2004)

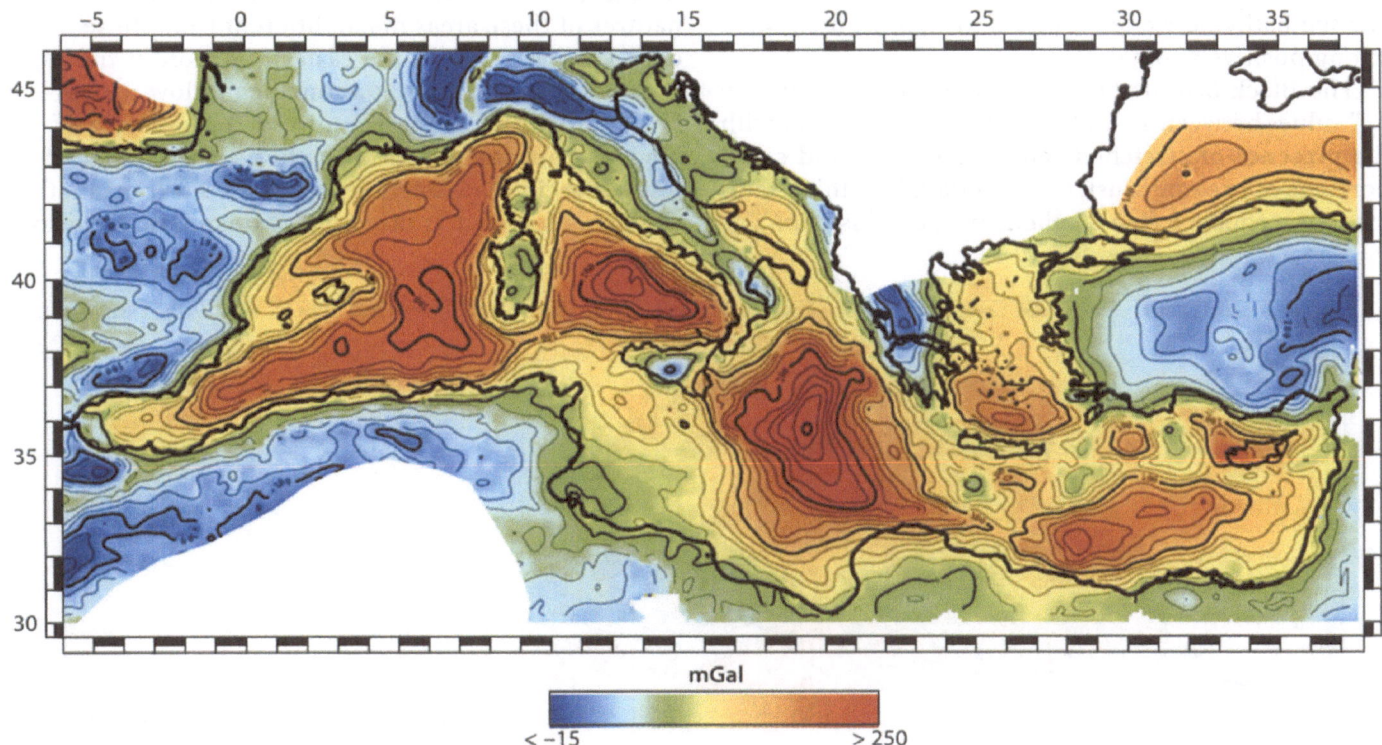

Fig. 1.5. Bouguer gravity anomalies (after Makris et al. 1998, with minor modifications)

bounded to the south by the northern margin of the African plate and by the strong negative anomalies along the Atlas Mountains. To the north, the thickened continental crust of the Betic Cordillera, the Pyrenees and the Alps mark areas of strong negative Bouguer anomalies (down to –180 mGal in the central Alps).

Bouguer gravity values in the Tyrrhenian Sea mimic closely the bathymetry and range between 180 mGal in the central part to about 20 mGal on the coasts of Italy and along the eastern margin of the Corsica-Sardinian block. As discussed in the text accompanying TRANS-MED Transect III, these gravity values are associated with stretched continental crust and denuded mantle, with volumetrically limited outpourings of basalts.

The highest values (ca. 300 mGal) of the Bouguer anomaly in the Mediterranean coincides with the deepest portion of the Ionian Sea, where the crust is thin and overlain only by a 4–5 km-thick sedimentary column (Makris et al. 1998).

The Aegean Sea is an area underlain by thinned continental crust and by relatively high values of $\Delta g''$, particularly in the Cretan Sea. The Aegean Sea is bounded by two areas of low Bouguer anomalies: the Hellenides to the west and Anatolia to the east. The Cretan arc has $\Delta g''$ values ranging between –10 and +30 mGal and separates the stretched Aegean Sea from the Ionian and Libyan Seas.

The eastern Mediterranean is dominated by a prominent ENE-WSW-trending positive anomaly culminating gravimetrically at about 220 mGal in the Herodotus abyssal plain. The anomalies decrease toward the east, mainly as the result of the increase in sediment thickness, reaching nearly 16 km in the Nile delta and offshore Israel. This broad anomaly is bounded to the north by a series of gravity highs and lows aligned east-west parallel the Anatolian margin; these highs and lows follow the bathymetric features. As to Cyprus, it shows a strong positive anomaly.

The Black Sea offshore Turkey is characterized by $\Delta g''$ values between 20 and 100 mGal, in agreement with the presence of stretched continental crust and small portions of oceanic crust, both overlain by a thick sedimentary sequence (see Stephenson et al., this publ.).

1.4.4 Magnetic Field

Figure 1.6 shows the magnetic anomalies in the Mediterranean basin at 50 nT intervals. This region is characterized both by very localized anomalies as well as by anomalies distributed throughout the area; seafloor magnetic lineations, as evident in the major oceans, are absent. Most local anomalies are the strong ones related to the fault systems along the continental slopes, particularly in the western Mediterranean, which is characterized by young, deep basins bounded by high-angle normal faults indicating very rapid subsidence. In the deep basinal areas of the western Mediterranean (i.e. west of Corsica and Sardinia) the depth of the top of the layer responsible for the magnetic anomalies is at 8–9 km (Le Borgne et al. 1971), which is the depth of the top of oceanic layer 2 as indicated by deep seismic soundings (Zanolla et al. 1998). This confirms that the crustal structure in the deep basinal area is typical of an oceanic crust. The nature of the crust in rifted areas such as the Alboran Sea (thinned continental crust, widespread yet areally discrete volcanism) and the Valencia trough (large volcanic body) makes the magnetic signatures of these areas much different from those of the deep basinal areas. In the eastern portion of the Algerian basin the magnetic anomalies follow a clear-cut N-S trend which is the southward continuation of the magnetic anomalies following the western Corsica-Sardinia margin. As elsewhere in the Mediterranean, they can be interpreted as the result of magmatic intrusions

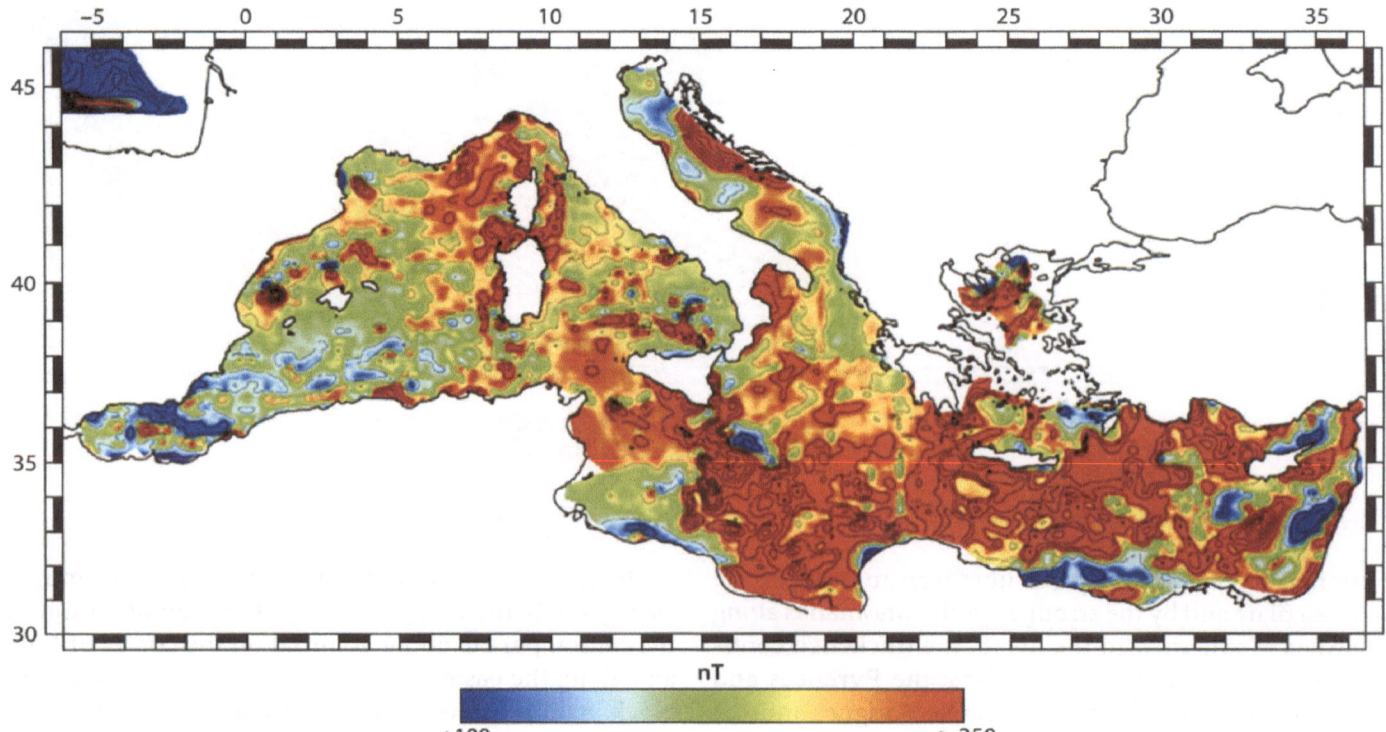

Fig. 1.6. Magnetic anomalies (after Zanolla et al. 1998, with minor modifications)

along extensional faults located at the transition between the bathyal plain and the continental margins. The magnetic anomalies along the western Sardinian margin thus confirm the results already obtained from seismic data: the continental crust of the Corsica-Sardinia block extends far offshore and oceanic crust is present only in the central part of the Algerian basin (see TRANSMED Transect II).

The eastern Ligurian Sea can be envisioned as the stretched southward continuation of the western Alps. Medium-high frequency magnetic anomalies, with approximately 100 nT maximum intensity and associated with 2–3 km deep, highly magnetized bodies, show an overall NNW-SSE trend connecting northeastern Corsica with the Alpine orogenic belt of western Liguria.

A complex array of local and regional magnetic anomalies characterizes the Tyrrhenian Sea, the result of the very young age and complexity of this peculiar basin born from the SE-ward drift of the Calabria-Peloritani terrane in latest Miocene-Pliocene times (see Bonardi et al. 2001 for a review). Rifting started along the Sardinian margin during the middle-late Miocene and the magnetic anomalies thus generated are evident along a N-S-trending strip east of Sardinia. The deep portion of the Tyrrhenian basin displays a magnetic fabric characterized by anomalies of medium wavelength (30 km) and high amplitude (> 30 nT) associated with a basic magnetic substratum (magnetic susceptibility $k = 0.020 - 0.045$). South of the 41st parallel magnetic anomalies trend approximately E-W. The calcalkaline and tholeiitic volcanic edifices of the Tyrrhenian bathyal plain are associated with very high-amplitude anomalies which mask the trend of the basic substratum as well as the weaker anomalies generated by small seamounts made of continental basement. Two clear-cut E-W-oriented alignments of local magnetic anomalies are present in the Tyrrhenian Sea at 41° and 39° latitude N: the northern one marks the boundary of the deepest part of the basin.

The foreland areas located on the Adriatic plate and genetically associated with the Apenninic and Dinaric fold-and-thrust belts are characterized by regional magnetic anomalies of long wavelength (60–100 km) and amplitudes ranging between 40 and 200 nT which are associated with structural highs in the basement. The very strong regional anomaly (> 320 nT) off the Dalmatian coast is probably caused by structural uplift of the basement or changes in its magnetic susceptibility (Zanolla et al. 1998).

Gravity anomalies (Fig. 1.5), seismic results (see TRANSMED Transects VII and VIII), the reduced magnetic anomalies in the deep basins (Fig. 1.6), and heat-flow data (Fig. 1.2) all suggest that the deep abyssal plains of the Eastern Mediterranean are underlain by a thick sedimentary cover lying on a thin crust of oceanic or intermediate nature which thickens to the south towards the African margin and becomes progressively more disrupted and tilted from east to west.

The entire easternmost Mediterranean, from the Libyan Sea to the Levant Sea, is characterized by incipient continental collision. This tectonically active area shows local magnetic anomalies: *positive* ones are related to igneous bodies (as for the continental margin of Israel and offshore Egypt between about 27° and 29° longitude); *negative* ones correspond to small trenches or sedimentary basins. The dipolar anomalies of the Cyprus arc can be attributed to the Cretaceous ophiolites.

As to regional magnetic anomalies, a large positive anomaly is centered south of Cyprus and corresponds to the Eratosthenes seamount (TRANSMED Transect VIII). Two large negative anomalies – interpreted by Zanolla et al. (1998) as the result of an oceanic crust of inverse polarity flooring the deepest parts of the E Mediterranean – are located off the coast of Egypt (between 25° and 27° longitude) and between the Eratosthenes seamount and Israel. The high amplitude anomalies of the Ionian Sea are associated with a semi-oceanic basement located at depths of 5–7 km b.s.l. and are probably linked to the rifting episode which formed the basin (Finetti 1982). In the rest of the eastern Mediterranean the magnetic field is mostly regular: this can be attributed to the thick pile of sediments covering the area and which becomes thicker in the Nile Cone (16 km) and the Eratosthenes trench (15 km).

1.4.5 Seismicity

Seismicity in the Mediterranean area is high both in terms of frequency and magnitude. Figure 1.7 depicts the geographic distribution of the epicenters of 1 112 seismic events with surface magnitudes $2.8 < M_s < 8$ for the period 1903–1999. Although rather diffuse, seismicity in the region is particularly concentrated in the Aegean and peri-Aegean area. Other areas of high seismicity include the Maghrebian orogen of northern Africa and Sicily, the Apennines, the Carnic Alps and the southern Dinarides. All these regions have been repeatedly affected in historical times by catastrophic earthquakes causing a high death toll (Giardini 1999).

Mediterranean seismicity distribution is related to the major known tectonic systems and clearly follows the boundary between the Eurasian and African plates. A few discrete seismogenic zones can be recognized, such as the subduction belt of the Hellenic arc and the geologically complex zone that includes the southern Dinarides, the northern Aegean Sea along the trace of the North Anatolian Fault system, and western Anatolia. Seismicity in this region is dominated by extensional focal mechanisms. Another significant seismogenic zone starts at the Azores, continues eastward and, after crossing Gibraltar, follows the Maghrebian orogen of northern

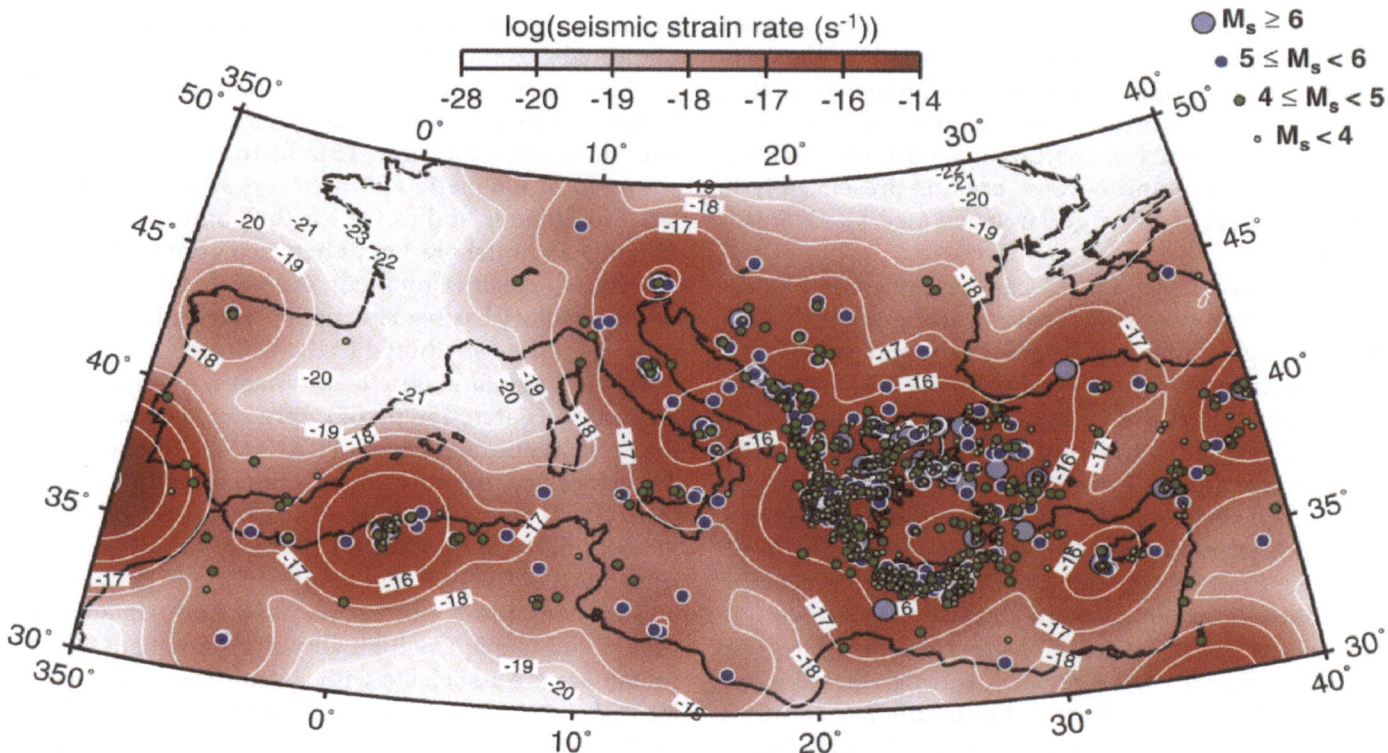

Fig. 1.7. Seismicity with surface magnitude (M_s) between 2.8 and 8 in the Mediterranean region, and calculated seismic strain. From Jiménez-Munt et al. (2003)

Africa (Rif and Tell Chains) and Sicily (Sicilian Maghrebides) as well as the Calabria-Peloritani terrane. This belt is considered representative for the collisional margin between the African and Eurasian plates, since along it there is a uniform stress field with a direction of relative movement in agreement with that expected from global plate-motion models. Conversely, in the remaining Mediterranean area, the situation is more complex.

In Europe there are other seismogenic areas where low-magnitude events ($M_s < 3$) are common and higher magnitude earthquakes occur sometimes. Among the more active zones there is the Rhine Graben, where, besides a frequent yet weak seismicity, there are also events of $M_s > 5$. The same pattern is observed in the Jura Mountains, in the Swiss-French Alps, in the Italian-French Alps and also in the lower Rhone and Durance valleys. Although less important, some seismic activity occurs also in the Central Massif of southern France.

Earthquake focal depths are generally restricted to crustal levels, except in areas where oceanic or continental lithosphere is subducted, either actively or passively. From west to east, this is the case in the Gibraltar region, beneath the Calabria-Peloritani terrane, along the Hellenic arc, in Vrancea along the SE Carpathians, and in the Antalya Gulf of southern Turkey.

Figure 1.7 also shows the associated seismic strain rates calculated by Jiménez-Munt et al. (2003). The results of such calculations indicate that the largest seismic strain rate release is of the order of 10^{-16}–10^{-15} s^{-1} occurring – from west to east – along the plate bound-ary in north Africa, in southern and northeastern Italy, in the eastern Alps, the southern Dinarides, and in the Aegean and peri-Aegean region, comprising the entire western Anatolia. Other areas of high seismic strain rates include easternmost Anatolia along the trace of the Assyrian-Zagros suture and Cyprus.

1.4.6 Geodetic Data

Figure 1.8 depicts a geodetic set of 190 vector velocities for the Mediterranean region with respect to a fixed Eurasia. Vector velocities were obtained and/or compiled by Jiménez-Munt et al. (2003) combining GPS, satellite laser ranging (SLR) and very long baseline interferometry (VLBI) data. The solution shown represents the residual velocity with respect to the Eurasian block obtained by subtracting the rigid motion of Eurasia expressed in the NUVEL-1A reference frame (see Devoti et al. 2002, for a discussion of the procedure).

In spite of areas such as the Iberian peninsula and the northern Adriatic block where large error ellipses result from the paucity of data, three coherent domains are present: (*i*) a generally SSE-ward direction in the Iberian peninsula and the Ligurian-Provençal coast; (*ii*) a generally northward direction for Italy, with a progressive clockwise rotation from Lampedusa Island (LAMP), through Calabria (COSE), to Matera (MATE); and (*iii*) a counterclockwise rotation from NW-ward to SSW-ward directions from eastern Anatolia to the south-

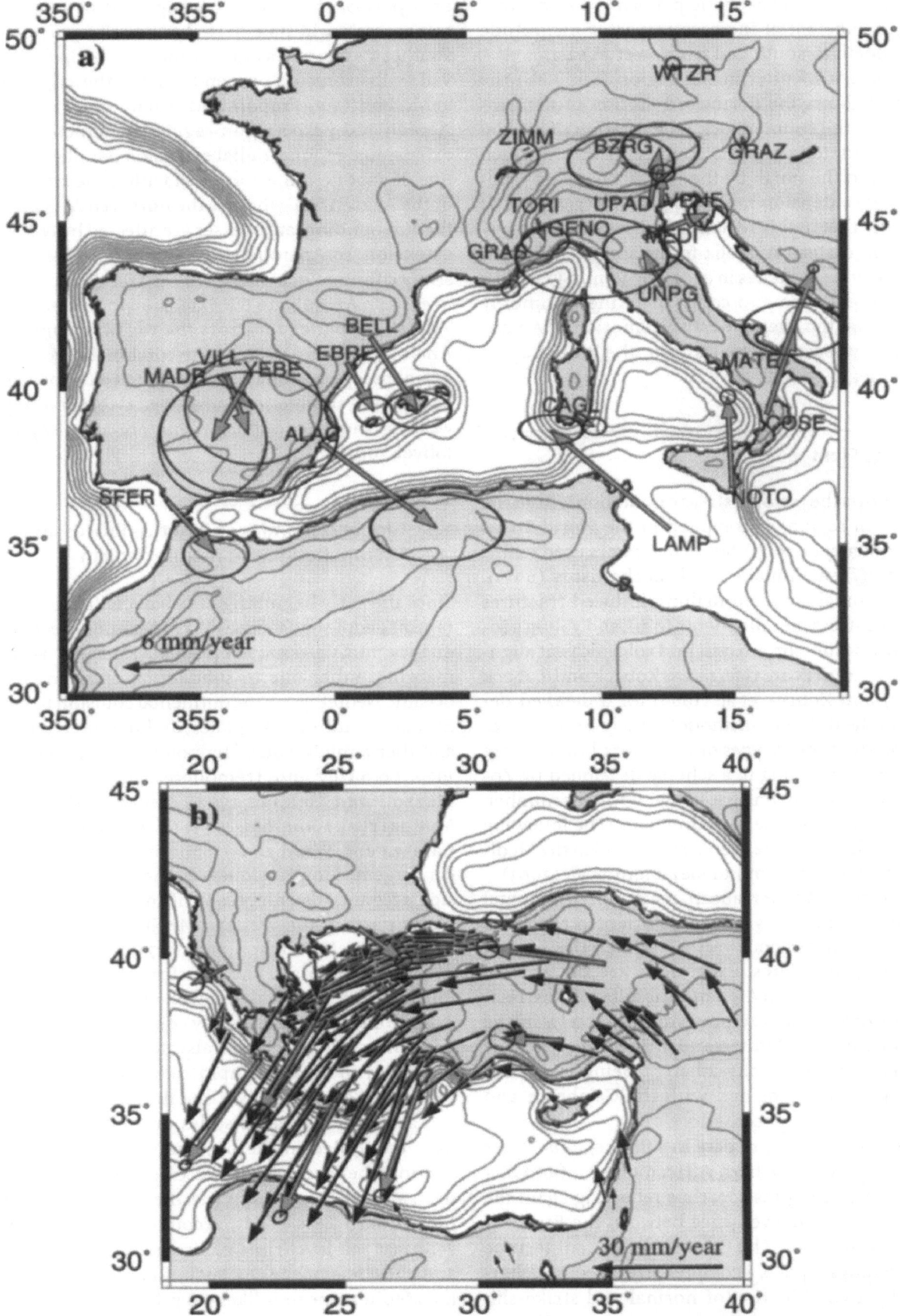

Fig. 1.8. Vector velocities in the western (**a**) and eastern (**b**) Mediterranean with respect to a fixed Eurasia (from Jiménez-Munt et al. 2003). *Gray arrows* combine data from GPS, satellite laser ranging (SLR) and very long baseline interferometry (VLBI); *black arrows* are from GPS data only

ern Aegean region. The geodetic pattern is also characterized by a substantial increase in vector magnitude both from north to south and from west to east.

In the eastern Mediterranean region (Fig. 1.8b) the velocity field shows the northward motion of the Arabian plate and the counterclockwise rotation of central and western Anatolia and the southern Aegean, which are bounded to the north by the North Anatolian Fault system and its extension in the northern Aegean Sea. It is noteworthy that in this region velocities progressively increase from eastern Anatolia to the southern Aegean Sea, where they reach values in excess of 30 mm yr^{-1}. This velocity gradient seems to contradicts the commonly held notion that the overall westward motion of Anatolia is driven by escape tectonics from the Bitlis-Zagros collision zone.

1.4.7 Stress Field

Figure 1.9 shows the tectonic stress orientation within the Mediterranean region, as compiled in the World Stress Map (Reinecker et al. 2003). Stress indicators used to this end include (i) earthquake focal mechanisms (69%), (ii) borehole breakouts and drilling-induced fractures (19%), in-situ stress measurements (8%), and young geologic data from fault-slip analysis and volcanic vent alignments (4%). [Readers are referred to Zoback and Zoback (1980, 1991) and Zoback et al. (1989) for a detailed description on the different methodologies to derive stress information from those types of indicators.] All data are quality ranked according to a scheme developed by Zoback and Zoback (1989) and based mainly on the number, accuracy and depth of the measurements. The following is a brief description of the overall stress field in the Mediterranean abstracted from Jiménez-Munt et al. (2003).

In the western Mediterranean, maximum horizontal compressional stress axis is roughly oriented NNW, i.e. parallel to the convergence vector between the African and the European plates. Exceptions are the arcuate structures of the western Alps and Gibraltar, where relatively small deviations are present. Conversely, in the central and eastern Mediterranean the stress field is rather variable yet regionally coherent, being associated with collisional orogens and with the Calabrian and Hellenic subduction zones.

In spite of the paucity of data in some regions, stress indicators along the northern African margin (and Sicily) indicate NW compression, thus reflecting again the overall direction of convergence between the Eurasian and the African plates. The Calabria-Peloritani terrane of southern Italy displays a complex stress regime, characterized by a combination of normal and strike-slip faulting which Rebai et al. (1992) ascribe to radial extension (Bonardi et al. 2001). The northern Apennines are dominated by extension in the internal (southwestern) part of the orogenic wedge and by compression-transpression in its external (Adriatic) portion. In the southern Apennines normal and strike-slip faulting prevail, with extension perpendicular to the axis of the orogenic belt (e.g. Frepoli and Amato 2000). The whole Aegean and peri-Aegean region is characterized by an extensional regime: radial extension parallel to the hinge line of subduction is localized within the southern part of the Aegean Sea, whereas the northern Aegean Sea and the surrounding landmasses are affected by N-S oriented extension. In Anatolia stress directions rotate progressively counterclockwise from NE-trending compression in eastern Anatolia to NE extension in western Anatolia. This stress pattern mirrors the westward movement of Anatolia, that is driven by the combination of syn-collisional escape tectonics (e.g. Kahle et al. 2000) and rollback of the subducting Ionian Sea slab, entailing southwestward advance of the Hellenic arc-trench system (e.g. Jolivet 2001).

1.5 Global Dynamics and Active Processes Exemplified in the Mediterranean

Since the early beginning of urban civilization, Mediterranean people had to face the devastating effects of earthquakes and volcanic eruptions associated with active margins, either due to subduction of remnants of the oceanic Neotethys or to continental collision between the European and African plates, and the intervening Adria and Iberia microplates (Tapponnier 1977; Channell et al. 1979; Letouzey and Trémolières 1980; Platt et al. 1989; Mazzoli and Helman 1994; Ziegler and Roure 1996; Muttoni et al. 2001; Nikishin et al. 2002; Ziegler et al. 2002; Cavazza and Wezel 2003). Superimposed on this overall convergent, compressional regime, local back-arc extension accounts also for recent rifting and seafloor spreading in parts of the Western Mediterranean and Tyrrhenian basins (Le Pichon et al. 1971; Durand et al. 1999; Ziegler et al. 2001a). However, current Mediterranean tectonic activity is not only controlled by compressional and extensional forces, as density contrasts, vertical compaction and gravity account also for active mud and salt diapirism in areas of rapid sedimentation such as the Nile delta, the Alboran Sea and the Mediterranean Ridge.

Lastly, successive episodes of uplift and subsidence have controlled in the past, and still control, the connections between the world ocean and the Mediterranean and adjacent basins (i.e. the former Paratethys and the modern Black Sea), accounting for either slow or very fast sea-level rises or drops during the Messinian and the Pleistocene. As such, the geological record of the Mediterranean is likely to provide key analogues for the study of global warming, greenhouse effects, related sea-level rise and its potential impact on densely populated coastal areas.

Fig. 1.9. Stress map for the Mediterranean region (after Reinecker et al. 2003). The map displays the orientation of the maximum horizontal stress SH. The length of the *stress symbols* represents the data quality, with *A* as the best quality category. A-quality data are believed to record the orientation of the horizontal tectonic stress field to within ± 10°–15°, B-quality data to within ± 15°–20° and C-quality data to within ± 25°. The tectonic regimes are: *NF* for normal faulting, *SS* for strike-slip faulting, *TF* for thrust faulting and *U* for an unknown regime

1.5.1 Subduction of the Eastern Mediterranean Lithosphere beneath the Calabrian and Aegean Arcs

Like the circum-Pacific rim, Greece and southern Italy display a pattern of deep focal earthquake mechanisms which delineates a Benioff plane (Wortel et al. 1990). However, these two Mediterranean subduction zones are far from typical. For instance, the nature of the East Mediterranean lithosphere currently subducting beneath both the Aegean and Calabrian arcs is still debated (Dixon and Robertson 1984; Casero and Roure 1994) although mounting geophysical evidence (e.g. De Voogd et al. 1992) points to its oceanic nature. The main question relates to the age of initial rifting and development of the rifted North African-Libyan-Sicilian passive margin, which locally displays marine Permian series onshore. Due to the huge thickness of Mesozoic and Cenozoic series deposited in the East Mediterranean basin, it is not yet known whether its substratum is formed by true Neotethyan oceanic crust of Permian or Triassic age, or rather by thinned continental crust of Gondwanan affinities. This issue is further discussed by Stampfli and Borel (this publ.).

Like in the Carpathian and Gibraltar oroclines, the Aegean and Calabrian subduction zones are not linear. Laterally, they grade into structural domains devoid of any deep focal mechanisms. It is now assumed that these Mediterranean subduction zones are the relics of a formerly more continuous active margin along the northern margin of the Neotethys. Rather than involving an efficient slab pull, most of the relative motion between the hangingwall and the subducted plates is to be attributed to a progressive retreat of the Ionian and East Mediterranean lithospheric slabs, which allows for southeastward and southwestward extension of the Tyrrhenian and Aegean plates, respectively (Royden et al. 1987; Royden 1993). The same process was recently proposed as the driving mechanism for the westward progradation of the Gibraltar arc (Gutscher et al. 2002). The lack of deep focal mechanisms, as well as the rapid uplift of the Apennines and Sicily, but also of the Dinarides-Albanides and the Maghrebides, have been recently attributed to slab detachment (Spakman 1990; Wortel and Spakman 2000; Spakman et al. 1993).

1.5.2 Rifting and Passive Margin Development in Back-arc Regions and Other Mediterranean Domains

In spite of prolonged indentation and mechanical coupling along the Alpine front, the Neogene geological history of the Mediterranean region is characteristically dominated by widespread extensional tectonism. A number of continental microterranes (Kabylies, Balearic Islands, Sardinia-Corsica, Calabria) rifted off the European-Iberian continental margin and drifted toward the south or southeast, leaving in their wake areas of thinned continental crust (e.g. Valencia trough) or small oceanic basins (Algerian, Provençal and Tyrrhenian basins) (Fig. 1.10). Back-arc extension and progressive southeastward drift of the Calabria-Peloritani terrane accounts for the successive opening of two neoformed oceanic domains in the Western Mediterranean, namely the Liguro-Provençal basin in Miocene time and the Tyrrhenian basin in Plio-Quaternary time (Burrus 1989; De Voogd et al. 1991; Vially and Trémolières 1996; Doglioni et al. 1997, 1999; Carminati et al. 1998a, b; Gueguen et al. 1998; Durand et al. 1999; Catalano et al. 2000; Roca 2001).

The rifted margins of Provence and western Corsica display the same Oligo-Aquitanian synrift sequences, being the conjugate margins of the same Neogene Provençal oceanic domain. Post-rift thermal subsidence commenced during the middle Miocene in the adjacent Provençal and Algerian basins, two areas where sedimentation was strongly affected by the Messinian salinity crisis. As a result, two sets of normal faults are usually evidenced in seismic profiles in the Western Mediterranean basins. The first one is made up of basement-involving high-angle normal faults that truly relate to rifting, and accounts for the deposition of syn-rift Oligocene and Early Miocene series, whereas a second, surficial set consists of listric normal faults that sole out within the Messinian salt horizon and account for a progressive gravitational gliding of the post-Messinian sedimentary successions.

Two episodes of rifting are recorded along the east Sardinian margin in the late Miocene and during the Pliocene, whereas two distinct oceanic domains developed in the adjacent Tyrrhenian basin during the Pliocene in the west, and only during the Pleistocene in the east, making it the youngest oceanic basin of the Mediterranean (Rehault et al. 1984a, b; Malinverno and Ryan 1986; Kastens and Mascle 1990; Patacca et al. 1993; Spadini et al. 1995; Mattei et al. 2002).

The East Mediterranean is similarly characterized by widespread Neogene extensional tectonism, as indicated by thinning of continental crust along low-angle detachment faults in the Aegean Sea and the peri-Aegean regions (see Durand et al. 1999, and references therein). Pliocene-Quaternary rifting in the hangingwall of the Aegean subduction zone accounts also for the opening of the Gulf of Corinth, which is currently the focus of much international research aiming at the study and monitoring of fracture development, pore-fluid pressure evolution in connection with seismically active fractures, and fluid transfer across or along fault planes (Corinth Natural Laboratory; Moretti et al. 2002).

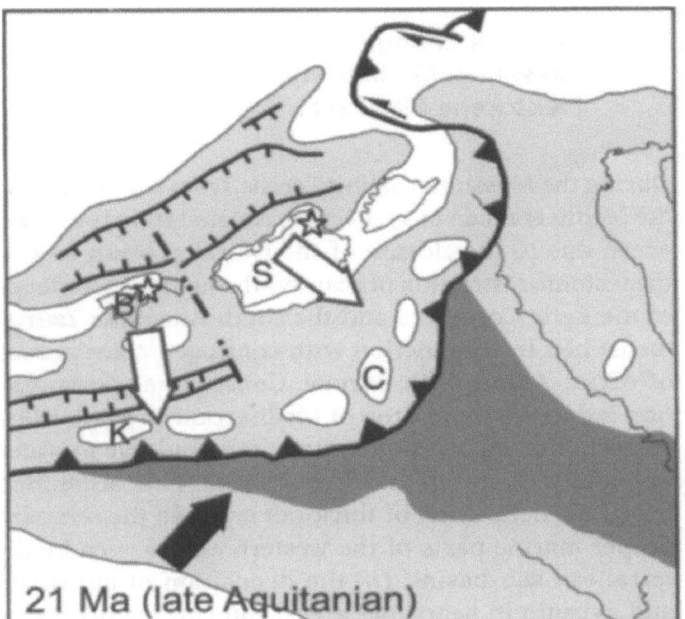
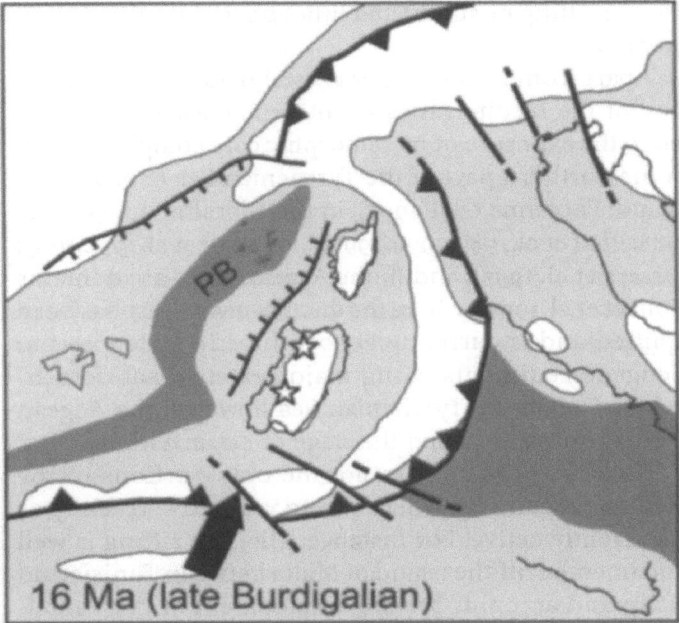
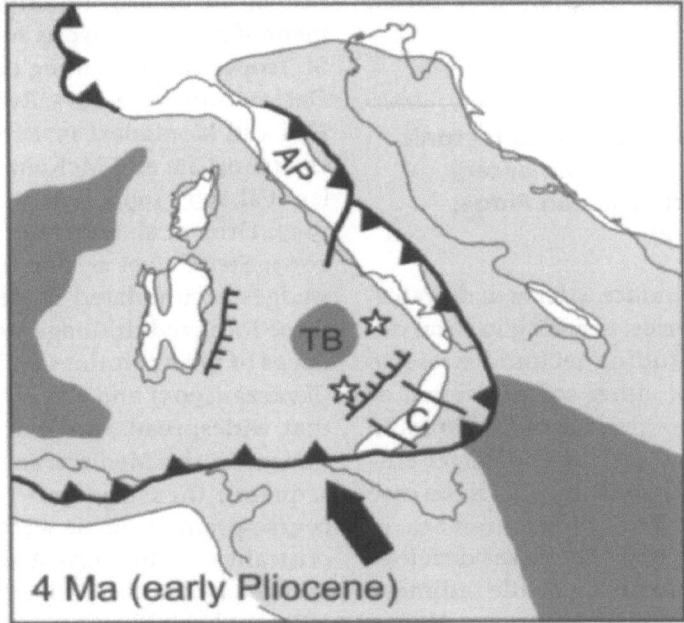

Fig. 1.10. Schematic maps showing the paleotectonic evolution of the W Mediterranean during Neogene time (modified after Bonardi et al. 2001, Roca 2001, and Cavazza and Wezel 2003). Only active tectonic elements are shown. *White*: exposed land; *light gray*: epicontinental sea; *darker gray*: oceanic crust. *Black arrows* indicate the direction of Africa's motion with respect to Europe (after Mazzoli and Helman 1994). *White arrows* indicate upper-plate direction of extension. *Stars* indicate subduction-related magmatism. *AP*, Apennines; *B*, Balearic block; *C*, Calabria-Peloritani terrane; *K*, Kabylies; *PB*, Provençal basin, *S*, Sardinia; *TB*, Tyrrhenian basin

Overall, Neogene extension in the Mediterranean can be explained as the result of roll-back of the subducting slabs of the Apennines-Maghrebian and Ionian-E Mediterranean subduction zones (e.g. Malinverno and Ryan 1986). As pointed out by Royden (1993), rapid extension of thickened crust in a convergent setting is a consequence of subduction roll-back. Neogene mountain belts throughout the Mediterranean region are characterized during their late orogenic stages by contemporaneous shortening in the external portion of the orogenic wedge and extension in its rear portions (e.g. Patacca et al. 1993).

Seismic tomographic images of the upper mantle velocity structure of the Mediterranean-Carpathian area (e.g. Wortel and Spakman 2000) point to the important role played by slab detachment, and its lateral migration along the plate boundaries, in the lithosphere dynamics of this region during the last 20–30 Myr. As such, mantle tomography provides a comprehensive explanation not only of arc-trench migration but also of along-strike variations in vertical motions, stress fields and magmatism. From this viewpoint, slab detachment represents the terminal phase in the gravitational settling of subducted lithosphere into the deeper mantle.

Apart from brittle, upper crustal deformation, extension of the Mediterranean continental crust has locally led to the exposure of metamorphic core complexes both in the northern part of the Tyrrhenian Sea (e.g. on Elba Island; Faccenna et al. 1997), in the Gibraltar arc (Dewey 1988; Platt et al. 1989, 1996, 1998; Cloetingh et al. 1992, 1995; Vissers et al. 1995), and in the Aegean back-arc domains (Jolivet et al. 1994), where the ductile lower crust has been uplifted and is currently almost directly juxtaposed to Neogene sediments along major extensional detachments. Unlike the Tyrrhenian Sea however, the Aegean Sea has not yet reached the stage of oceanization.

Back-arc domains are not the only portions of the Mediterranean area where rifting was active recently or is currently active. For instance, Pliocene rifting is well documented off the island of Malta between Tunisia and Sicily, and accounts for the development of the Pantelleria, Malta and Linosa grabens (Argnani and Torelli 2001).

1.5.3 Mud and Salt Diapirism Related to Tectonic Wedges, Tilted Blocks and Sedimentary Loading (Eastern Mediterranean Ridge, Alboran Sea, Nile Delta)

High sedimentation rates can induce a delay in dewatering processes in siliciclastic series, resulting in local development of overpressure. Adding tectonic stress to undercompacted sediments, in either compressional or extensional regimes, is likely to produce mud diapirs in areas where the overburden and seals are no longer efficient enough to balance the overpressure. This is the case for instance in the Neogene series of the Alboran Sea, in association with transtension and the coeval development of tilted blocks within underlying brittle sedimentary units and basement (Comas et al. 1990, 1999; Alonso and Maldonado 1992; Perez-Belzuz et al. 1997). Mud diapirs occur also at the toe of the Mediterranean Ridge, a major accretionary wedge associated with the Hellenic arc-trench system (Kopf and Behrmann 2003).

Salt diapirs and salt tectonics with development of listric faults and turtle-back features have been widely described in the Western Mediterranean, both in the Gulf of Lion and in the Algerian basin, in direct connection with the regional distribution of the Messinian salt. Mud and salt diapirs occur also in the Eastern Mediterranean basin in the area of the Herodotus abyssal plain, where rapid loading of shale and salt horizons by the clastics provided by the Nile River resulted in a progressive gravitational gliding of the sedimentary wedge toward the north, coeval with the development of listric faults and diapirism (Sage and Letouzey 1990; Mascle et al. 2002, 2003; Loncke et al. 2003).

1.5.4 Sea-level Changes, Salinity Crisis, Flooding (Messinian Mediterranean versus Pleistocene Black Sea)

During the Messinian salinity crisis, from 5.6 to 5.32 Ma, the Mediterranean basin became isolated from the world ocean due to the closure of the Gibraltar strait and to the tectonic restriction of seaways through the foredeeps of the Betic Cordillera and the south Anatolian Taurus thrust belt in conjunction with continued convergence of Africa-Arabia with Europe. Consequently, evaporation induced a rapid drop of Mediterranean sea level by at least 1 000 m, causing isostatic rebound due to water unloading (Norman and Chase 1986). This accounted for (i) the deposition of thick salt layers in the remnant deeper marine parts of the western and eastern Mediterranean sub-basins, (ii) the deposition of anhydrite and gypsum in nearshore areas, and (iii) the deep erosion of the emergent margins, resulting in the development of narrow canyons such as the Stoechades, Var and St. Tropez canyons along the Provençal coast (Hsü 1972; Decima and Wezel 1973; Ryan 1976; Montadert et al. 1977; Hsü and Montadert 1978; Cita and Colombo 1979; Rouchy 1980; Cita and McKenzie 1986; Gelati et al. 1987; Butler et al. 1995, 1999; Rabineau et al. 1998; Krigsman et al. 1999; Droz et al. 2001; Bernier et al. 2004; Gorini et al. 2003; Steckler et al. 2003). Thick clastic sedimentary wedges accumulated in deep-sea fans associated with these localized drainage systems, as well as in the foredeeps of the Apennines and the Albanides. DeCelles and Cavazza (1995) and Cavazza and DeCelles (1998) argue that widespread and short-lived intra-Messinian tectonism in the Mediterranean region, including out-of-sequence thrusting and the associated deposition of coarse-grained clastic wedges, may be the result of subcriticality of the peri-Mediterranean orogenic wedges induced by isostatic adjustment forced by unloading of adjacent basinal areas.

Like the Messinian Mediterranean, the Black Sea was temporarily isolated from the world ocean during the Pleistocene, due to a combination of tectonic activity and sea-level oscillations (Ryan et al. 1997, 2003; Ryan and Pitman 1998; Ballard et al. 2000; Clauzon et al. 2004). Consequently, its sea level and salinity have fluctuated much, as indicated by the changes recorded from fresh water to brakish or marine faunas, and the recent discovery of archeologic remnants at a water depth of −100 m. However, as for the Mediterranean, it is still debated whether post-Messinian flooding of the Mediterranean, and historic flooding of the Black Sea, now referred to as the Noah's flooding in popular literature, were as rapid and catastrophic as postulated by some authors, who envision the Messinian Gibraltar and Pleistocene Bosphorus gates as Niagara-like waterfalls (Ballard et al. 2000; Aksu et al. 2002 a, b).

1.6 Record of Ancient Dynamics of the Tethyan Oceans, Ophiolitic Sutures, Mantle Tomography versus Paleogeography of the Mediterranean Realm

In addition to being a natural laboratory for the study of active tectonics, the Mediterranean region is also a proxy of the past evolution of the former Tethyan oceans. Since the onset of geological prospecting and geophysical imaging, hundreds of geologists and geophysicists have progressively unraveled the records of past tectonics, mountain-building processes and basin development in this region. As with archeologic remains however, uncertainties increase when moving from the present to the past. As such, we can consider the Cenozoic and Mesozoic history of the Mediterranean domain relatively well constrained, whereas its Paleozoic evolution still remains highly conjectural. Hereafter we review a few salient aspects of the geological evolution of the Mediterranean domain.

1.6.1 Collisional versus Intracontinental Thrust Belts and Oceanic Sutures

Ophiolitic sutures have been studied for a long time along the true collisional belts of the Mediterranean such as the Alps, the Dinarides-Hellenides, and in Anatolia. Obducted ophiolites indeed constitute the only relics of the complex array of former Tethyan and associated oceanic domains, most of which have been subducted (see Stampfli and Borel, this publ.). Alternatively, deep seismic imagery in the Alps provide direct control of the present, post-collisional architecture of the orogen (Pfiffner et al. 1988; Roure et al. 1990, 1996; Pfiffner 1996; Schmid et al. 1996 and this publ.).

Other thrust belts, such as the Outer Carpathians and the Neogene Apennines, are almost devoid of ophiolitic sutures. Therefore, their development was presumably controlled by different dynamic processes in which slab retreat and lateral block escape in the hangingwall of the subduction zone played a dominant role. As such, they differ from true frontal collisional orogens involving more rigid continental lithosphere, as seen in the Alps (Royden et al. 1987; Royden 1993; Linzer 1996).

Unlike in the Alps, no ophiolite has ever been described in the Pyrenees. Albian meta-sediments associated with lherzolite outcrops and granulitic units of the Axial Zone are the only evidence for local thinning of the former continental lithosphere during the mid-Cretaceous separation of Iberia from the European plate (see TRANSMED Transect II). However, crustal stretching accounted only locally for denudation of the mantle-lithosphere, consisting of lherzolite-type peridotite, without any oceanic accretion. Albian to Turonian rotational translation of Iberia in conjunction with opening of the Bay of Biscay and the Valais ocean occurred in a dominantly transtensional regime. Subsequent Senonian to Paleogene crustal shortening in the Pyrenees, amounting to some 150 km and involving basin inversion and crustal imbrication (Vergés and Garcia-Senz 2001), occurred under a partly transpressional regime, as evidenced by strike-slip mechanisms along the North Pyrenean Fault (Choukroune and Ecors Team 1989).

1.6.2 Plate Dynamics and Palinspastic Restorations: Demise of the Concept of a Single Tethys

Mantle tomography provides ways to trace the dense and cold lithospheric slabs currently still sinking into the asthenosphere, no matter whether or not slab detachment has already occurred (Spakman 1990; DeJong et al. 1993; Spakman et al. 1993; Carminati et al. 1998; Wortel and Spakman 2000). As documented by W. Spakman in this publication, the current depth and lateral extent of the subducted lithospheric material can be traced over the entire Mediterranean domain and its surroundings, i.e. beneath the Alps, the Carpathians, the Aegean and the Gibraltar arcs, thus providing first-order constraints to restore the former evolution of retreating slabs and the lateral connections of former Tethyan subduction zones.

Balanced cross-section techniques applied to the external parts of fold-and-thrust belts such as the Alps, Pyrenees, Carpathians, Apennines, Albanides and Maghrebides, are useful to palinspastically restore the former continental margins of the Tethyan oceans, now forming part of the respective orogenic wedges, to their initial configuration (Bally et al. 1988; Roure et al. 1989, 1993, 2004; Ellouz and Roca 1994; Frizon de Lamotte et al. 2000; Ziegler et al. 2001b). However, this cannot be applied to former intervening oceanic domains, which have been entirely consumed in subduction zones, except for the few remnants still preserved along the ophiolitic sutures. Hence, paleomagnetic data become useful to trace the north-south displacement and rotations of Adria, Iberia and Africa with respect to Europe through Mesozoic and Cenozoic times.

Lastly, biostratigraphy and lithofacies can be used also to validate palinspastic restoration and to better constrain paleogeography and ancient plate dynamics. Oceanic domains are likely to isolate distinct benthic faunas on their conjugate continental margins, whereas they can help homogenizing pelagic fossils across wide areas. Fossil contents and sedimentologic trends, together with carbonate vs. silica contents of sedimentary layers are also used to obtain paleobathymetric estimates, which in turn provide information on tectonic subsidence history, all processes relevant to rifting, continental crustal stretching and oceanic accretion.

Numerous attempts have been made to propose palinspastic reconstructions of the entire Tethyan-Mediterranean domain since the Permo-Triassic (Sengor 1979, 1984; Ziegler 1988; Dercourt et al. 1993, 2000; Roure 1994; Yilmaz et al. 1996; Stampfli et al. 2001a and b, 2002a). The latest of these attempts is presented by G. Stampfli and G. Borel in this publication. A discussion of the various hypotheses proposed for the evolution of the western Tethyan domain goes beyond the purpose of this contribution. We provide here a brief summary of the post-Variscan evolution of the Mediterranean domain following the paleogeographic reconstructions presented in Chapt. 3 and refer readers to the abundant literature available on this subject.

Following the late Carboniferous-early Permian assemblage of Pangea along the Variscan-Appalachian-Mauritanian-Ouachita-Marathon and Uralian sutures, a wedge-shaped ocean basin widening to the east – the Paleotethys – was comprised between Eurasia and Africa-Arabia (Fig. 3.5). At this time, a global plate reorganization induced the collapse of the Variscan orogen and continued northward subduction of Paleotethys beneath the Eurasian continent (e.g. Vai 2003). A new oceanic basin – the Neotethys – began to form along the Gondwanan margin due to the rifting and NNE-ward drifting of an elongate block of continental lithosphere, the Cimmerian composite terrane (Sengor 1979, 1984). The Cimmerian continent progressively drifted to the northeast, leaving in its wake the Neotethys (Fig. 3.6). The Permo-Triassic history of this part of the world is hence characterized by progressive widening of Neotethys and contemporaneous narrowing of Paleotethys, culminating in the late Triassic docking of the Cimmerian terrane along the Eurasian continental margin (although portions of the Paleotethys closed as early as the late Permian) (Figs. 3.7 and 3.8). The Cimmerian collisional deformation affected a long yet relatively narrow belt extending from the Far East to SE Europe (see Sengor 1984, for a discussion). Cimmerian tectonic elements are clearly distinguishable from the Far East to Iran, whereas they are more difficult to recognize across Turkey and SE Europe, where they were overprinted during later orogenic pulses. The picture is complicated by back-arc oceanic basins (Hallstatt-Meliata, Maliac, Pindos, Crimea-Svanetia and Karakaya-Küre) which opened along the southern margin of Eurasia during subduction of Paleotethys and which were mostly destroyed during the docking of the Cimmerian continental terranes.

The multi-phased Cimmerian collisional orogeny marked the maximum width of the Neotethyan ocean, which during Jurassic-Paleogene times was progressively consumed by northward subduction along the southern margin of the Eurasian plate (Figs. 3.8–3.14). Whereas the Paleotethys was completely subducted or incorporated in very minor quantities in the paleotethyan suture, remnants of the Neotethys are still preserved in the Ionian Sea and the Eastern Mediterranean. Throughout the Mesozoic new back-arc marginal basins developed along the active Eurasian margin. Some of these basins are still preserved today (Black Sea and Caspian Sea) though most of them were closed (e.g. Vardar, Izmir-Ankara) with the resulting sutures masking the older suture zones of the Paleotethyan and Neotethyan oceanic domains.

The picture is further complicated by the mid-Jurassic opening of the Ligurian-Piedmont-south Penninic ocean which resulted in the development of a new set of passive margins which were traditionally considered as segments of the northern margin of a single "Tethyan Ocean" stretching from the Caribbean to the Far East. It is somehow a paradox that the Alps – which for almost a century served as an orogenic model for the entire Tethyan region – are actually related neither to the evolution of Paleotethys nor to Neotethys evolution and instead have their origin in a branch of the Atlantic ocean that was closed by late Eocene times to form the Alps-Carpathians orogenic system (Fig. 3.19) (Stampfli et al. 2002). Furthermore, development of the Pyrenean rift zone, which was activated during the Triassic at the same time as the North Atlantic rift system, culminated in the mid-Cretaceous detachment of Iberia from Europe (Fig. 3.12) and the opening of the oceanic Bay of Biscay and the Valais trough. The Pyrenean rift and the Valais trough were closed during the Eocene.

Paleogene collision of the evolving Alpine orogenic wedge (the leading edge of Adria) with its foreland was accompanied by their progressive collisional coupling, inducing intraplate deformation in the foreland (Ziegler et al. 2002), as well as lateral block-escape and oblique motions within the orogen. For example, eastward directed orogenic transport from the Alpine into the Carpathian domain during the Oligo-Miocene was interpreted as a direct consequence of the deep indentation of Adria into Europe (Ratschbacher et al. 1991). From a wider perspective, strain partitioning clearly played a major role in the development of most of the Mediterranean orogenic wedges as major external thrust belts parallel to the former active plate boundaries coexist with sub-vertical, intra-wedge strike-slip faults which seem to have accommodated oblique convergence components (e.g. Insubric line of the Alps, intra-Dinarides peri-Adriatic line).

1.6.3 Cenozoic Magmatism in the Mediterranean Region (*with the contribution of C. Savelli*)

Figures 1.11 and 1.12 depict the time-space distribution of the puzzling mosaic of magmatic associations occur-

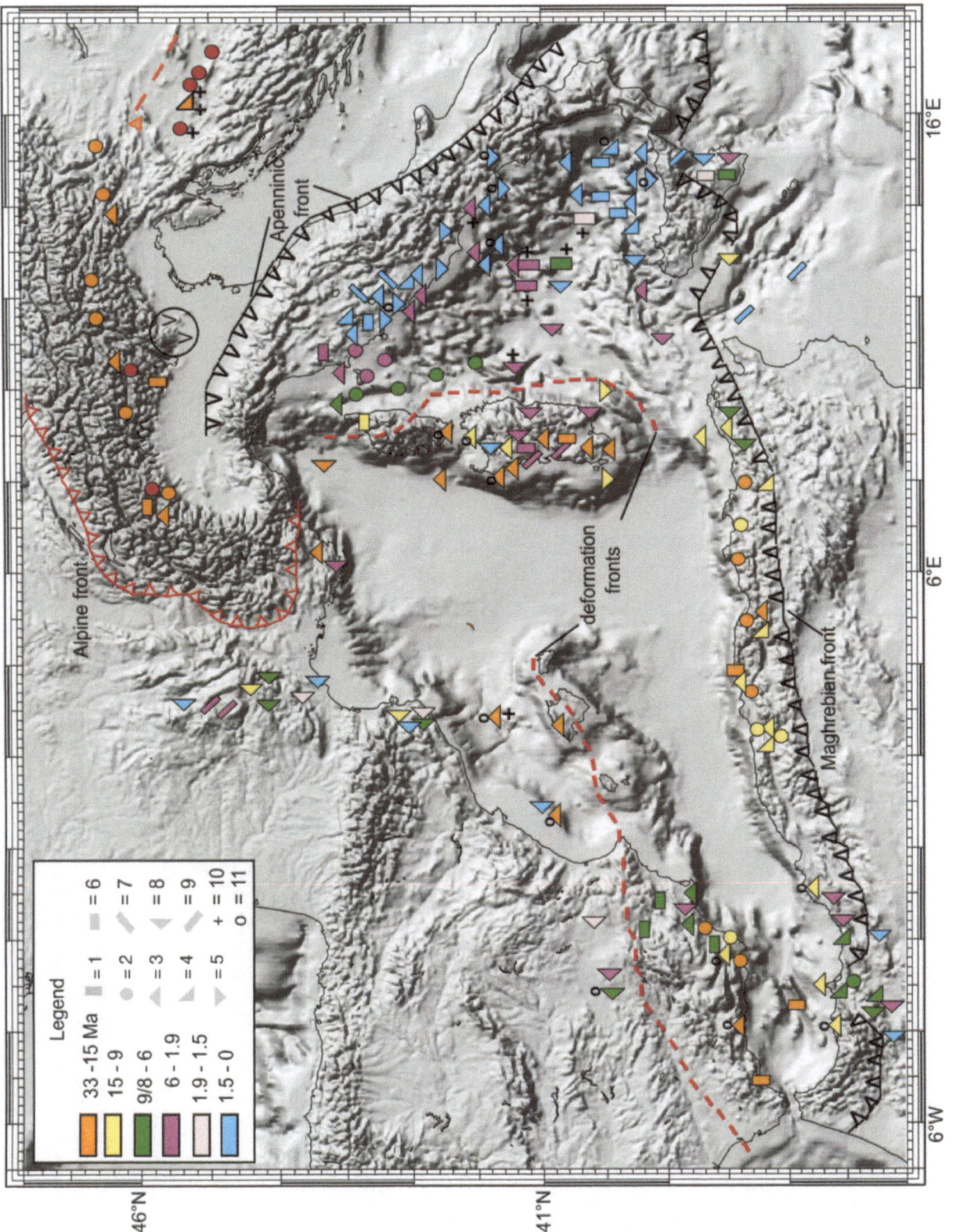

Fig. 1.11. Digital relief-shaded image of time-space-composition characteristics of magmatic rocks of the western Mediterranean and peripheral orogens (modified after Savelli 2002). Legend: rock types: *1* = tholeiites; *2* = medium-K and high-K calcalkaline plutons (*red circles* = Eocene plutons from the Alps and adjacent Sava zone); *3* = medium-K and high-K calcalkaline volcanics; *4* = shoshonites; *5* = ultrapotassic volcanics; *6* = lamproites; *7* = carbonatites; *8* = intraplate volcanics; *9* = rocks from intraplate, large central volcanoes; *10* = volcanics from deep drillings (DSDP and ODP sites, and Neapolitan area); *11* = pyroclastic and/or ignimbritic rocks. *VV* = Veneto volcanic district

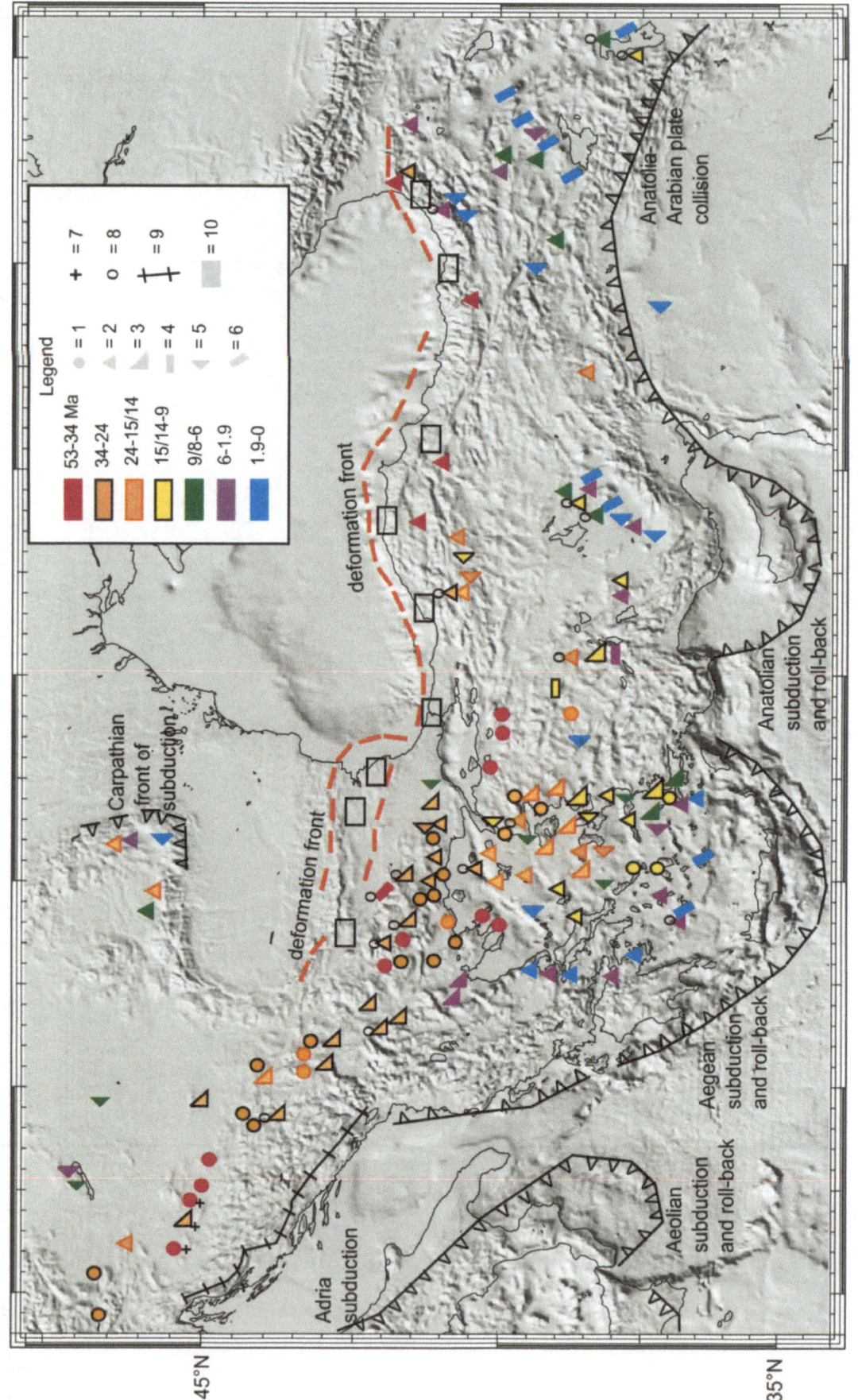

Fig. 1.12. Digital relief-shaded image and distribution of magmatic rocks in regions bordering the eastern Mediterranean. Legend, rock types: *1* = high-K and medium-K calcalkaline acidic plutons; *2* = high-K and medium-K calcalkaline volcanics; *3* = shoshonites; *4* = lamproites and ultrapotassic rocks; *5* = intraplate volcanics; *6* = large central volcanoes; *7* = igneous rocks from deep drillings; *8* = ignimbritic and/or pyroclastic rocks; *9* = Adria-Dinarides boundary; *10* = late Cretaceous-Paleocene plutonic and volcanic rocks (courtesy of C. Savelli, ISMAR-CNR, Bologna)

ring in and around the western and eastern Mediterranean Sea, respectively. Some imprecisions most likely affect large compilations of analytical data that were generated by different laboratories with different methods. Notwithstanding this drawback, such data sets are useful to study the possible relationhips between magmatic activity and dynamic processes in the Mediterranean domain across which convergence of the African-Arabian and European plates led to major shortening, from the Betic Cordillera of southern Spain to Anatolia.

Mediterranean magmatic associations occur within (*i*) the Hercynian European lithosphere, (*ii*) the Alpine and Dinarides orogenic wedges, (*iii*) the internal parts of the Apennines where they are associated with a NW- to SW-dipping subduction system, (*iv*) the internal parts of the Hellenic-Aegean orogen which is dominated by a generally N-dipping subduction zone, (*v*) several back-arc basins, and (*vi*) the collision zone of the Arabian shield with the East Anatolia orogen (Bitlis suture). Such diversity in tectonic settings was somehow reduced to three distinct associations by Wilson and Bianchini (1999):

1. orogenic, subduction/post-collisional magmatism related to plate convergence characterized by a spectrum of calcalkaline, high-K calcalkaline and potassium magma series (shoshonites and lamproites), including relatively primitive mafic magmas and their differentiates;
2. extension-related intra-plate magmatism, typically alkali basalts, basanites and their differentiates, but locally including subalkaline (tholeiitic) basalts and rare potassic magma types (leucitites and leucite nephelites);
3. localized oceanic spreading centers, erupting subalkaline basalts akin to ocean ridge basalts (MORB).

Magmatic rocks of calcalkaline (orogenic) affinity are most abundant. Overall, prior to about 15 Ma the calcalkaline rocks have a dominant acidic K-rich composition. Calcalkaline volcanism of the Aegean and Tyrrhenian basins (e.g. Santorini volcano of Greece, Aeolian Islands of Italy) clearly relates to the Aegean and Calabrian subduction systems, its current location being directly controlled by the dip of the slab. This igneous activity was accompanied by processes of crustal attenuation, stretching and slab break-off. It appears that the post-15 Ma calcalkaline volcanics exhibit mainly intermediate-basic compositions with their age decreasing towards the fronts of subduction and roll-back. The islands of Stromboli and Vulcano in the Aeolian archipelago are volcanically active.

Volcanic signatures can provide also key informations on the current state of the underlying lithosphere at the time of lava emplacement. For instance, Pleistocene calcalkaline volcanism of the Aegean and Tyrrhenian Seas was used in the TRANSMED sections, in addition to focal mechanisms and mantle tomography, to delineate the current attitude of the East Mediterranean and Ionian slabs beneath the Hellenic and Calabrian arcs, respectively.

Of special interest is the occurrence of high-K calcalkaline and shoshonitic magmas which is interpreted as the magmatic signature of lithospheric extension and slab detachment (Davies and von Blankenburg 1995). For instance, the latest Eocene-Oligocene magmatism of the Alps is interpreted as marking detachment of the south-dipping Alpine Tethys slab from the lithosphere (Schmid et al. 1996). Similarly, the Oligocene magmatism of the Dinarides is related to detachment of the east-dipping Vardar(?) slab (Pamic et al. 2002), whereas the late Miocene magmatism of the Maghrebides reflects detachment of the north-dipping Maghrebian Tethys slab (Wilson and Bianchini 1999).

Besides the migrating calcalkaline magmatism, also anorogenic and tholeiitic volcanics provide important information on Mediterranean geodynamic processes. Intraplate basic lavas with Na-alkaline and subalkaline compositions are present in the upper plate as well as in the peripheral foreland domain. The upper plate anorogenic volcanics appear to be located at larger distance from the roll-back fronts if compared with the distribution of coeval orogenic products. From the late Eocene to the present, anorogenic magmatism has occurred throughout the European and African margins and within the Mediterranean Tethyan domain. Major anorogenic-type volcanic activity occurred in the early Miocene and particularly in the late Miocene-Pliocene. Such volcanic flare up may reflect a reorganization of the upper mantle convection system following the Alpine collisions (Wilson and Bianchini 1999).

The West European rift system has been the site of rift-related volcanism since the late Eocene in the Massif Central (Limagnes, Chaîne des Puys) as well as in the Rhine Graben. Rift-related volcanism with significant subduction-related geochemical character has been also well documented in the vicinity of the Tyrrhenian Sea, i.e. in the Latium province near Rome.

Some magmatic bodies in the Mediterranean area have rather peculiar settings. For instance, Mt. Etna is located very close to the Neogene thrust front in NE Sicily, and it is associated with a long-lasting north-trending crustal lineament which extends to the south along the Malta escarpment and accounts for Mesozoic and Cenozoic volcanic episodes on the Ragusa plateau of SE Sicily. Mt. Vesuvius, like other K-rich volcanoes from the Latium-Campania area, is likely related to back-arc opening and foundering of the Tyrrhenian Sea and coeval upwelling of the asthenosphere.

1.7 Conclusions

The Mediterranean basin and the surrounding regions constitute a natural laboratory for studying active geodynamic processes related to the final stages of continent-continent collisions, such as passive subduction of oceanic lithosphere, microplate development, back-arc rifting and subduction-related volcanism. The Mediterranean basins constitute also modern analogues for former active margins. Areas flanking the Mediterranean basins comprise almost continuous Late Cretaceous to Neogene fold-and-thrust belts both on its northern (Betics, Pyrenees, Alps, Carpathians, Apennines, Dinarides, Albanides, Hellenides, Pontides, Taurides) and southern margins (Maghrebides, Atlas). These contain remnants of preexisting oceanic basins and their Mesozoic passive margin sedimentary prisms that have been tectonically accreted and can be studied by field geologists and used for palinspastic reconstructions.

The last twenty-five years of geological investigation of the Mediterranean region have disproved the traditional notion that the Alpine-Himalayan mountain ranges originated from the closure of a single, albeit complex, oceanic domain – the Tethys. Instead, the present-day geological configuration of the Mediterranean region is the result of the opening and subsequent consumption of two major oceanic basins – the Paleotethys and the Neotethys – and of additional smaller oceanic basins within an overall regime of prolonged interaction between the Eurasian and the African-Arabian plates. In greater detail, there is still some debate about exactly which "Tethys" existed at what time. There is consensus on the presence of (*i*) a mainly Paleozoic Paleotethyan ocean north of the Cimmerian continental terrane(s), (*ii*) a younger late Paleozoic-Mesozoic Neotethyan ocean that had opened south of these terranes, and finally (*iii*) a middle Jurassic ocean, the Alpine Tethys, an extension of the central Atlantic ocean into the western Tethyan domain. Additional late Paleozoic to Mesozoic back-arc marginal basins along the active Eurasian margin complicated somewhat this basic scheme. The closure of these heterogenous oceanic domains produced a system of connected yet discrete orogenic belts which vary in terms of timing of their main deformation, tectonic setting and internal architecture. As such, they cannot be interpreted as the end product of a single "Alpine" orogenic cycle

Progress in deep seismic imagery and mantle tomography have considerably improved our understanding of the crustal-lithospheric architecture and overall evolution of the Mediterranean margins and adjacent fold-and-thrust belts, making possible the present compilation of the TRANSMED transects. However, in some areas (e.g. Anatolia and Macedonia) subsurface geophysical constraints are still limited. Correspondingly, the TRANSMED transects presented for these areas must be considered as tentative, leaving space for alternative interpretations.

Reflection-seismic data acquired by the industry in its search for hydrocarbons has considerably increased our knowledge on the sedimentary fill of Mediterranean basins and margins. New developments of deep-water prospecting have effectively reached the Mediterranean with a renewed interest of the petroleum industry for areas which hitherto have not yet been explored. Accurate three-dimensional description of sedimentary lithosomes is also locally required for water management, because water resources are frequently restricted to fossil aquifers in onshore parts of the Mediterranean domain, where they should be managed with caution.

Apart from being of interest to academic research, the Neogene-to-Recent geological record of the Mediterranean region has also direct and major societal implications. For instance, the sedimentary fill of the Mediterranean basins is likely to provide key information on recent vertical motions and climate, the combination of which accounts for local relative sea-level changes which are superimposed on truly eustatic sea-level oscillations. A good knowledge of the interactions between deep processes (geodynamics) and surface processes (climate, erosion, transport and deposition of sediments, as well as biosphere) is a prerequisite for risk assessment in such highly populated area as the Mediterranean shores and their hinterlands. Major hazards that are controlled by global dynamics comprise earthquakes and volcanoes, but flooding and landslides result also from direct interactions between local topography, vegetation, rainfall and river drainages.

The Mediterranean Sea is a relatively small, partly oceanic basin that is enclosed by orogenic systems. In view of this, the environment and overall equilibrium of the entire Mediterranean region is highly susceptible to external factor. This was the case during the Messinian crisis that affected both eastern and western Mediterranean basins, and that was controlled by the tectonic closure of the seaways connecting them to the world ocean. Similarly the Black Sea was intermittently connected with the Mediterranean during the Pleistocene, as recorded by major fluctuations in its salinity and sea level. Direct and indirect consequences of CO_2 increase in the atmosphere and coexisting greenhouse effects will be amplified in a confined domain such as the Mediterranean. Deforestation has been continuous since the onset of urban development in Roman times and the associated agricultural boom. Efforts to reconstitute the Mediterranean forest are continuously facing the devastating effects of artificial fires and desertification. Lastly, global warming has been also assumed to account for the recent increase of catastrophic flooding and forest fires in the Mediterranean, hot and dry summers inducing an increase in average temperatures of the sea water on one hand, and destruction of the terrestrial plants on the other hand. The reduced vegetation cover cannot absorb the excess of precipitation and the associated soil erosion during increasingly wetter winters.

Acknowledgements

A. Camerlenghi, I. Jimenez-Munt, R. Sabadini and C. Zanolla kindly provided the original files of some of the figures reproduced here. C. Savelli (ISMAR-CNR, Bologna, Italy) contributed Figs. 1.11 and 1.12 and provided expertise for the discussion of Mediterranean magmatism.

Chapter 2

A Tomographic View on Western Mediterranean Geodynamics

Wim Spakman · Rinus Wortel

Abstract

During the Cenozoic, the Western Mediterranean region has experienced a complex subduction history which involved the destruction of the Late Triassic/Jurassic Ligurian ocean and the West Alpine-Tethys. Lithosphere remnants of this evolution have been detected in the upper mantle by seismic tomography imaging. However, no general consensus exists on the interpretation of these remnants/slabs in the context of Ligurian ocean and West Alpine-Tethys subduction. In this paper we search for subduction remnants of the entire Cenozoic evolution in the recent global tomography model of Bijwaard and Spakman (2000) and compare these tomography results and our interpretations with those obtained in previous studies. Next, we present an analysis of imaged mantle structure in the context of the tectonic evolution of the Western Mediterranean during the Cenozoic. Our analysis leads to the following main results:

1. The identification of the remnant of the West Alpine-Tethys (Piedmont ocean) found at the bottom of the upper mantle under the Alps and northern Apennines region.
2. A surface reconstruction of the Ligurian ocean from subduction remnants found in the upper mantle under the Western Mediterranean.
3. The confirmation of the earlier propositions by Lonergan and White (1997) concerning slab roll-back and lithosphere tearing which led to two dominant Ligurian subduction systems: the Betic-Alboran subduction and the Apennines-Calabria subduction.
4. Propositions of a short (300–400 km) continuous north-Apennines slab and of slab detachment beneath the central-southern Apennines.
5. Slab detachment and lithosphere tearing are considered crucial processes for facilitating slab roll-back in the Western Mediterranean region.
6. A new kinematic model for slab roll-back in the Betic-Rif-Alboran region which involves slab detachment under the Betics, lithosphere tearing along the African margin, and which explains both the inferred slab geometry and the arcuate geometry of the Betic-Rif orogen.

2.1 Introduction

Seismic travel-time tomography is an imaging method which allows to construct three-dimensional (3-D) models of the Earth's internal structure from observations of seismic travel times (e.g. Spakman et al. 1993). Earth structure is obtained in terms of the propagation speed of seismic P- and/or S-waves. Seismic-wave speed is a material parameter which depends on local properties as temperature, pressure, and composition (e.g. Trampert et al. 2001). Thus, implicitly tomography delivers a snapshot of mantle dynamics. We note, however, that tomographic imaging generally provides a blurred view on Earth structure as a result of the combined influence of lack of observations, data errors, and theoretical and numerical approximations. Image blurring complicates the interpretation of a tomogram. Other factors that complicate interpretation are introduced by incomplete knowledge of how seismic wave speed relates to temperature and composition and by occasional degrees of freedom to explain the same image with different types of mantle processes. The latter basically derive from insufficient knowledge of, and/or constraints on, mantle dynamics and the crustal response to lithosphere processes (e.g. subduction, lithosphere delamination).

For the Alpine-Mediterranean region, seismic tomography has considerably narrowed the range of possible scenarios for the geodynamic evolution of the region. The first mantle models revealed a complex pattern of upper mantle heterogeneity underlying the entire Alpine belt which was interpreted as subducted remnants of Tethys lithosphere (Spakman 1986a, 1990). Subsequent tomographic studies of the Mediterranean, generated by predominantly Dutch and Italian groups, have considerably focussed the image of mantle structure and revealed, for example, flat-lying slabs under the Western Mediterranean (Lucente et al. 1999; Piromallo and Morelli 1997, 2003; Wortel and Spakman 2003) or subduction beneath the Aegean to depths of 1 500 km (Bijwaard et al. 1998). In this paper we discuss recently obtained travel-time tomography results of the 3-D structure of the Western Mediterranean mantle (Fig. 2.1). The results are extracted from the global-mantle model BS2000 of Bijwaard and

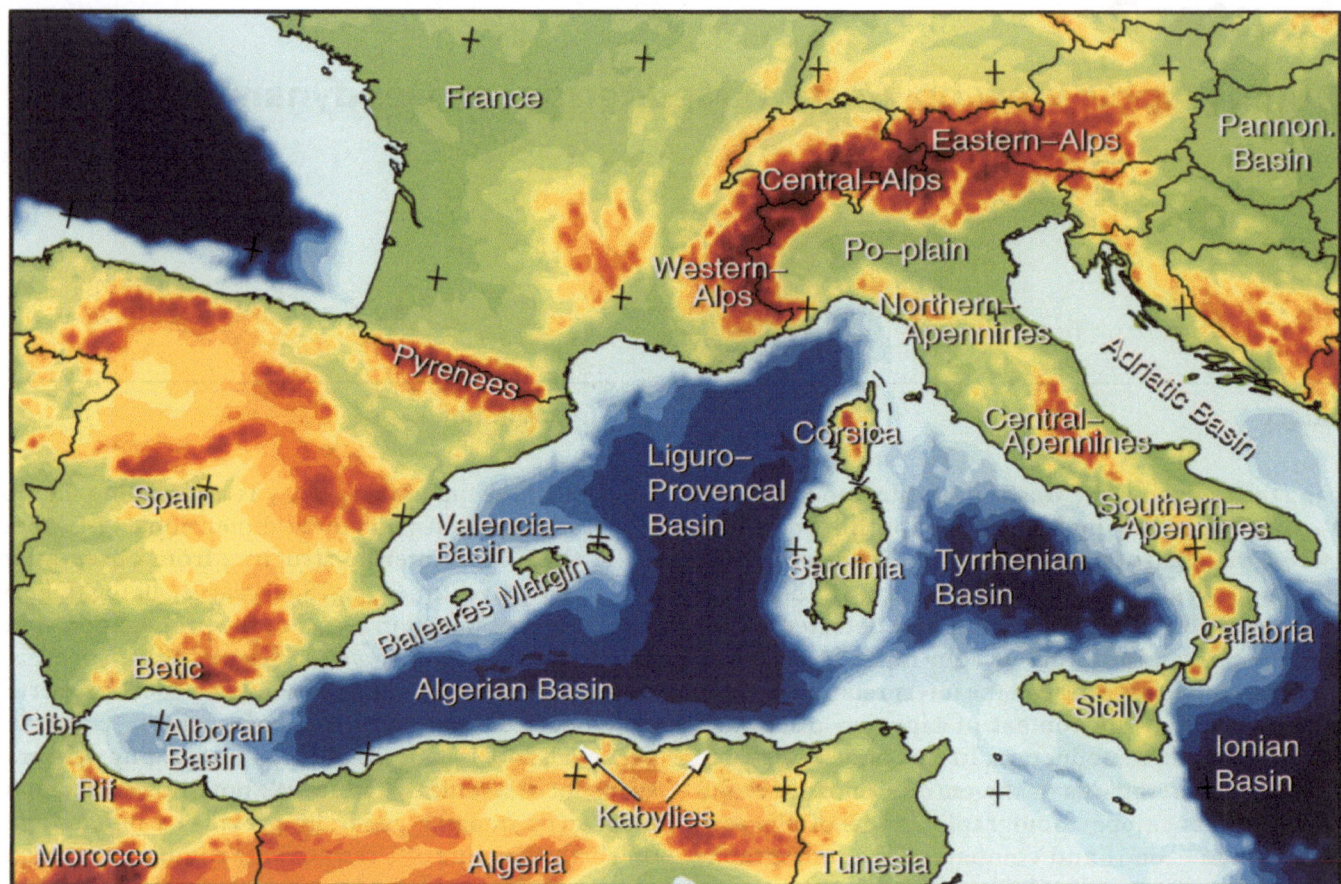

Fig. 2.1. The Western Mediterranean and surrounding regions. Indicated names are used in the text. For an introduction to the main geological features of the region, see Cavazza et al., this publication

Spakman (2000). Details on data analysis and tomographic method can be found in Spakman (1993), Spakman et al. (1993), Bijwaard et al. (1998), and Bijwaard and Spakman (2000). Some results of BS2000 concerning Mediterranean mantle structure were presented earlier (Wortel and Spakman 2000, 2001 (Erratum), Gutscher et al. 2002). We will analyse in detail what can be learned from imaged mantle structure to help unravel the complex geodynamics of the Western Mediterranean. Based on this new tomographic information we adapted some of our earlier interpretations. Furthermore, we can confirm, and in some cases refine, existing interpretations and we will also add some new results such as the identification of the West Alpine-Tethys slab (Piedmont ocean slab), a surface reconstruction of the Ligurian ocean, and a new qualitative reconstruction of the Betic-Alboran subduction history. In our analysis, we include inferences drawn from tectonic reconstructions of the Western Mediterranean, but stay relatively away from detailed comparisons with geology, in particular, timing of events. Rather we like to present, as the title of this contribution states, a tomographic view on Western Mediterranean mantle dynamics, leaving the test of our propositions to future research.

2.2 The Global Tomography Model BS2000

Model BS2000 (Bijwaard and Spakman 2000) is the successor of the BSE-model (Bijwaard et al. 1998). Both models result from a tomographic analysis of the global data set of 7.5 million P- and pP-phase delay times of Engdahl et al. (1998). They are determined as seismic wave-speed anomaly models with respect to the 1-D global reference model ak135 (Kennett et al. 1995). Furthermore, the models are based on exactly the same model parameterization. The basic difference results from the application of 3-D ray tracing and nonlinear inversion by a step-wise linearized approach (Bijwaard and Spakman 1999, 2000). Nonlinear inversion is performed to account for the effects of ray bending caused by 3-D wave speed heterogeneity in the mantle and for the effects of an inadequate reference model. BS2000 is a model of the entire mantle based on a special cell-parameterization technique in which cell size is adaptive to the local data density (Spakman and Bijwaard 2001). Small cells are constructed in regions of high data density and larger cells elsewhere. This effectively allows for solving a tomographic inverse problem for the entire mantle while retaining resolution for the relatively small structural detail known from regional mantle studies (e.g. Spakman et al. 1993; Piromallo and Morelli 1997, 2003). Figure 2.2 shows examples of the spatially variable cell parameterization of the Western Mediterranean mantle.

BS2000 results from a very large-size inverse problem (about 400 000 model parameters) for which formal resolution analysis is computationally not feasible. An alternative and approximate method is to assess spatial resolution through sensitivity analysis with synthetic wave speed models (Spakman and Nolet 1988). Synthetic data are first computed by integrating a synthetic wavespeed anomaly model along the seismic ray paths of the real data. Usually some Gaussian noise is added and finally the synthetic data are inverted in the same way as the actual data. Qualitative inferences about model resolution are derived from the comparison between the synthetic model and its tomographic image. Examples are given in Fig. 2.3 showing results of so-called spike- or block tests for synthetic blocks of 1.2 to 2.4 degrees in size, at different depths. Synthetic spike tests are conducted for a wide variety of block sizes ranging from 0.6 degrees to 6.0 degrees. This allows for investigating the potential for resolving both small and larger scale structure at specific locations in the tomographic model and reduces the risk of misinterpreting sensitivity test results (e.g. good recovery of detail does not imply good recovery of larger structures; see Leveque et al. 1993). Many examples of sensitivity tests are given in Appendix 1 (CD-ROM). The degree of recovery of a variety of synthetic patterns is used as a measure of confidence to engage interpretation of the actual tomogram.

Model BS2000 shows a complex and heterogeneous pattern of seismic wave-speed anomalies across the entire upper mantle beneath the Western Mediterranean. Generally, the imaged wave-speed heterogeneity becomes more smooth from the top of the mantle downward (see map view images of Fig. 2.4). Amplitudes decrease from 3–4% in the upper 200 km to 1–1.5% at the base of the upper mantle. We deduce from sensitivity test results that the noted amplitude decrease with depth is real, although amplitudes may be underestimated on average. This was also found in other studies (e.g. Spakman et al. 1993; Piromallo and Morelli 2003). Sensitivity tests also indicate a general decrease in spatial resolution from, locally at best, 50 km in the first 100 km to 100–200 km in the mantle transition zone (410–660 km). For reasons of model comparison later in this paper, we note that, from a technical point of view, the regional mantle models of Spakman et al. (1993) and Piromallo and Morelli (1997, 2003) are closest to BS2000. The former work employs a spatially uniform parameterization with 0.8 degree cells and cell thicknesses similar to BS2000, whereas the latter uses a detailed node parameterization with a node distance of 0.5 degrees and

50 km node spacing in depth. This is comparable to using 0.25 degrees cells with thicknesses of 25 km (Spakman and Bijwaard 2001).

2.3 Interpretation of Model BS2000 for the Western Mediterranean Mantle

Our interpretations will concentrate on subduction of lithosphere. Previous tomographic studies have contributed strongly to unravelling the 3-D geometry of subducted slab in the region (e.g. Spakman 1986a, 1990, 1991; Spakman et al. 1993; Amato et al. 1993, 1998; Cimini and De Gori 1997; Piromallo and Morelli 1997, 2003; Lucente et al. 1999). Model BS2000, however, allows to make new inferences on Mediterranean subduction systems. In tomography, subducted lithosphere translates into a positive (fast) wave-speed anomaly, predominantly, due to the temperature contrast between the cold slab and the warmer ambient mantle, whereas relatively warmer mantle regions lead to a negative (slow) wave-speed anomaly (e.g. De Jonge et al. 1994; Goes et al. 2000). We first discuss mantle structure in the eastern part of the region (Alps-Apennines-Calabria) and separately discuss imaged structure of the Betic-Alboran mantle. In the last part of this paper we analyse our findings in the context of the geodynamic evolution of the entire Western Mediterranean.

2.3.1 Alps, Apennines, and the Western Mediterranean

The tomographic image of the upper mantle of this region is very complex. The 3-D cartoon displayed in Fig. 2.5 shows our schematization of the imaged positive mantle anomalies and summarizes our geodynamic interpretation. This cartoon can be verified against the imaged positive anomaly patterns as extensively displayed in map view sections (Fig. 2.4), selected cross sections (Figs. 2.6–2.8) and cross sections presented in Appendix 2 (CD-ROM). To justify our choices leading to the geometric and geodynamic interpretation depicted in Fig. 2.5, we start with a brief description (with some interpretation) of the larger scale patterns, followed by a comparison with related work.

Imaged Upper Mantle Structure

The positive anomalies found in the upper 150–200 km beneath the Alps and northern Po plain are dipping to the south, beneath the central Alps, and to the ESE below the western Alps. This is best shown in cross sections (Fig. 2.8b and Appendix 2). This pattern is consist-

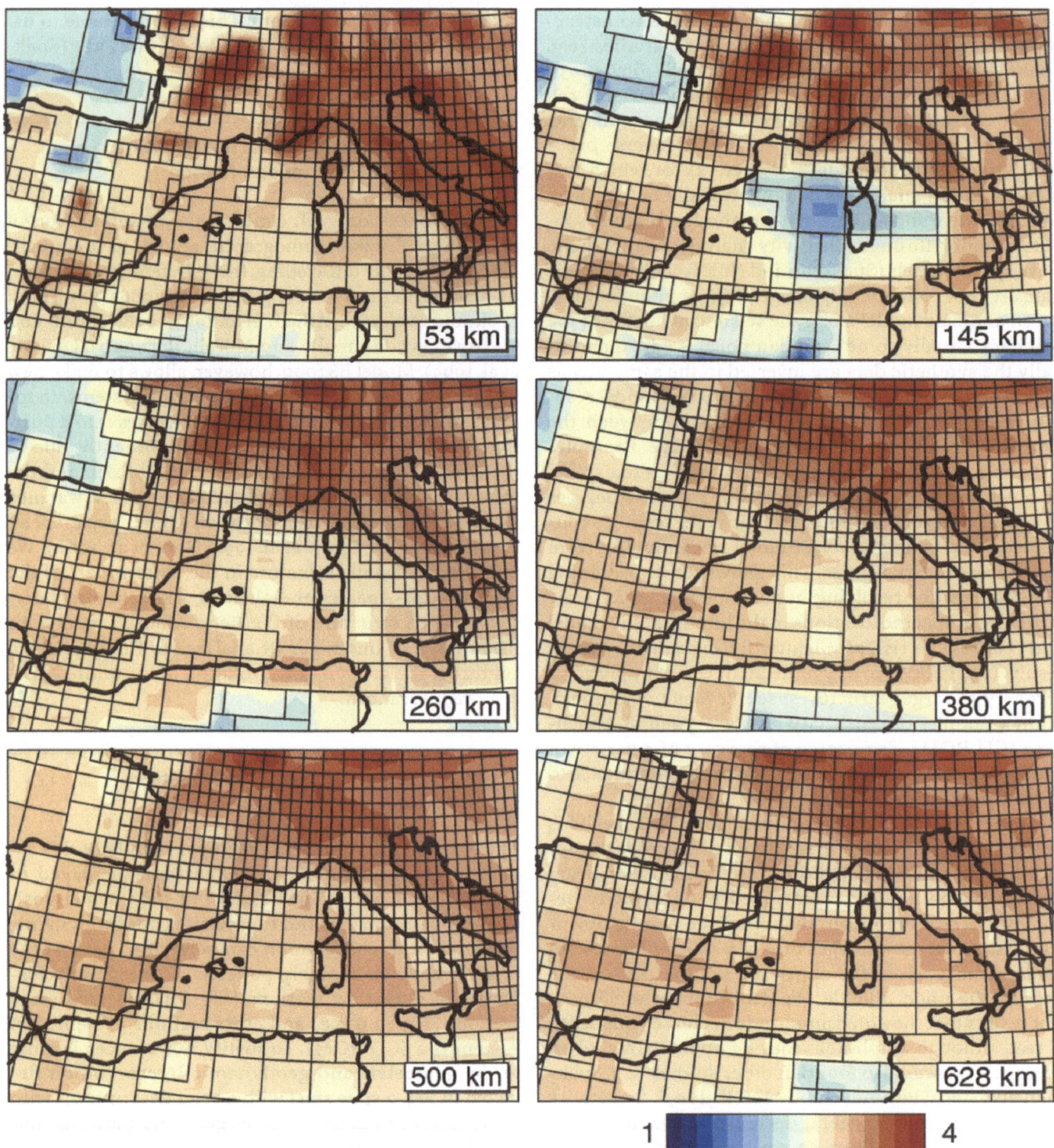

Fig. 2.2. Cell hit count and model cell-parameterization at selected depths beneath the Mediterranean region. The smallest cells have lateral dimensions of 0.6 degrees. Lateral dimensions of larger cells are a multiple of 0.6 degrees. The smallest cells have lateral dimensions of 0.6 degrees and thickness of 35 to 65 km increasing with depth, respectively. Larger cells are found below the sea areas and northern Africa where station and earthquake density is strongly reduced. Still, the 1.8 × 1.8 degree cells constructed at a depth of 628 km in the Western Mediterranen have a thickness of only 65 km. The *color contouring* quantifies the cell hit count, i.e. the number of rays passing through each cell. The *color coding* uses a 10-logarithm scale. The irregular cell grid is specifically designed to minimize hit-count differences between adjacent cells as well as possible. This is done in a self-adaptive way using a target hitcount of about 500 rays per cell in the upper mantle (see Spakman and Bijwaard 2001, for details). This explains why cells of different volume have similar hit count. The smallest cells used in the grid construction procedure have dimensions of 0.6 degrees therefore their hit count can exceed the target value

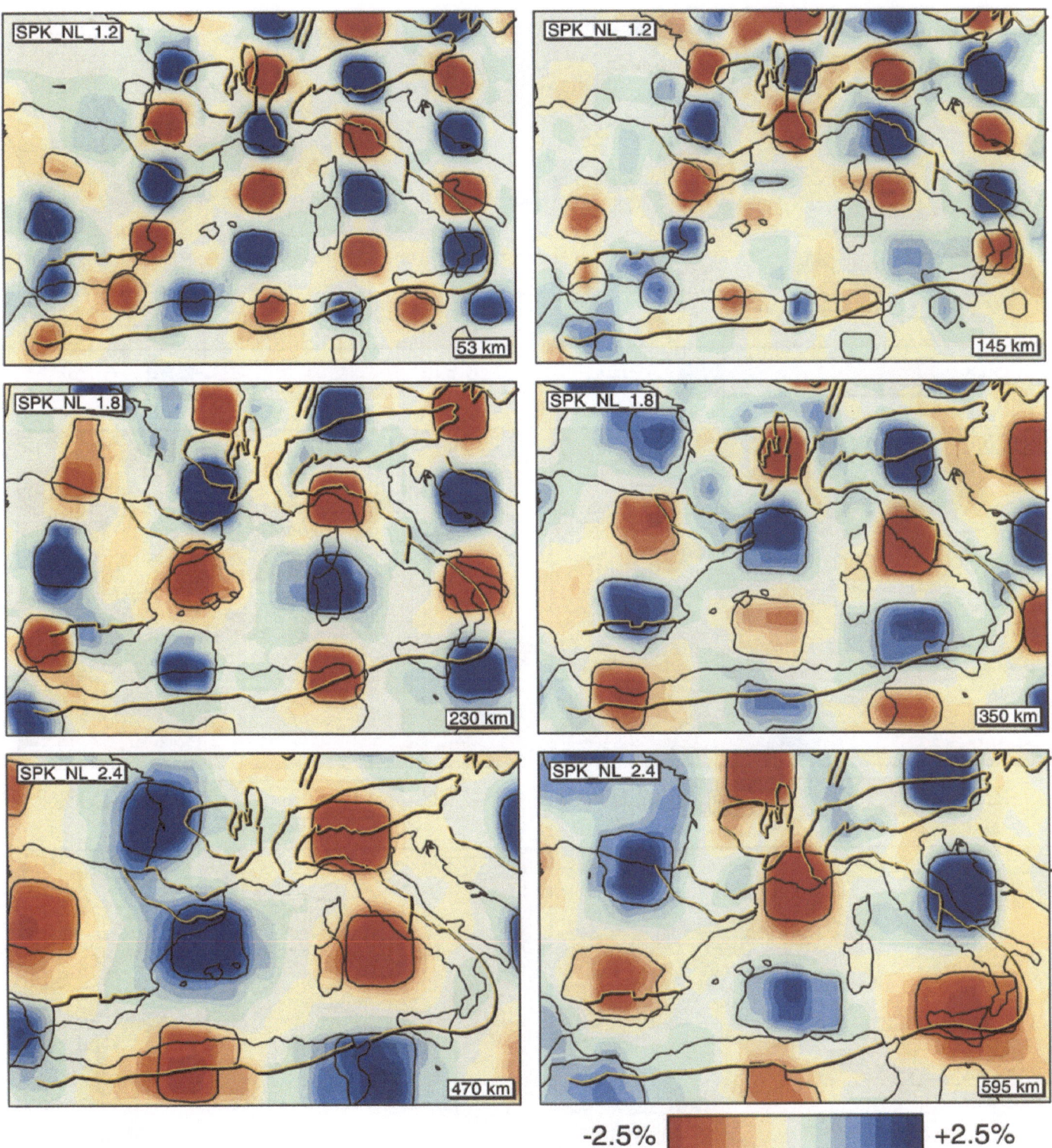

Fig. 2.3. Results of a sensitivity test for the recovery of a regular patterns of 1.2, 1.8, and 2.4 degree blocks at different depths in the upper mantle. *Circular lines* denote the location of the isolated synthetic blocks with a seismic wave-speed anomaly amplitude of +5% or –5% with respect to the 1-D reference model ak135 of Kennett et al. (1995). Between these blocks the synthetic anomaly is 0%. The *colors* denote the tomographic image of the synthetic block models. Comparison of "input" and "output" model leads to qualitative assessment of spatial resolution. The figure shows results for the Western Mediterranean taken from a global synthetic mantle model (model BS2000 is a *global* mantle model). The synthetic models are only constructed in regions where cell sizes of 1.2 (or 0.6), 1.8, or 2.4 degrees were permitted. This explains, e.g., the absence of synthetic blocks of 1.2 degree below the Western Mediterranean at 145 km depth. Lack of resolution can be detected where significant amplitudes occur between the blocks where block anomalies smear into the model. Generally the synthetic patterns are well recovered although with systematically smaller amplitudes than the input values

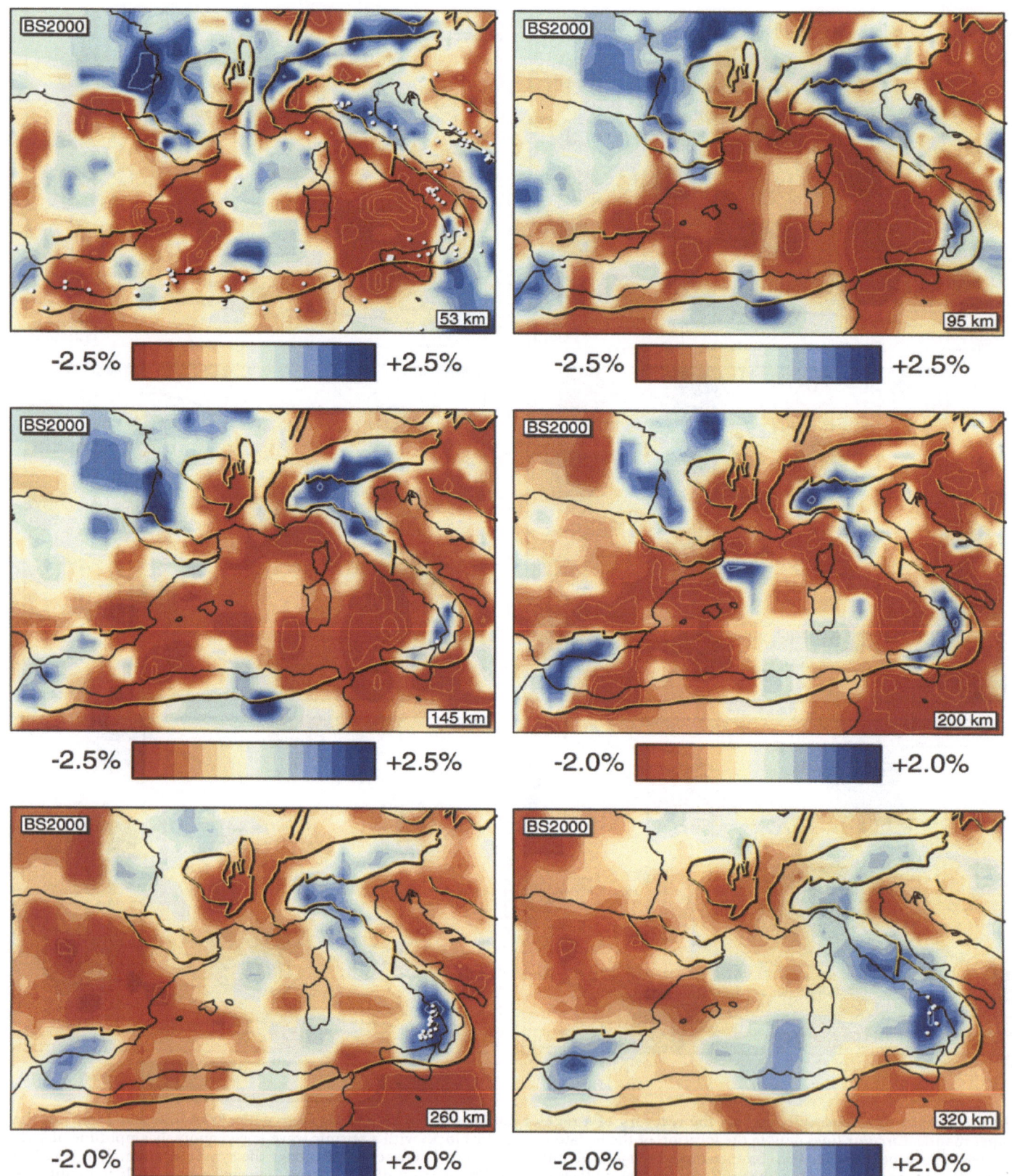

Fig. 2.4. Tomographic map view images at twelve selected depths for the upper 710 km of the Western Mediterranean and surrounding regions. *Colors* display the percentage deviation of seismic wave speed with respect to the 1-D reference model ak135 (Kennett et al. 1995). Negative (positive) anomalies represent slower (faster) than average wave speed at depth. Reference model values are different for each depth. Negative (positive) wave speed anomalies likely represent predominantly higher (lower) temperatures than average (Goes et al. 2000). Temperature anomalies can be as large as 10%–20%. In regions where imaged amplitudes are larger than the limits of the contouring scale, *additional line-contours* are plotted for every step in 1%-anomaly value. *Shaded yellow lines* indicate outlines of major tectonic features for reference. *Capital letters A, B, C, D,* at depths of 500 km, or larger, label individual anomaly patterns for discussion in the text

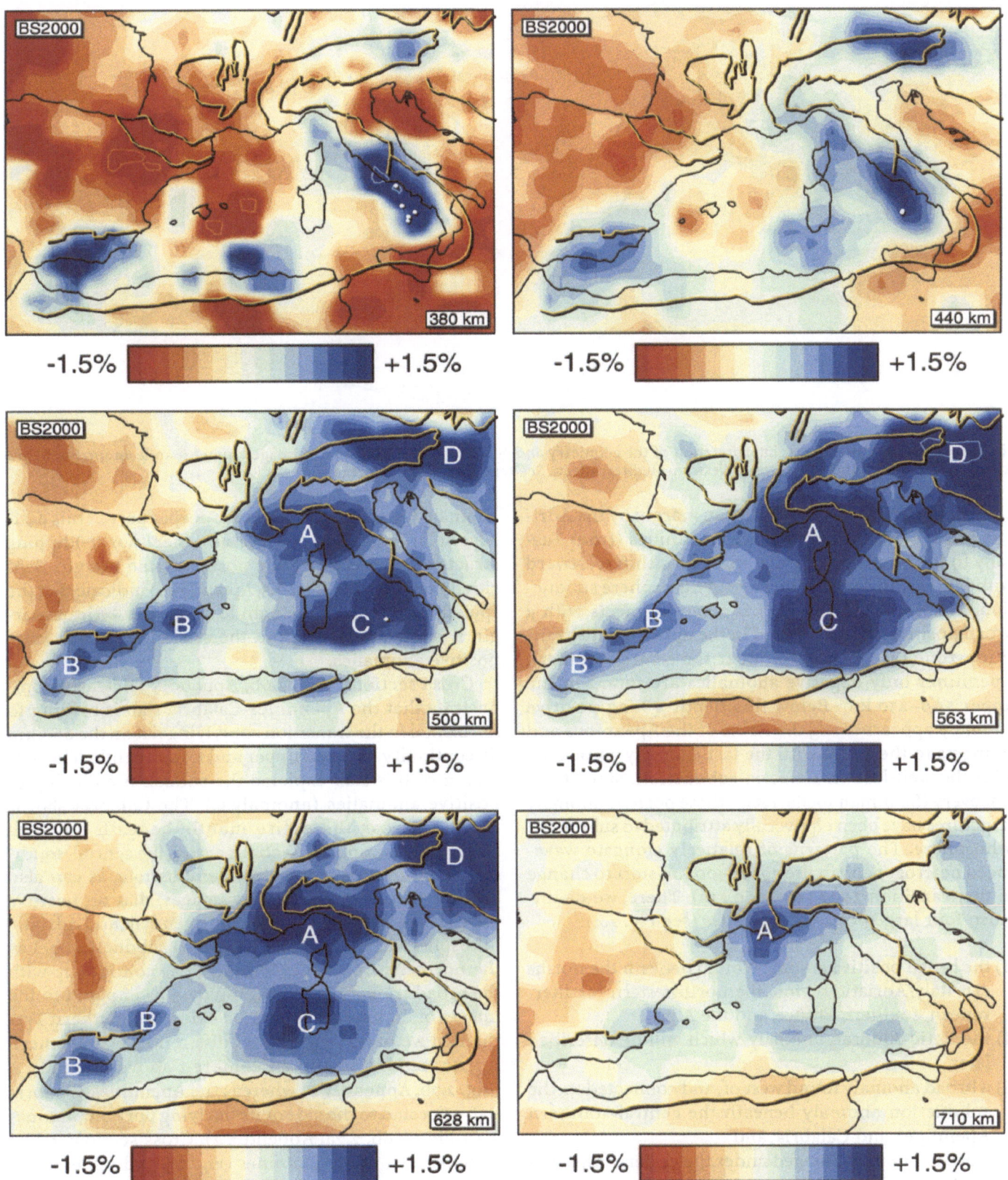

Fig. 2.4. *Continued*

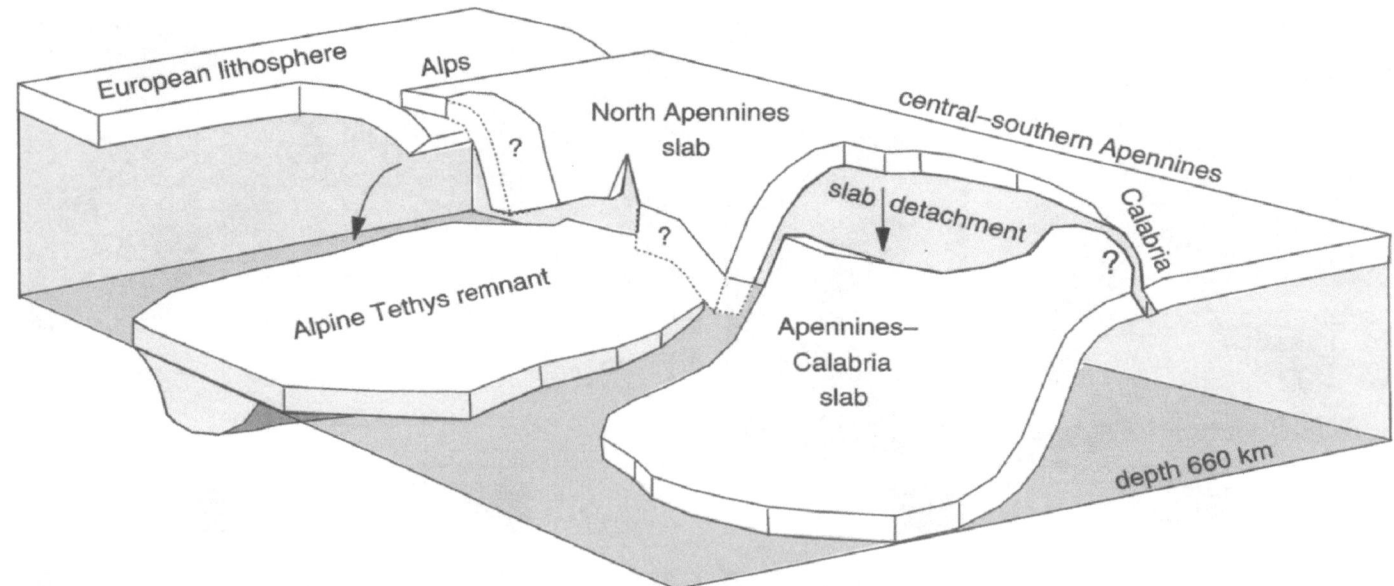

Fig. 2.5. Cartoon depicting our interpretation of the geometry and geodynamic significance of imaged upper mantle anomalies below the Alps, Apennines and the western Western-Mediterranean. See text for further information

ent with overriding of the European margin by the Adriatic plate. Beneath the northern Apennines (from Tuscany to the Po plain) positive anomalies shift westward with increasing depth, indicative of lithosphere subduction. Below 400 km this slab anomaly connects with a broad positive anomaly found in the upper mantle transition-zone (410–660 km). Under the central-southern Apennines only negative anomalies are found in the upper 200–250 km. Below this depth a long positive anomaly appears along the strike of the Apennines and connects to the south with the fast anomaly associated with the well-known Calabria subduction. Since their discovery (Spakman 1986a, 1990), these positive anomaly structures have been equivocally attributed to subducted lithosphere. The patterns of relatively elongate wave-speed heterogeneity in the upper 400 km start to change in the transition zone (410–660 km). There, we distinguish four large positive anomalies (see Fig. 2.4):

A) the broad positive anomaly found beneath the northern Italy/Adriatic region, the northwestern Mediterranean, southern France, and the Alps,
B) the Betic-Alboran anomaly which will be described later,
C) a broad anomaly found west of, and connected to, the subduction anomaly beneath the central-southern Apennines and Calabria, and
D) a broad anomaly imaged under the eastern-Alps and the Pannonian basin.

Note that anomaly A also attains significant amplitudes at the top of the lower mantle (710–810 km) while the other anomalies fade away. Under east Algeria we find a positive anomaly of varying amplitude which trends northward with increasing depth. In the transition zone,

anomaly C also encompasses the deeper part of this east-Algeria anomaly. We observe no other (north-)dipping structures connected to the north-African margin. Sensitivity tests results (Fig. 2.3, Appendix 1) indicate that the spatial resolution in the mantle under the African margin is sufficient to rule out the existence of such large positive anomalies.

Cross sections (Fig. 2.6b, Appendix 2) clearly demonstrate that the Apennines-Calabria slab is turning to horizontal in the transition zone, lying flat on the 660 km discontinuity between upper and lower mantle. The flattening of this slab explains the sudden broadening of positive anomalies (anomaly C). The Calabria slab is imaged across the entire mantle beneath Calabria (Fig. 2.6b). In contrast, sections across the central-southern Apennines (e.g. Fig. 2.7a) demonstrate, as can also be observed in the map view images, that no positive wave speed connection exists between slab and Adriatic lithosphere at the surface. Westward subduction below the northern Apennines is imaged as continuous with the broad transition zone anomaly A. Compared to the images of flattening subduction below the southern Tyrrhenian we observe a distinct difference: The northern Apennine slab is almost centered above anomaly A (Fig. 2.7b, Appendix 2) whereas the Apennines-Calabria slab is located to the side of its flat lying portion (anomaly C; Figs. 2.6b, 2.7a, Appendix 2). Cross sections taken along strike of the Apennines (Fig. 2.8 and Appendix 2) lend additional support for large structural differences between the Apennines-Calabria slab (to the right of the 8-degree tick mark), the northern Apennines slab (to the left of the 8-degree tick mark and found in the upper 300 km), and the broad anomaly A in the transition zone, suggesting that these anomalies result from (at least partly) a difference in geodynamic evolution.

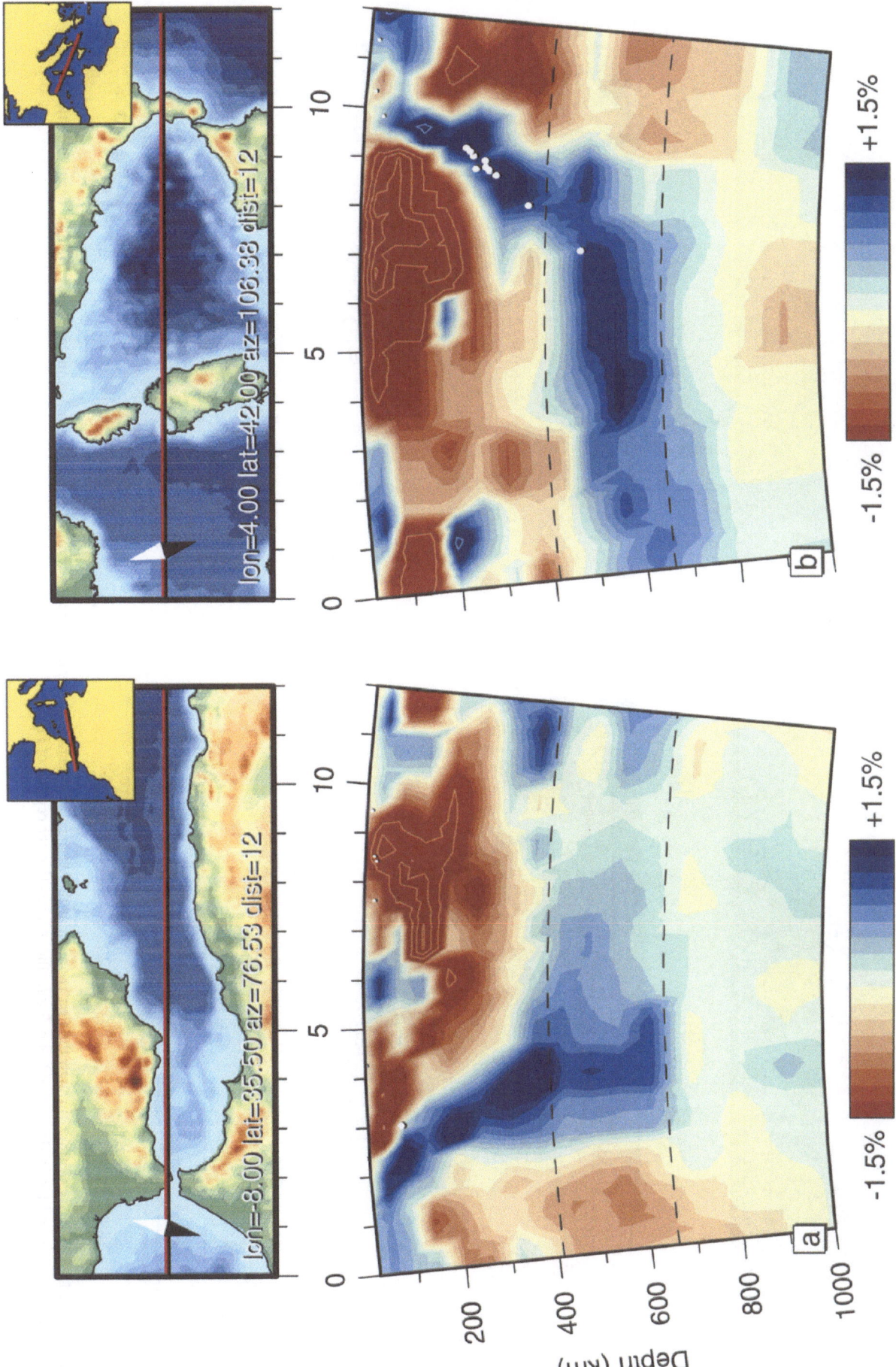

Fig. 2.6. Two BS2000 cross sections through the first 1 000 km of the Western Mediterranean mantle. **a** Section through the Betic-Alboran region and Algerian basin; **b** section through the Tyrrhenian mantle and Calabria. The sections are computed along a great circle segment indicated by a *straight red line* in the center of the map above each mantle section. *Great circle coordinates* are printed in the map. Lateral units are in degrees measured from the start of the section (*left*); 1 degree = 110 km; all dimensions are plotted to scale. *White dots* indicate major (magnitude > 4.8) earthquakes which occur within 25 km distance of the vertical section. The *diamond symbol* to the left in the map indicates a compass needle (white pointing north). The small map-inset shows a larger map of the region with the great circle segment indicated as a red line. For *color coding* (and *line contours*) of the tomographic image see caption of Fig. 2.4. *Dashed lines* in the section represent the mantle discontinuities at 410 and 660 km depth

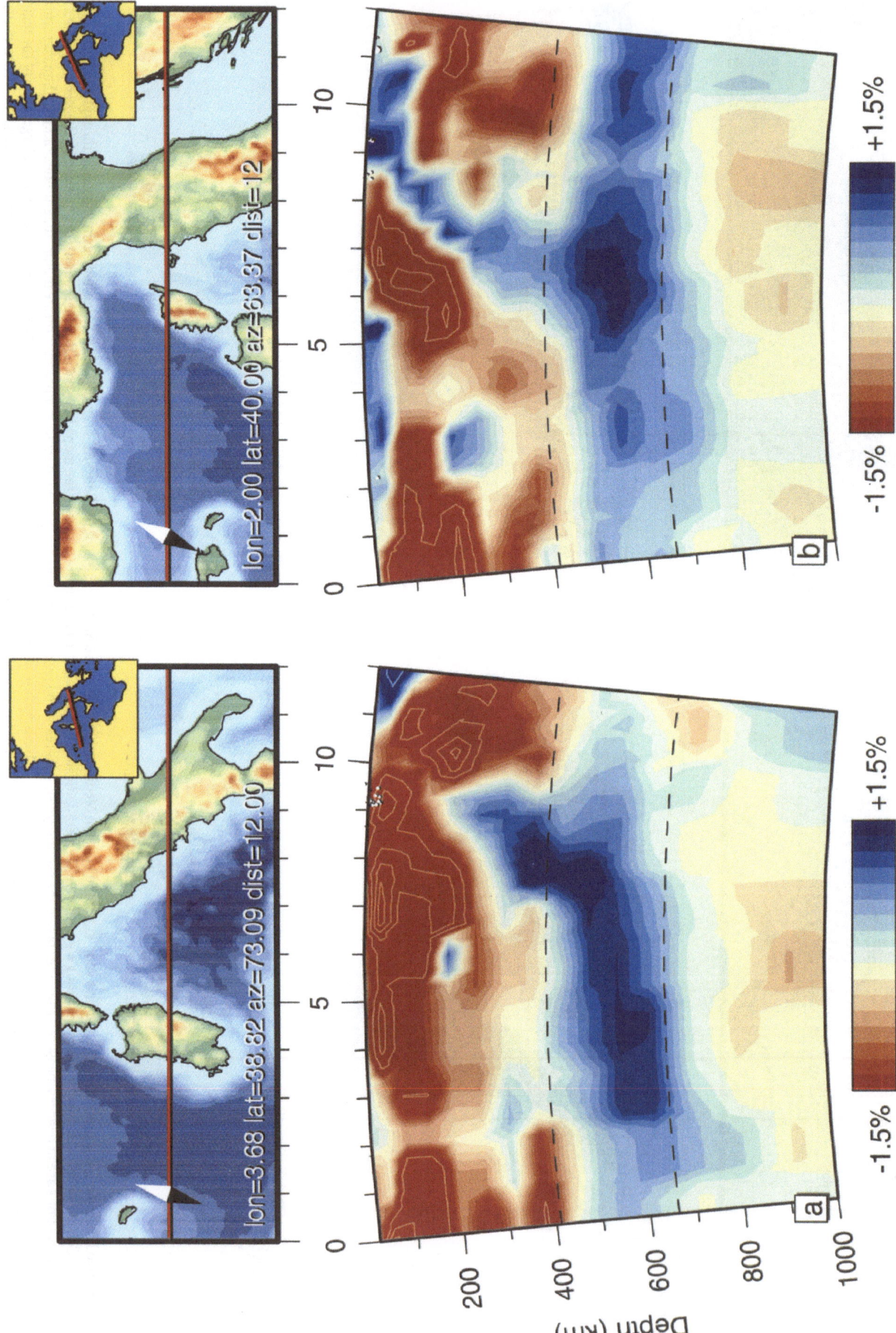

Fig. 2.7. Two BS2000 cross sections through the first 1000 km of the Western Mediterranean mantle. **a** Section through the southern Apennines and Tyrrhenian basin; **b** section through the northern Apennines and Liguro-Provençal basin. For further explanation see caption of Fig. 2.6

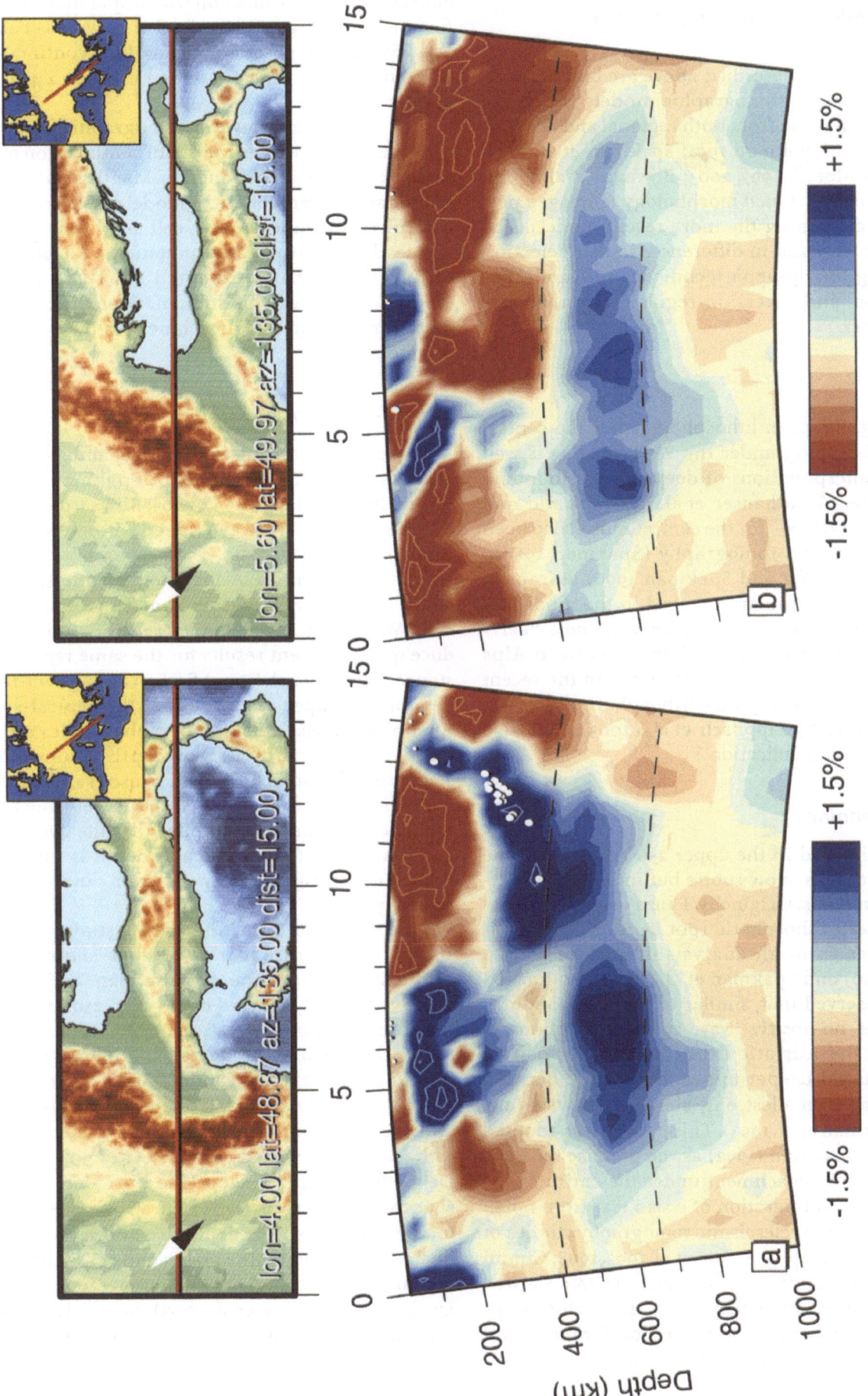

Fig. 2.8. Two BS2000 cross sections through the first 1000 km of the Western Mediterranean mantle taken along strike of the Italian peninsula. **a** Section through western Italy and the central Alps; **b** section through the Adriatic basin and the central-eastern Alps. For further explanation see caption of Fig. 2.6

Comparison with Other Tomographic Work

The general patterns described above are broadly consistent with previous tomographic models which are based on inverting data from both regional and teleseismic earthquakes (Spakman 1991; Spakman et al. 1993; Piromallo and Morelli 1997, 2003; Bijwaard et al. 1998). However, the more detailed morphology of imaged patterns still differs among the more recent models. The reason should be found in differences in data selection and processing, tomographic techniques and model parameterization, and inversion regularization, as discussed hereafter.

Alps

The S-dipping European lithosphere below the central Alps and ESE-dipping under the Western Alps is well known from interpretations of deep seismic sounding and geology (e.g Waldhauser et al. 1998; Schmid and Kissling 2000) and has – for the central Alps – also been inferred from mantle tomography (Spakman 1986b; Kissling 1993; Kissling and Spakman 1996). European plate geometry was similarly imaged in the models of Piromallo and Morelli (1997, 2003). We also find a clearly S-dipping European margin below the eastern Alps (Fig. 2.8b, Appendix 2) which contrasts with the recent proposition of north-dipping lithosphere along the TransAlp transect by Lippitsch et al. (2003). (See also Transect VI of this publication.)

Northern Apennines

The patterns imaged in the upper 250 km beneath the northern Apennines show subtle but important differences among models. Originally, Panza et al. (1980) discovered the deep lithospheric root beneath the northern Apennines from an analysis of surface waves. Spakman (1990) and Spakman et al. (1993) also imaged a root and observed that, similar to the central-southern Apennines, no positive wave speed connection exists between the Adriatic lithosphere and positive anomalies imaged deeper under this region. Beneath Tuscany the gap is smallest, only about 50 km wide (e.g. Fig. 16 of Spakman et al. 1993). This pattern was adopted by Wortel and Spakman (1992) as the premise for their hypothesis of slab detachment under the entire Apennines and of lateral migration of a slab tear from north to south. Subsequent teleseismic tomography studies of the Italian peninsula (e.g. Amato et al. 1993, 1998; Cimini and De Gori 1997; Lucente et al. 1999), however, showed continuous positive anomalies under the northern Apennines. The regional mantle models of Piromallo and Morelli (1997, 2003) exhibit local evidence for a continuous slab at the same location where Spakman et al. (1993) find the smallest gap. The model of Bijwaard et al. (1998) and BS2000 also possess evidence for a continuous positive anomaly in the upper 200 km (Figs. 2.4, 2.7b, Appendix 2).

The differences between tomographic models concern the 50-km detail in the depth continuation of north-Adriatic lithosphere subduction. Mantle structure of this scale is at the current limit of model resolution and, even if resolved in some models, cell amplitude errors may still lead to equivocal interpretations. All these factors of uncertainty have prompted Wortel and Spakman (2000, 2001) to put a question mark on their earlier suggestion regarding slab detachment in the northern Apennines.

Central-southern Apennines

A very consistent feature of regional and global tomography studies (Spakman 1991; Spakman et al. 1993; Piromallo and Morelli 1997, 2003; Bijwaard et al. 1998) is a zone of negative wave-speed anomalies which is imaged in the top 200–250 km under the central-southern Apennines. A local teleseismic-tomography study (Cimini and De Gori 2001) corroborate this result. We note, however, that teleseismic studies can produce quite different results for the same region. For instance, Lucente et al. (1999) find on average zero anomalies for the upper 170 km under the central-southern Apennines. Teleseismic tomography studies are intrinsically hampered by a poor depth resolution (only steeply emergent seismic rays are used) and, for the Italian region, by the fact that *relative* travel-time residuals are inverted. The latter implies that the average wave-speed anomaly for a particular depth is invisible, and thus a low or high wave-speed layer cannot be imaged properly.

Overall, results from independent studies mostly lend support for the case that no positive-anomaly connection exists between the Adriatic plate and the deeper positive anomaly imaged below 200–250 km. Proposed interpretations of this feature include (*i*) a slab window (absence of slab or subducted continental lithosphere; Amato et al. 1993); (*ii*) subduction of a promontory of continental lithosphere (Lucente et al. 1999; Lucente and Speranza 2001); (*iii*) lithosphere attenuated by a hot asthenospheric wedge (Amato et al. 1993; Cimini and De Gori, 2001); (*iv*) deep (250 km) slab detachment (Lucente and Speranza 2001), in which case the low velocities above 250 km depth are still assigned to subducted continental lithosphere; and (*v*) shallow slab detachment (Wortel and Spakman 1992, 2000; Cimini and De Gori 2001; Panza et al. 2003), in which case the low wave-speed anomaly represents mobile (asthenospheric) mantle.

Calabria

The Calabria slab is imaged across the entire upper mantle in regional/global mantle models (e.g. Spakman et al. 1993; Amato et al. 1993; Cimini and De Gori 1997; Piromallo and Morelli 1997, 2003; Bijwaard et al. 1998; BS2000) and detailed local tomography models (e.g. Selvaggi and Chiarabba 1995). Compared to the studies of the eighties and early nineties, the broad positive anomalies in the transition zone, particularly the flat lying portion of the Calabria slab (anomaly C in Fig. 2.4), are imaged more clearly in the recent tomographic models. These flat anomalies, or parts of them, in the transition zone of the western Mediterranean were already present in the models of Cimini and De Gori (1997), Piromallo and Morelli (1997, 2003), and Lucente et al. (1999) although with a different morphology and depth extent as compared to our models BSE (Bijwaard et al. 1998) and BS2000.

2.3.2 The Betic-Rif and Alboran Region

Beneath the Betic-Rif and Alboran region a positive anomaly is found from the base of the crust across the entire upper mantle (Fig. 2.4). The deeper part of the anomaly extends more to the ENE of the Alboran region. At the base of the upper mantle it underlies a large part of the east Iberian margin and the Valencia basin. Figure 2.6a displays a W-E cross section through the anomaly which clearly shows an eastward dip and confinement of the anomaly to the upper mantle. Cross sections with a more N-S orientation exhibit no dip. Figures of Appendix 2 allow to follow this anomaly along many W-E directed slices.

Previous tomographic studies, based on different tomographic approaches and data sets, have already imaged a positive anomaly in the upper mantle of the Betic-Rif/Alboran region. Blanco and Spakman (1993) and Spakman et al. (1993) find the anomaly from a depth of 200 km downward, whereas Seber et al. (1996), Bijwaard et al. (1998), and Calvert et al. (2000) image the anomaly locally from sub-Moho depth downward. The latter authors prefer a division into two separate bodies which however are not imaged in BS2000. Piromallo and Morelli (1997, 2003) find a positive anomaly connection below northern Morocco, but most of the positive anomaly is overlain by low wave-speed anomalies.

Figure 2.9 shows a 3-D cartoon schematically depicting the geometry of the positive anomaly in model BS2000. Several options exist for its interpretation, each associated with a different geodynamic process: (*i*) delamination of the lithospheric mantle (Seber et al. 1996; Calvert et al. 2000), (*ii*) removal of thickened continental lithosphere (Platt and Vissers 1989), and (*iii*) subducted lithosphere (Blanco and Spakman 1993; Spakman et al. 1993). Gutscher et al. (2002) combined the positive anomaly of model BS2000 with marine-seismic observations of a deforming fore-arc west of Gibraltar. They propose a still active eastward-dipping subduction system involving a continuous slab. Blanco and Spakman (1993) and Spakman et al. (1993) proposed an interpretation of a completely detached slab because only low velocities were imaged above 200 km in their models. The recent tomography results, which are based on much more accurate data, weaken the basis for their detachment interpretation, at least where complete slab break-off is concerned.

Fig. 2.9.
Cartoon displaying our interpretation of the 3-D geometry of imaged structure below the Betic-Rif-Alboran region. For more explanation see text

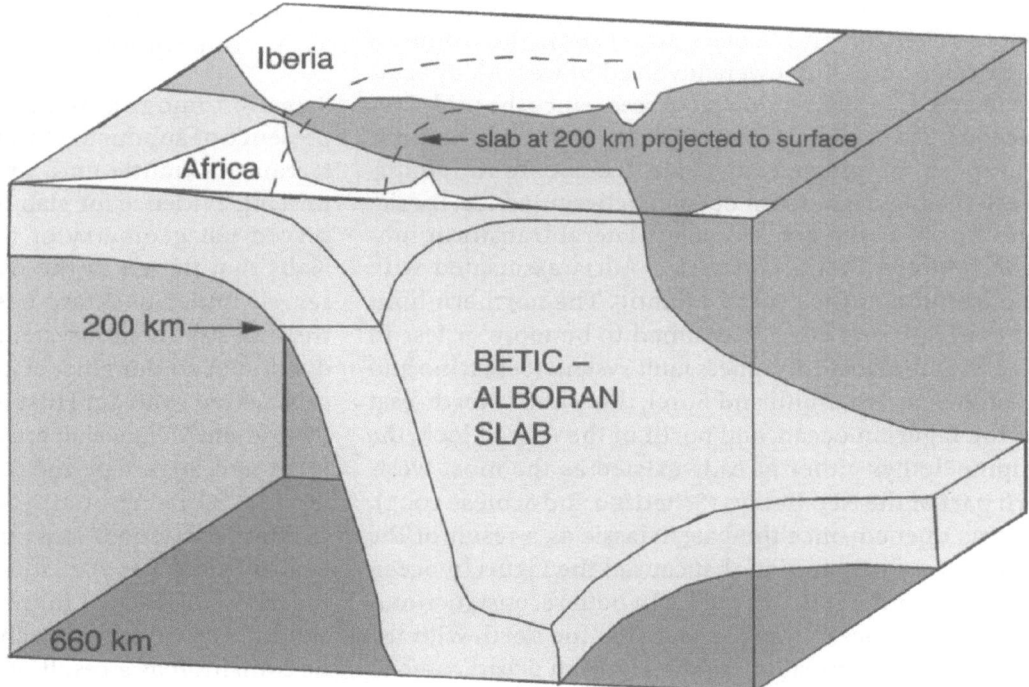

2.4   Analysis: The Geodynamic Evolution of the Western Mediterranean

Several detailed reconstructions of the tectonic evolution of the Western Mediterranean s.l. have been published which are based on interpretations of geology, magnetic anomalies, and marine seismics in the overall context of Africa-Europe convergence (e.g. Dercourt et al. 1986; Dewey et al. 1989; Lonergan and White 1997; Gueguen et al. 1998; Jolivet and Faccenna 2000; Gelabert et al. 2002; Rosenbaum et al. 2002a; Frizon de Lamotte et al. 2000; Mantovani et al. 2002; Cavazza and Wezel 2003). Also attempts were made to combine tectonic reconstructions with inferences made from seismic tomography (e.g. Wortel and Spakman 1992; De Jonge et al. 1994; Carminati et al. 1998; Wortel and Spakman 2000; Faccenna et al. 2001a, 2001b, 2003). These kinematic and geodynamic models differ considerably in detail, basically because the scarcity of data allows for degrees of freedom in their interpretation. But they all share the notion that slab roll-back is invoked as the most prominent process for reshaping the region in the past 25–30 Ma. Roll-back started in the northwest, along the Oligocene Iberianmargin, and progressed outward to the southwest, south, and southeast. As a result the Liguro-Provençal basin was opened in the central Western Mediterranean, the Alboran-Algerian basin in the south and, as a second phase, the Tyrrhenian basin in the southeast (see Transect III, this publication). Details of the roll-back evolution are still poorly known.

Most of the above cited kinematic and geodynamic models focus on the Cenozoic evolution. Reconstructions of the pre-Cenozoic evolution (e.g Stampfli and Borel 2002; Stampfli et al. 2002; Stampfli and Borel, this publ.; Schettino and Scotese 2002) show the origin of the oceanic areas that were involved in west Mediterranean and Alps subduction processess since the mid-Cretaceous. Two small oceans bordered the south Iberian-European margin since the Late Triassic. To the southwest, the Ligurian ocean opened between Africa, Iberia, and Adria, as the result of a left-lateral transform motion between Iberia and Africa-Adria associated with the opening of the central Atlantic. The northern limit of the Ligurian ocean is assumed to be more or less in line with the North-Pyrenees fault system (Schettino and Scotese 2002; Stampfli and Borel, this publ.). North-east of the Ligurian ocean, and north of the Adria block, the Alpine Tethys either already existed as the most western part of the Neo-Tethys (Schettino and Scotese 2002), or was opened since the Late Triassic as a result of the same transform motion that caused the Ligurian ocean (Stampfli and Borel, this publ.). In both reconstructions, the Ligurian ocean was connected in the north with the Alpine Tethys. Spreading in the Ligurian ocean came to a halt during the Early/mid-Cretaceous and the basin

became incorporated as a passive embayment in the Iberia-Africa-Adria domain. Since the mid-Cretaceous, this entire domain experienced a counterclockwise rotation with respect to central Europe associated with the opening of the South Atlantic (e.g. Stampfli and Borel, this publ.). As a result of this motion the western Alpine-Tethys (Piemont/Valais ocean) was overrriden by the Adria block and perhaps also by the northern part of the Ligurian ocean. This convergence culminated in the closure of the Pyrenees-Valais basin and western/central Alps orogeny during the Eocene-Oligocene period when Adria collided with Europe.

Destruction of the western Alpine-Tethys effectively locked the Ligurian ocean between Africa, Iberia, central Europe, and Adria. Continuing convergence between the African and European plates may have caused the onset of NW-directed subduction of the Ligurian ocean along the east Iberian margin. According to some studies, initiation of this subduction system may have commenced earlier (Late Creteaceous; Schettino and Scotese 2002; Faccenna et al. 2001a) whereas many other reconstructions assume initiation of subduction during the Tertiary. The latter timing is related to the Late Oligocene opening of the Valencia basin. Because the Ligurian ocean got trapped (land-locked) during the Eocene between the slowly converging African and European plates, roll-back of the gravitationally unstable Ligurian ocean lithosphere eventually took over as the dominant mode of subduction (Le Pichon 1982; Wortel and Spakman 1992, 2000; Jolivet and Faccenna 2000). In the following, the remnants of subducted lithosphere found in the mantle under the Western Mediterranean and Alps region will be identified as parts of the Ligurian ocean and of West Alpine-Tethys.

2.4.1   Tomographic Evidence for Slab Roll-back

Seismic tomography has basically substantiated the presence of subducted lithosphere in the Western Mediterranean mantle and, importantly, has provided the missing evidence for slab roll-back by means of the observed flat geometry of the Apennines-Calabria slab. Slabs that flatten in the transition zone have been observed under back-arc basin behind several other retreating subduction systems such as the Izu-Bonin subduction (Van der Hilst et al. 1991), the Tonga-Kermadec subduction (Van der Hilst 1995; Bijwaard et al. 1998), and behind the Melanesian arc in the region east of Australia (Hall and Spakman 2002). Convection modeling studies (e.g. Olbertz et al. 1997; Christensen 1995, 2001; Cizkova et al. 2002) and tank experiments (e.g. Griffitts et al. 1995; Becker et al. 1999) conclusively demonstrated that slab roll-back of more than a few cm yr^{-1} may cause subducted slab to flatten above the upper-to-lower mantle transition as a result of encountering (initial) resistance against lower-mantle penetration by the slab.

2.4.2 Northern Apennines and Alpine-Tethys Subduction

Below the Po plain and the northernmost Apennines, the southward underthrusting Eurasian plate meets with the westward dipping north-Apennines slab. The European margin may reach a depth of 200 km but is tomographically difficult to distinguish from the Adriatic plate under the Po plain. However, for the steeply dipping anomalies reaching a depth of 300 km beneath this region, close inspection of cross sections (Appendix 2) suggests a geometry of subducted/overridden Ligurian-Adriatic lithosphere which is strongly curving to the west below the Po plain and continues up to the Western Alps. In our cartoon (Fig. 2.5) we indicate this interpretation with a question mark.

Slightly to the south-east, Lucente et al. (1999) and Lucente and Speranza (2001) propose the presence of a long (700 km) westward subducted slab below the northern Apennines (Tuscany). The Oligocene-to-present extension in the northern Liguro-Provençal basin amounts to only about 250 km (Faccenna et al. 2001a) and thus cannot explain the inferred slab length. Faccenna et al. (2001a) combine these inferences with the geological development of the northeastern Iberian margin and propose that initiation of Ligurian ocean subduction dates back to about 80 Myr. The tomography results presented here allow for different interpretations of the positive anomalies in the northern Apennines mantle. As noted in the first part of Sect. 2.3.1, the top-400 km of the north-Apennines slab is nearly centered above the flat lying transition-zone anomaly A (Fig. 2.4, 500 km depth). This peculiar geometry is well resolved (Appendix 1) and may suggest a lithosphere slab plunging into, and "feeding", a broad transition-zone anomaly where lithosphere material spreads in many directions. For several reasons, this interpretion is rather unrealistic. First, the rheological feasibility of such lateral flow in the transition zone is highly questionable. Secondly, the amount of subduction implied by the geometry of the slab and transition zone anomaly is much larger than 1 000 km which by far exceeds the E-W extent of the northern part of the Ligurian ocean between Iberia and continental Adria in tectonic reconstructions of the past 100 Myr (e.g. Stampfi and Borel, this publ.; Schettino and Scotese 2002).

An alternative explanation, which we prefer, is that the slab anomaly above the transition zone is actually unrelated to transition zone anomaly A described above. This interpretation allows for eastward overriding of the Ligurian-Adriatic lithosphere by only a few hundred km corresponding to a similar amount of opening of the northern Liguro-Provençal basin. We propose that anomaly A represents the remnant of the West Alpine-Tethys (or Piedmont ocean) and thus – as introduced earlier – results from S-dipping subduction below the Ligurian-

Adria domain created by northward overriding of the Piedmont ocean by the Ligurian-Adria domain. This interpretation is consistent with the fact that anomaly A is partly found in the lower mantle which suggests a longer history than – for instance – the Apennines-Calabria subduction. Oligocene detachment of the West Alpine-Tethys slab under the Alps (Davis and Von Blanckenburg 1995) allowed it to sink further into the upper mantle and drape itself on the 660 km discontinuity. We note that the geometry of flat lying slabs can also result from a situation in which an oceanic basin is being actively overriden. We estimate the NS extent of Piedmont ocean to be between 500 and 600 km. Its EW extent has been a few hundred km larger. This interpretation of anomaly A implies that Europe has hardly moved with respect to the mantle during the second half of the Cenozoic and that flat-lying remnants of subducted slab can reside for 30 Myr or more in the upper mantle transition-zone. According to its position in the mantle, the Piedmont ocean may have existed in the region east of the Pyrenees up to and including the central Alps. We remark that remnants of subducted lithosphere with a coeval or even older subduction history have also been identified elsewhere in the mantle (e.g. Van der Hilst et al. 1997; Bijwaard et al. 1998; Van der Voo et al. 1999a,b; Hall and Spakman 2002).

In line with our interpretation of anomaly A, we interpret anomaly D (Fig. 2.4) as the remnant of the east Alpine-Tethys of which the subduction history is quite different from the Piedmont ocean and specifically related to the evolution of the Austro-Alpine-Carpathian-Pannonian domain (e.g. Stampfli and Borel, this publ.). Last remnants of the East Alpine-Tethys were subducted during the Neogene (e.g. Wortel and Spakman 2000).

2.4.3 Slab Detachment beneath the Central-southern Apennines

Following our earlier propositions (Wortel and Spakman 1992, 2000), we interpret the absence of cold (seismically fast) Adriatic lithosphere beneath the central and southern Apennines as a result of slab detachment. The lower seismic velocities imaged in the detachment zone result from inflow of hot mobile asthenosphere. Other interpretations have as common factor the subduction of the Adriatic continental lithosphere, or more specifically, the subduction of a promontory of continental lithosphere (Lucente et al. 1999; Lucente and Speranza 2001). We see, however, no compelling reasons why such a promontory would exist or why subducted Adriatic continental/transitional lithosphere beneath the south-central Apennines should image as a slow wave-speed anomaly. In the north, continental Adriatic lithosphere images as fast at the surface and as a fast slab anomaly. Similarly, the overriden European continental lithosphere below the Alps is seismically fast.

The detachment hypothesis provides many testable predictions for surface development above the detachment zone and thus can be used as a working hypothesis to understand geological evolution (Wortel and Spakman 1992, 2000; Davies and Von Blanckenburg 1995). These predictions concern for example vertical motions, and sedimentary depocenter formation and migration in central-southern Italy for which laterally migrating slab detachment provides a reasonable explanation (Van der Meulen et al. 1998, 2000). Furthermore, numerical modelling (Yoshioka and Wortel 1995; Wong a Ton and Wortel 1997) demonstrates that slab detachment and its lateral migration is a feasible process, particularly, in the late stage of subduction when continental lithosphere enters the trench. Modelling results by Van der Zedde and Wortel (2001) show that slab detachment can occur at shallow levels (e.g. Moho depth) and that inflow of hot asthenosphere and subduction wedge material can raise the geotherm considerably (creating slow seismic anomalies) to the point that the mantle and lower crust can start to melt. As noted by Wortel and Spakman (1992, 2000) and Davies and Von Blanckenburg (1995), slab detachment may cause a specific change in character of the geochemistry of volcano products from more subduction-related calc-alkaline to more alkaline (potassium enriched). Because of the geochemical complexity of volcanic rock in the region (e.g. Serri et al. 2001) interpretation is still controversial. For example, Doglioni et al. (1999) relate the high-potassium content to the subduction of Adriatic continental lithosphere (see Transect III, this publ.).

2.4.4 Calabria Subduction

The detachment gap becomes smaller toward the Calabria arc. Whether the Calabria slab is still attached to the Ionian basin (Neo-Tethys) lithosphere is questionable. Although tomographic mantle models mostly show a continuous slab up to the crust, none of these models possesses the spatial resolution to exclude a small detachment gap as would result from shallow and recent (e.g., past million years) slab detachment. Indicative for a continuous slab would be progressing slab roll-back which can be expected because Ionian oceanic lithosphere is still found east of the trench. Wortel and Spakman (2000) argue on the basis of space geodetic observations of crustal motion that at present evidence for strong roll-back is absent. These observations show a distinct contrast between the speed of Aegean roll-back (about 3 cm yr^{-1}, Noomen et al. 1996; McClusky et al. 2000) and southern Apennines-Calabria-Sicily motion which more closely follows the motion of Africa (< 1 cm yr^{-1}) relative to the European plate (Noomen et al. 1996; Oldow et al. 2002). The observed small ($\ll 1$ cm yr^{-1}) eastward component of southern Apen-

nines motion (Oldow et al. 2002) cannot be uniquely attributed to roll back since it can also be explained by extrusion of a crustal block, as a result of continuing Africa-Europe convergence, above a detached-slab setting.

2.4.5 The North African Margin

Below northern Sicily we find the free edge of the Apennines-Calabria slab. Tomography does not show subducted slab beneath the African margin except for a zone beneath east Algeria. This corresponds to the location where the Kabylies (derived from the Iberian margin) accreted around 15 Ma ago with the African margin (e.g. Frizon de Lamotte et al. 2000). Carminati et al. (1998) proposed that vertical slab tearing, east of the Kabylies, initiated the last phase of slab roll-back which led to the opening of the Tyrrhenian basin. Detachment tearing of the slab along the African margin may have turned into a lithosphere (surface) tear-fault along the African margin propagating toward Sicily (Carminati et al. 1998; Wortel and Spakman 2000). Faccenna et al. (2001b) observed in tank experiments an acceleration of slab roll back after the slab starts to interact with the 660 km discontinuity. They propose that this interaction can explain the accelerated opening of the Tyrrhenian basin after an apparent stalling of the roll-back process. A combination of slab interaction with the 660 km discontinuity and slab rupture processes should be considered possible. Eastward migration of slab detachment[1] and lithosphere tearing explains why tomography shows no north-dipping slab between east Algeria and Sicily. This development along the African margin is in line with the original suggestion made by Lonergan and White (1997): they propose that southward roll back of Ligurian ocean toward the African margin came to a halt in the middle Miocene after which plate rupture initiated below the African margin and progressed in two directions: to the east facilitating slab roll-back toward present-day Calabria (creating the Tyrrhenian basin), and to the west facilitating slab roll back toward Gibraltar (creating the Alboran-Algerian back-arc basin) (Fig. 2.11).

Westward migrating slab-detachment below Algeria is proposed by Coulon et al. (2002) who observe a distinct change from calc-alkaline subduction-related

---

[1] With slab detachment we mean lateral tearing of the subducted slab beneath depths of say 40–50 km along a more or less horizontal fault plane cutting the slab (see Wortel and Spakman 2000). Lithosphere tearing cuts the surface and propagates along a more vertical fault plane. The horizontal slab-detachment fault can turn into a vertical lithosphere fault and viceversa. This depends for instance on the development of the angle between the strike of the roll-back system and the local strike of the continental margin: perpendicular strikes would favor lithosphere tearing whereas a parallel strike would favor detachment.

arc volcanism (12–9 Ma) to more alkaline volcanism (4–0.8 Ma) with a distinct asthenospheric/plume contribution. Slab detachment along the entire African margin is in accord with the observation that this margin is under compression since the Pleistocene as a result of continuing Africa-Europe convergence (Frizon de Lamotte et al. 2000; Bracene and Frizon de Lamotte 2002). All of these observations are consistent with the fact that no north-dipping slab is found under the African margin (except locally below the east Algerian margin). Importantly, all geological interpretations invoke the presence of lithosphere slab below this margin at some period in the past. The subducted lithosphere, however, is now found at the base of the upper mantle and as dipping slabs at either extremity of the present Western Mediterranean.

2.4.6 Betic-Rif and Alboran Region: I. Subduction and Roll-back of Predominantly Oceanic Lithosphere

In line with Gutscher et al. (2002) we prefer to explain the Betic-Alboran BS2000 anomaly by subduction of (mostly) oceanic lithosphere. The alternatives of convective removal of thickened lithosphere and of delamination of the continental lithospheric mantle are attractive processes, and perhaps may have contributed to the mantle anomaly, but fail to explain the origin of the largely oceanic Neogene Alboran-Algerian basin. More promising in this respect is a westward roll-back model (Royden 1993; Lonergan and White 1997) in which the Alboran-Algerian basin can develop as a back-arc basin (which may include thinning of the continental Alboran microplate). Roll-back should have involved predominantly oceanic lithosphere because lithosphere delamination (or a convective removal mechanism) would have left at least the upper crust at the surface in the entire Alboran-Algeria basin, which is not observed. A slab retreating southward toward the African margin would provide an alternative for creating the Algerian basin. However, no north-dipping slab is observed with tomography. Instead the prime candidate to fuel regional geodynamics is found below the Betic-Rif region, which in fact implies that a northward dipping slab, creating the Algerian basin, may have eventually rolled westward.

For the sake of clarity we emphasize that, as for the Apennines subduction systems, the Betic-Alboran slab does not result from west-east convergence between plates but should be viewed as unstable lithosphere sinking *passively* into the mantle (with possible south- and westward components of retreat) and being replaced at the surface by a back-arc basin which may contain continental fragments (Alboran microplate). Delamination of continental mantle (Seber et al. 1996; Calvert et al. 2000)

may start to get involved in the subduction process in its advanced stage when slab roll-back has reached the present Betic-Rif region. This depends on the distribution of oceanic and continental lithosphere in the region between Africa and Iberia. Recent paleogeographic reconstructions (Rosenbaum et al. 2002b; Schettino and Scotese 2002; Stampfli and Borel, this publ.) all suggest that, since about 150 Myr, at least a small stripe (less than 200 km) of Ligurian ocean lithosphere separated Iberia and Africa until it became involved in the Betic-Alboran subduction. Still, at present continental lithosphere may also be involved by delamination in the roll-back process.

Many arguments for slab roll-back in the Alboran region are given by Lonergan and White (1997) and will not be repeated here. We like to add the following inferences which are all in line with a model of WSW-directed slab retreat and consistent with observed mantle structure:

1. The conspicuous arcuate geometry of the slab anomaly in the upper few hundred kilometers (particularly around a depth of 200 km) is quite similar to that of the retreating Calabria slab.
2. Platt et al. (2003) demonstrate that, in the Middle Miocene, the Alboran domain was at least about 250 km to the ESE of its present position with respect to Iberia and 200 km to the ENE of its present position with respect to the African margin. They suggest that convergence of the Alboran domain with Iberia and Africa results from lithosphere roll-back in a westerly direction (coeval with Africa-Iberia convergence in a roughly NNW direction). The WNW convergence of the Alboran domain with Iberia proves to be about perpendicular to the arcuate geometry of the slab anomaly imaged at a depth of 200 km (Fig. 2.4).
3. Duggen et al. (2003) combine the evidence for an east-dipping Betic-Alboran slab (Gutscher et al. 2002) with the spatial-temporal evolution of calc-alkaline to alkaline volcanism in the Betic-Rif and Alboran regions. They propose to explain their observations by a model of westward slab retreat which invokes lithosphere rupture along the north African and Betic-Balearic margin allowing for inflow of asthenospheric mantle in the lithosphere tear-zones.
4. From detailed marine-seismic observations and the geology of the Balearic margin, Acosta et al. (2002) infer southeast migration of the Alboran microplate during the Late Oligocene and Miocene and opening of the Alboran-Algeria basin behind it, while in the north the Valencia basin is opening and the Balearic Islands undergo clockwise rotation. Also, these authors report a transition from calc-alkaline to alkaline volcanism pointing at early slab detachment or southwestward lithosphere tearing along the Balearic margin.

Tectonic reconstructions place Betic-Alboran fragments in the Oligocene Balearic margin (e.g. Lonergan and White 1997; Rosenbaum et al. 2002a; Stampfli and Borel, this publ.) implying at least 400 km of westward transport to the present Betic-Rif system. The Early Miocene orogenic collapse event of Platt and Vissers (1989) occurred along the Balearic margin where it evolved in a tectonic environment of WSW-directed slab roll-back which was likely accommodated by SW-ward propagating slab detachment or lithosphere tearing. Although more research is needed, lithosphere tearing may cause considerable subsidence (juxtaposed to uplift at the other side of the lithosphere tear) and may allow for exhumation of mantle rock (Ronda peridotite) along the lithosphere tear (Duggen et al. 2003). These observations along the Balearic margin are consistent with the absence of a north (or south) dipping slab under this margin, and with the observed east-dipping Betic-Alboran slab found to the southwest.

2.4.7 Betic-Rif and Alboran Region: II. Development of Arc Geometry and Subduction Roll-back

The shape of the Betic-Alboran slab, e.g. at a depth of 200 km, correlates quite conspicuously with the asymmetry in arcuate shape between the Betic and Rif orogens – the Betic orogen being much larger than the Rif orogen – and with the spatial distribution of Neogene volcanism (e.g. Lonergan and White 1997). The reconstruction of Betic-Rif shortening by Platt et al. (2003) leads to shortening directions roughly perpendicular to the slab (as imaged at 200 km depth). These observations point at a possible relation between the shape of the slab and the (creation of) arcuate geometry of the Betic-Rif orogen. Duggen et al. (2003) propose that part of the Iberian continental lithosphere, attached to the Ligurian ocean, delaminated in the roll-back process in the Late Miocene. In this case, at least the top part of the slab (e.g. from 145 to 250–300 km) could consist of delaminated lithosphere. In map view (Fig. 2.4), the arcuate geometry of the slab can more or less be followed with increasing depth from which we infer that it rather is a deep feature of the slab, independent of lithosphere delamination. In the following, we will relate this feature to the evolution of the roll-back process.

Slab roll-back can transport continental fragments over large distances (e.g. Betics, Corsica-Sardinia, Kabylies, Calabria, Alboran microplate) but does not necessarily imply the building of mountain belts. Special circumstances are needed, for instance effects of continental margin geometry or continental margins entering the trench, to lead to orogenic activity. The BS2000 results show that the slab is located only slightly under the Rif margin whereas it largely underlies the Betics

(e.g. at a depth of 200 km). The Rif orogen is created at the location where the African margin turns toward a NW strike (Fig. 2.1), thus, where westward slab roll-back (accommodated by E-W lithosphere tearing) would start to involve the continental margin (and possibly lithosphere delamination). The slab's north-south extent is narrowing in the top 200 km toward Gibraltar. At 200 km depth, the slab width along strike is 400 ± 50 km whereas Africa and Iberia have been separated by a much smaller distance. We suggest that the strong curvature of the slab at 200 km depth is the result of the narrow(ing) corridor between the African and Iberian margins. Along the African margin roll-back is more advanced toward the west, as reflected by the position of the slab under the Rif orogen compared to that below the Betic Cordillera. Lithosphere tearing has accommodated faster westward roll-back along the African margin whereas roll-back/ lithosphere tearing was inhibited or stalled temporarily against the Betic margin. This leads to an evolution of trench positions as depicted qualitatively in Fig. 2.10. In the late Miocene, detachment below the eastern Betics, possibly involving continental lithosphere, must be invoked to allow the slab to deflect to its position under the Betic region. Slab detachment under the Betics is consistent with the noted trends from calc-alkaline to alkaline volcanism and the recent uplift of Neogene basins along the Betic margin (e.g. Lonergan and White 1997; Duggen et al. 2003). The position of the slab under the African margin is consistent with small, but continuing, NNW-directed motion of Africa.

Lastly, we remark that, in map view, the Betic-Rif slab geometry is like a mirror-image of the Calabria slab (e.g. at a depth of 200 km). Also for the Calabria subduction the corridor for slab roll-back narrowed between the Adriatic and African margin. In this comparison, the Apennines are in a similar position as the Betic orogen (including slab detachment) whereas the free end of the Calabria slab below Sicily compares well with that of the Betic-Alboran slab under the Rif orogen (including the lithosphere tearing along the African margin). The angle between the continental margins of Africa and Iberia is however much smaller than the angle between the margins of Africa and Adria which may entail a different evolution of slab geometry and crustal response in both regions.

2.4.8 Synthesis of Tomographic Constraints on the Geodynamic Evolution of the Western Mediterranean Region

The surface area occupied by the Ligurian ocean can be reconstructed by restoring the Betic-Alboran and Apennines slabs to their former position at the surface (Fig. 2.11a). We estimate from the E-W tomographic cross sections a length of about 700–800 km for the Betic-

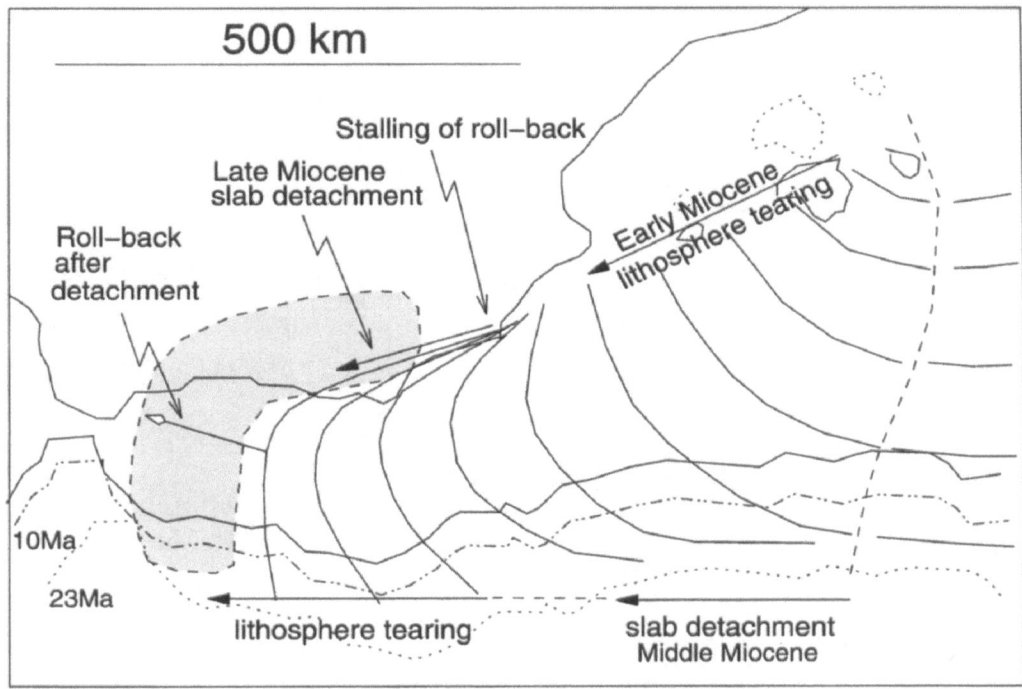

Fig. 2.10. Kinematic evolution of slab roll-back in the Betic-Rif-Alboran region. *Grey shaded area* gives the present location of the Betic-Alboran slab at a depth of 200 km (cf. Fig. 2.4). The location of the African margin is denoted by *dashed/dotted lines* for 10 and 23 Ma ago (after Gueguen et al. 1998). The *dashed line on the right* indicates the outline of that part of the Ligurian ocean that rolled-back to form the Betic-Alboran slab (see Fig. 2.11). *Curved lines* denote the location of the subduction trench through time, starting at the Balearic Islands and ending at the time of slab detachment under the Betic (Late Miocene). Initially, south- and southwest-ward roll-back moved the trench in a roughly SW direction while slab bending progressed. Lithosphere tearing along the Balearic margin in a WSW direction occurred already in the Early Miocene and facilitated slab roll-back. During the Middle Miocene the trench reached the African margin in the south after which slab detachment initiated and migrated to the west along the African continental margin. This lithosphere rupture facilitated the westward turning of the roll-back system. When the angle between trench and margin became large enough, slab detachment evolved into lithosphere tearing along the African margin eventually leading to a margin-perpendicular trench (slab). Along the Balearic margin, lithosphere tearing came to a halt at some time during the Middle-Early Miocene perhaps as the combined effect of (*i*) encountering a southward-directed turn in the continental Iberian-Balearic margin, i.e. a strong narrowing of the Africa-Iberia corridor (see Bos et al. 2003 for an analogous situation in the Taiwan region), and (*ii*) of faster slab roll-back to the south. The trench rotated and turned parallel to the Iberian margin while lithosphere tearing and faster roll-back along the African margin continued. Such a slab-bending process seems necessary to facilitate subduction to pass through the narrow corridor between Africa and Iberia and basically caused the present-day geometry of the slab as imaged at 200 km depth. When the trench became more or less parallel to the Iberian margin, Early-Miocene slab detachment allowed the slab to tear away and deflect under the Betic by which it facilitated the last phase of roll-back to the west, leading to the geometry as imaged by tomography

Alboran slab which defines its extent along the African margin. Because the anomaly broadens with depth toward the NE we expect that the part of the Ligurian ocean associated with this subduction extended more to the NE (along the Balearic margin). For the Calabrian subduction we estimate a slab length between 1 000 and 1 100 km in a NW direction. This is shorter than the estimate of 1 200 km of Faccenna et al. (2001a) which is based on the tomographic model of Lucente et al. (1999). The difference is related to our interpretation of the flat lying anomaly A (West Alpine-Tethys) which we do not incorporate in the length estimate. With these estimates, space is left in the surface reconstruction for the east Algeria slab. The length of the northern Apennines slab we estimate to be between 300 and 400 km. Figure 2.11a shows the entire area of the Ligurian ocean region affected by subduction in the past 30 Ma. The three slab surfaces of the southern systems meet at the surface between the Balearic Islands and Sardinia. This configu-

ration of the Ligurian ocean is in agreement with the starting geometry of many tectonic reconstructions. We note that the transition zone anomaly A is not needed for this reconstruction. This also attests for its different origin independent of Ligurian ocean subduction. We also note that part of the Calabria slab may in fact consist of Neo-Tethys ocean. Plate tectonic reconstructions (e.g. Stampfli and Borel, this publ.; Schettino and Scotese 2002; Rosenbaum et al. 2002) do not agree on how the Ligurian ocean was connected in the SE to the Neo-Tethys ocean which makes it difficult to distinguish between Ligurian and Neo-Tethys contributions to the Calabria slab.

In Fig. 2.11a we assume that initiation of Ligurian ocean subduction occurred along the Sardinia-Corsica segment which we consider to be the zone of greatest lithosphere weakness, considering its proximity to the Pyrenees orogeny (up to Eocene) and the Pyrenees-to-Alps suture left after Alpine collision s.s. in the Eocene.

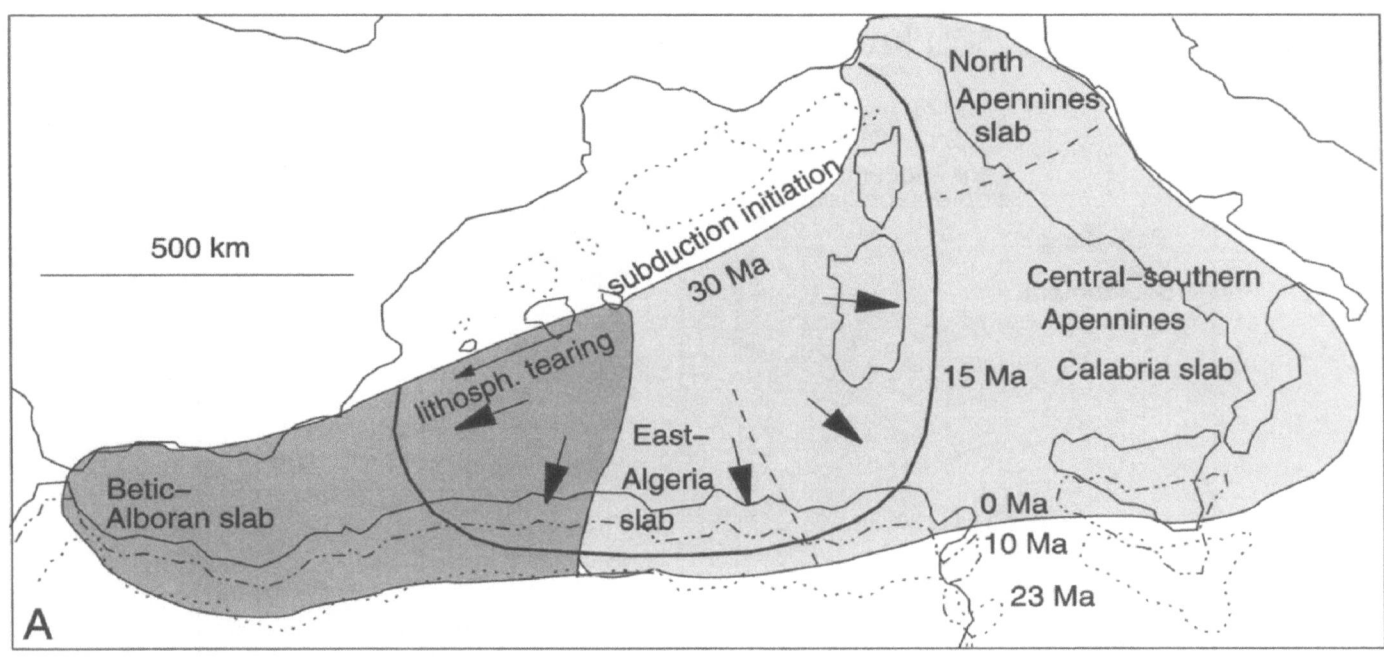

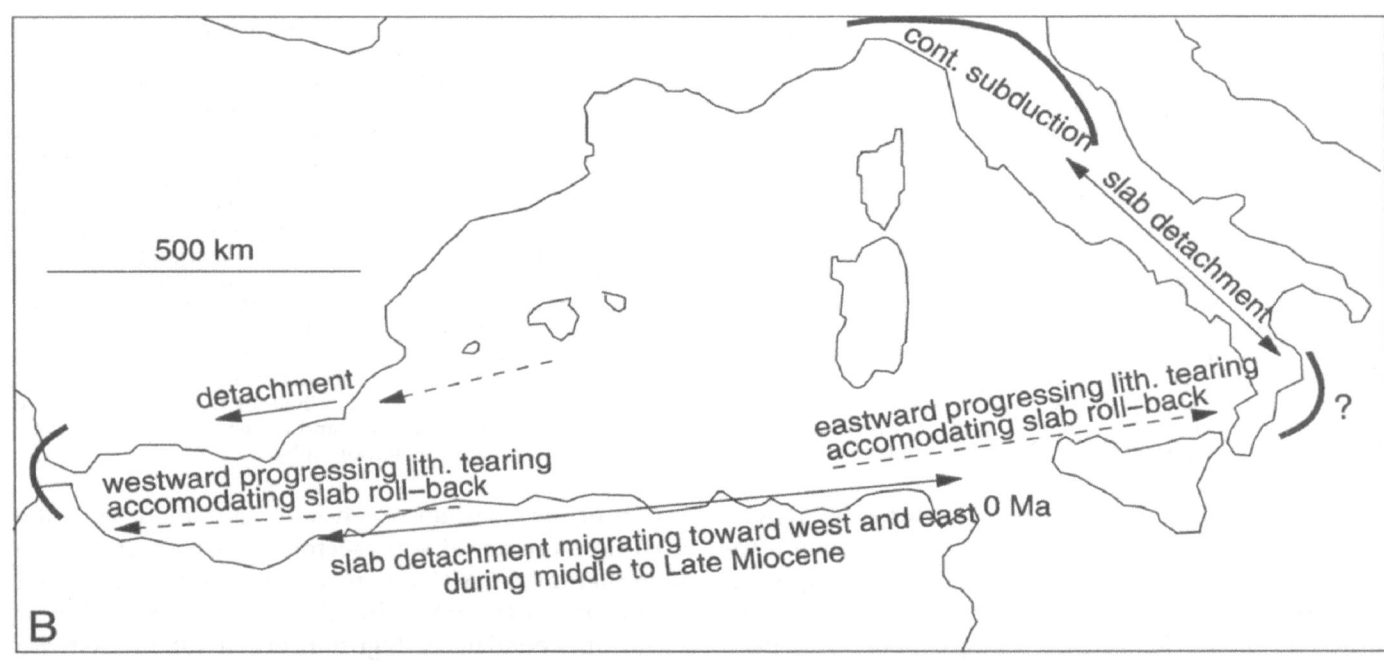

Fig. 2.11. a Surface reconstruction of the Ligurian ocean based on the amount and geometry of subducted slab as estimated from the BS2000 tomography model. *Dash-dotted lines* denote the African margin at about 10 Ma, while *dotted lines* denote the location of the Balearic Islands, Corsica and Sardinia, and the African margin in the Late Oligocene/Earliest Miocene (after Gueguen et al. 1998). The reconstruction assumes a short 300–400 km north-Apennines slab, a 1000–1100 km Calabrian slab (measured along a NW-SE line) and a 700–800 km long Betic-Alboran slab. This leaves space for the East-Algeria slab as imaged by tomography. We note that part of the Calabria slab may actually derive from Ionian (Neo-Tethys) lithosphere; by how much is unknown. The remnant of the West Alpine-Tethys (anomaly A; Fig. 2.4) is not used for this reconstruction. The *thick solid line* gives an impression of the trench location at about 15 Ma, after which slab detachment initiated along the African margin. **b** Schematic indication of where we propose that slab detachment and lithosphere tearing occurred to facilitate the overall development of roll-back of the Ligurian ocean. Continuous slab is assumed below Gibraltar and the northern Apennines. Continuity of the Calabria slab is doubtful

Dissociation of the Betic-Alboran slab, the east Algerian slab, and the Apennines slab is a necessary result of accumulating tensional stresses due to surface enlargement during subduction roll-back. It may have occurred along preexisting weakness zones and even before the late Miocene opening of the Tyrrhenian basin. The counterclockwise rotation of Corsica and Sardinia is much larger than the clockwise rotation of the Balearic margin. This suggests much larger initial roll back in the former region which points at a quite early decoupling between the two major subduction systems. Particularly, to accommodate the southeastward roll-back of narrow Betic-Alboran slab it seems geometrically necessary to initiate early tearing of the Ligurian ocean lithosphere along the Balearic margin. The work of Acosta et al. (2002) suggests that lithosphere tearing may have commenced already in the Early Miocene, coeval with the extensional event of Platt and Vissers (1989). A transpressive stress regime associated with tear propagation – due to continuing convergence between Africa and Europe – would give an explanation for the observed compression in Majorca coeval with Valencia basin extension (Gelabert et al. 1992).

In Fig. 2.11b, we schematically indicate the consequences of slab retreat for the development of slab detachment below the north African margin developing into propagating lithosphere tearing in the west and east (as suggested earlier by Carminati et al. 1998 for the eastern segment and by Duggen et al. 2003 for the Betic-Alboran region). Beneath Gibraltar and the northern Apennines the slabs are assumed to be continuous as evidenced by continuing fore-arc deformation (Gutscher et al. 2002; Lucente and Speranza 2001; Carminati et al. 2003).

2.5 Summary

Our interpretation of Western Mediterranean mantle structure leads to a surface reconstruction of the Ligurian ocean prior to its subduction which is in good agreement with the starting configuration of many independent tectonic reconstructions of the Western Mediterranean. The Betic-Alboran and Apennines-Calabria subduction are the two governing systems in the geodynamic evolution of the Western Mediterranean during the last 30 Ma. An unknown part of the Calabria slab may derive from subducted Ionian (Neo-Tethys) lithosphere.

Large scale roll-back, as evidenced by tomography, must be invoked to explain the present location of the subducted slabs. Southward roll-back of the Ligurian ocean reached the African-Maghrebian margin, at least in its central portion (Kabylies in Algeria). Because no N-dipping slabs are imaged beneath either side of this region, slab detachment and/or lithosphere tearing along the African margin must have assisted the slabs in retreating toward their present position. This independently confirms the original suggestion by Lonergan and White (1997).

Following our earlier work (Carminati et al. 1998) we stress the important role of lithosphere rupture (i.e. slab detachment or lithosphere tearing) in the more detailed, late stage, development of slab roll-back in the Western Mediterranean. In particular, we propose that lithosphere rupture is a key element in the evolution of Betic-Alboran slab roll-back toward the present-day Betic-Rif

region and in linking slab geometry with the geometry of the Betic-Rif orogenic belt. Lithosphere delamination may also have played a role in the latest development of this region, but this process is not considered important for the overall evolution of the westernmost Mediterranean at large.

We consider the large transition zone anomaly beneath the Alps, southern France and northern Italy as the remnant of the West Alpine-Tethys (Piedmont ocean) and the transition anomaly under the Eastern Alps and Pannonian basin as the remnant of the East Alpine-Tethys. This interpretation implies that during the second half of the Cenozoic the European plate was relatively stationary with respect to the base of the upper mantle. Importantly, the West Alpine-Tethys anomaly is not required for the surface reconstruction of the Ligurian ocean.

We propose a relatively short (300–400 km) and continuous northern Apennines slab which may strongly curve to the west beneath the Po plain. With this interpretation we depart from our earlier suggestion of slab detachment below the northern Apennines (Wortel and Spakman 1992), but are in contrast with the long slab suggested by other researchers (e.g. Lucente et al. 1999; Piromallo and Morelli 2003). A short north-Apennines slab is sufficient to explain the opening history of the northern Liguro-Provençal basin.

The negative anomalies found at depths of 200 km under the central-southern Apennines are attributed to slab detachment, which remains in line with our earlier propositions (Wortel and Spakman 1992, 2000). The Calabria slab, although imaged up to the crust, may have recently detached. We have refined the proposition of Blanco and Spakman (1993) and Spakman et al. (1993) for complete slab detachment of the Betic-Alboran slab: a continuous slab is proposed below the Gibraltar region (Gutscher et al. 2002) and slab detachment is confined to the central-eastern Betic region and probably occurred during the Late Miocene (Duggen et al. 2003).

Acknowledgements

We are grateful for the constructive comments on this paper provided by William Cavazza, and two anonymous reviewers. The work described in this paper was conducted under research programs of the Vening Meinesz School of Geodynamics (VMSG) and the Netherlands Centre for Integrated Solid Earth Science (ISES).

Appendix 1 (CD-ROM)

The appendix contains seven figures summarizing the results of sensitivity tests for different synthetic models. See Sect. 2.2 and the caption of Fig. 2.3 for general information. Inspect Fig. 2.4 for cell dimensions as constructed for model BS2000 in the upper mantle below the Western Mediterranean. See Spakman and Nolet (1988), Spakman et al. (1993), and Leveque et al. (1993) for more information on sensitivity analysis with synthetic models.

Fig. 2.A1.1. Spike test for block size of 0.6 degrees and about 50 km thickness in the upper 300 km. Note that the BS2000 model does not possess these cell dimensions directly below the Western Mediterranean basins. Generally the sensitivity for 60 km-detail is good below the land areas. The amplitude response decreases with increasing depth but note that the pattern is still reasonably well recovered at 320 km although the individual blocks are below the limit of being resolved

Fig. 2.A1.2. Spike test for block size of 1.2 degrees and about 50 km thickness in the upper 710 km. As for the 0.6 degree cells, the 1.2 degree cells are not constructed everywhere in the BS2000 model. Note that where 0.6 degree cells are used in the BS2000 model, we can perform a sensitivity test for 1.2 degree structure. With increasing depth, the sensitivity for 1.2 degree (50 km thick) detail decreases, although it stays good below the European land areas and the Adriatic. Below the Western Mediterranean detection of this detail is highly variable with position in the mantle. Specifically, this detail is difficult to image in the westernmost Mediterranean and under Iberia

Fig. 2.A1.3. Spike test for block size of 1.8 degrees and about 100 km thickness in the upper 710 km. This detail is very well recoverable in the synthetic test, although not perfect. But it is sufficient to interpret mantle anomalies such as the Betic-Alboran slab

Fig. 2.A1.4. Spike test for block size of 2.4 degrees and about 120 km thickness in the upper 710 km. This detail is generally well recovered. Some anomaly smearing occurs, particularly, in locations where the sensitivity for smaller scale detail is reduced

Fig. 2.A1.5. Spike test for block size of 3.0 degrees and about 120 km thickness in the upper 710 km. Similar to the 2.4-degree blocks, 3.0 degree structures are generally well recovered

Fig. 2.A1.6. The sensitivity for different scales of detail at a depth of 710 km (uppermost lower mantle). Note that the penetration of the West Alpine-Tethys anomaly (anomaly A) is well detectable

Appendix 2 (CD-ROM)

Three sets of upper-mantle cross sections (Figs. 2.A2.2 to 2.A2.4) are presented, each sweeping in detail across the BS2000 model (see Fig. 2.A2.1 for general locations). For description of the layout of cross sections see the caption of Fig. 2.6. All cross section are plotted to scale. Small ticks along the vertical axis are plotted at every 100 km of depth increase; large ticks at 500 km. Along the horizontal axis small ticks are placed at every 1 degree (110 km) and large ticks at every 5 degrees. Above each of the cross sections a small map is plotted for geographical reference with Fig. 2.A2.1. Anomalies are labelled with the following letters: B = Betic-Alboran slab; Al = East Algeria slab; C = Calabria slab; A = West Alpine-Tethys remnant; Ad = northern Apennines subduction of oceanic-to-continental lithosphere; D = positive anomaly related to the Eastern-Alps-Dinarides-Pannonian system, interpreted as a remnant of the East Alpine-Tethys (not discussed in detail in this paper); Eu = European continental lithosphere

Fig. 2.A2.1. Map indicating the location of the three cross section sweeps displayed in Figs. 2.A2.2–2.A2.4

Fig. 2.A2.2. A sweep of 20 west-east parallel sections across the Western Mediterranean from the African margin (*a*) to southern France-Alps (*t*). See for reference the labeling in Fig. 2.A2.1

Fig. 2.A2.3. A sweep of 10 NW-SE parallel sections across the Alps-Apennines-Tyrrhenean region. See for reference the labeling in Fig. 2.A2.1

Fig. 2.A2.4. Ten sections across the Alps, Po plain and northernmost Apennines. See Fig. 2.A2.1 for labeling

Chapter 3

The TRANSMED Transects in Space and Time: Constraints on the Paleotectonic Evolution of the Mediterranean Domain

Gérard M. Stampfli · Gilles D. Borel

Abstract

The Phanerozoic evolution of the western Tethyan region was dominated by terrane collisions and accretions, during the Variscan, Cimmerian and Alpine cycles. Most terranes were derived from Gondwana and present a similar early Palaeozoic evolution. Subsequently, they were detached from Gondwana and affected by different deformation and metamorphic events, which permit to decipher their geodynamic history. Lithospheric scale peri-Mediterranean transects show the present-day juxtaposition of these terranes, but do not allow to unravel their exotic nature or their duplication. To create a reliable palinspastic model around these transects, plate tectonics constraints must be taken into consideration in order to assess the magnitude of lateral displacements. For most of the transects and their different segments, thousand km scale differential transport can be demonstrated.

3.1 Introduction

The geodynamic evolution of the Tethyan domain as presented in this paper is based on the studies of our research team during the last fifteen years, integrating numerous recent and older publications as well as the results obtained by IGCP Project 369 "Comparative Evolution of PeriTethyan Rift/Wrench Basins" (Stampfli et al. 2001a, 2001b; Ziegler et al. 2001a). In papers to be published in conjunction with the 32nd International Geological Congress we defined the main elements of the Cimmerian cycle in the western Tethyan domain (Stampfli et al., in press) and presented a revised plate tectonic model for Adria-Apulia (Stampfli, in press). The conclusions of these studies are incorporated in the first part of this paper together with revised plate tectonic reconstructions.

In Stampfli and Borel (2002) we developed a new plate reconstruction method which represents a distinct departure from classical continental drift models. These new plate models for Palaeozoic and Mesozoic times (Ordovician to Cretaceous) integrate dynamic plate boundaries, plate buoyancy factors, ocean spreading rates, subsidence patterns, and new stratigraphic and palaeobiogeographic data, as well as major tectonic and magmatic events. Lithospheric plates were constructed through time by adding/removing oceanic material (symbolized by synthetic isochrones) to major continents and terranes. This approach offers a good control on plate kinematics and provides new constraints for plate tectonic scenarios in the Tethyan realm. A revised set of reconstructions is partly presented in Figs. 3.2–3.14, accompanied by extended captions providing an overview of the Tethyan region evolution in space and time. The complete set is found in the accompanying CD-ROM. The major changes in respect with the previous publications (Stampfli et al. 2001a; Stampfli and Borel 2002) concern mainly the West Tethys domain, such as a revised Permian Pangea fit allowing a better placing of Iberia and Apulia. From our plate reconstructions we could develop a coherent scheme of birth and death and genealogy of the many oceans whose history contributed to the make-up of the Tethyan realm. This scheme, modified from Ziegler et al. (2001b), is presented in Fig. 3.1.

For a long time, the Tethys was considered as a large and single oceanic space, mostly of Mesozoic age, located between Gondwana and Eurasia. Already in the 1940s and 1950s, a distinction between a Palaeo- and a Neo-Tethys appeared, and it was recognized that the latter comprised marine Permian and younger strata, whereas the former opened during the Early Paleozoic. Stöcklin (1968, 1974), following extensive field work in Iran, gave a formal definition of these two large oceanic entities, the Neotethys becoming a Permian to Cretaceous peri-Gondwanan ocean (whose suture was between Iran and Arabia), whereas the Paleotethys suture was located just north of Iran, thus, between the Cimmerian terranes (Sengör 1979) and Eurasia. In that sense, Paleotethys separates the Variscan domain from Gondwana-derived terranes. We follow here Stöcklin's definition, adding to it that most Variscan terranes were also derived from Gondwana, but, they drifted away from Gondwana during the opening of Paleotethys, and, in contrast with the Cimmerian terranes, they have been strongly affected by Variscan metamorphism and magmatism.

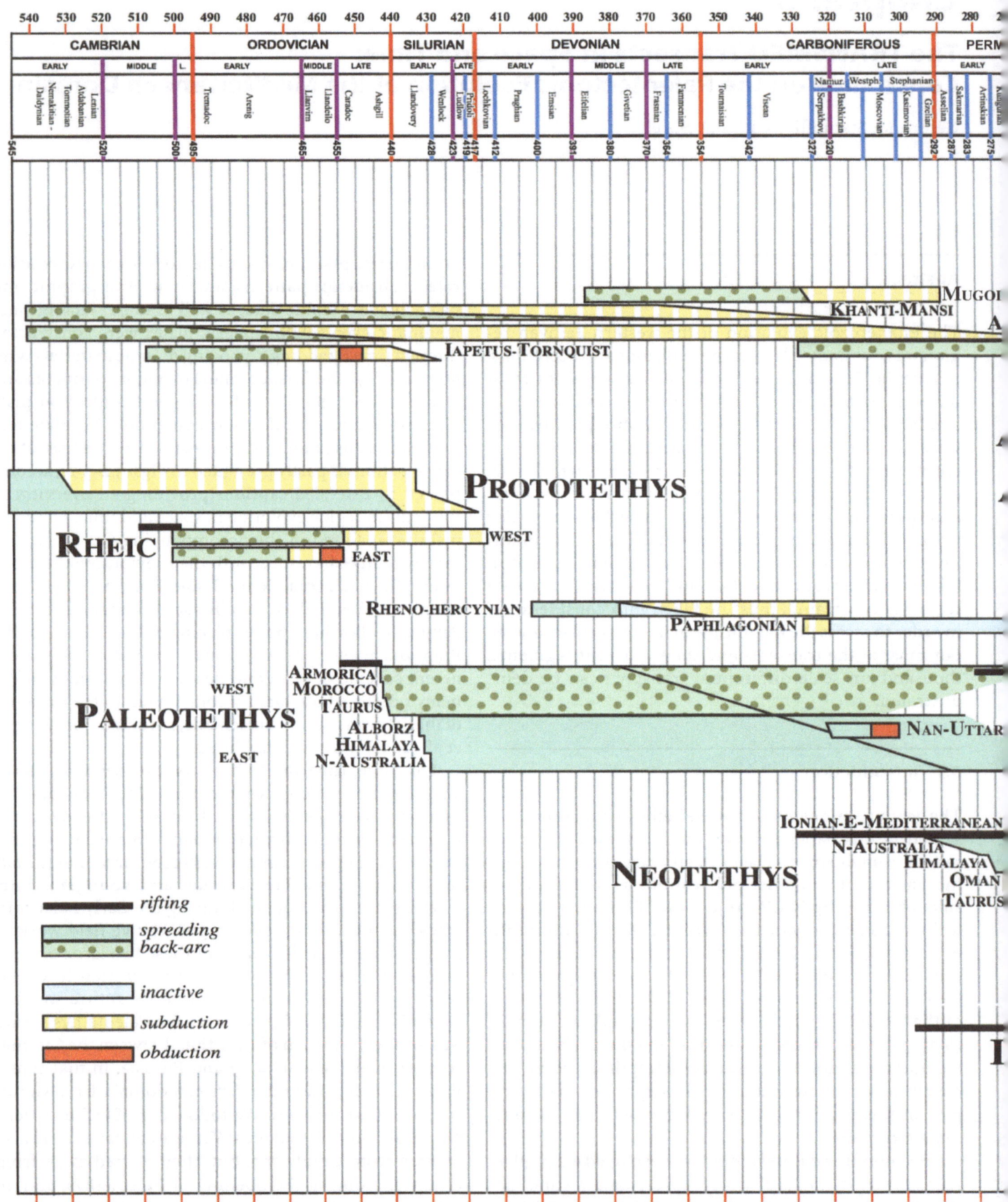

Fig. 3.1. Time table of opening and closure of Tethyan and related oceanic basins, modified from Ziegler et al. (2001). Time scale Unesco (2000), modified for the Permian stages

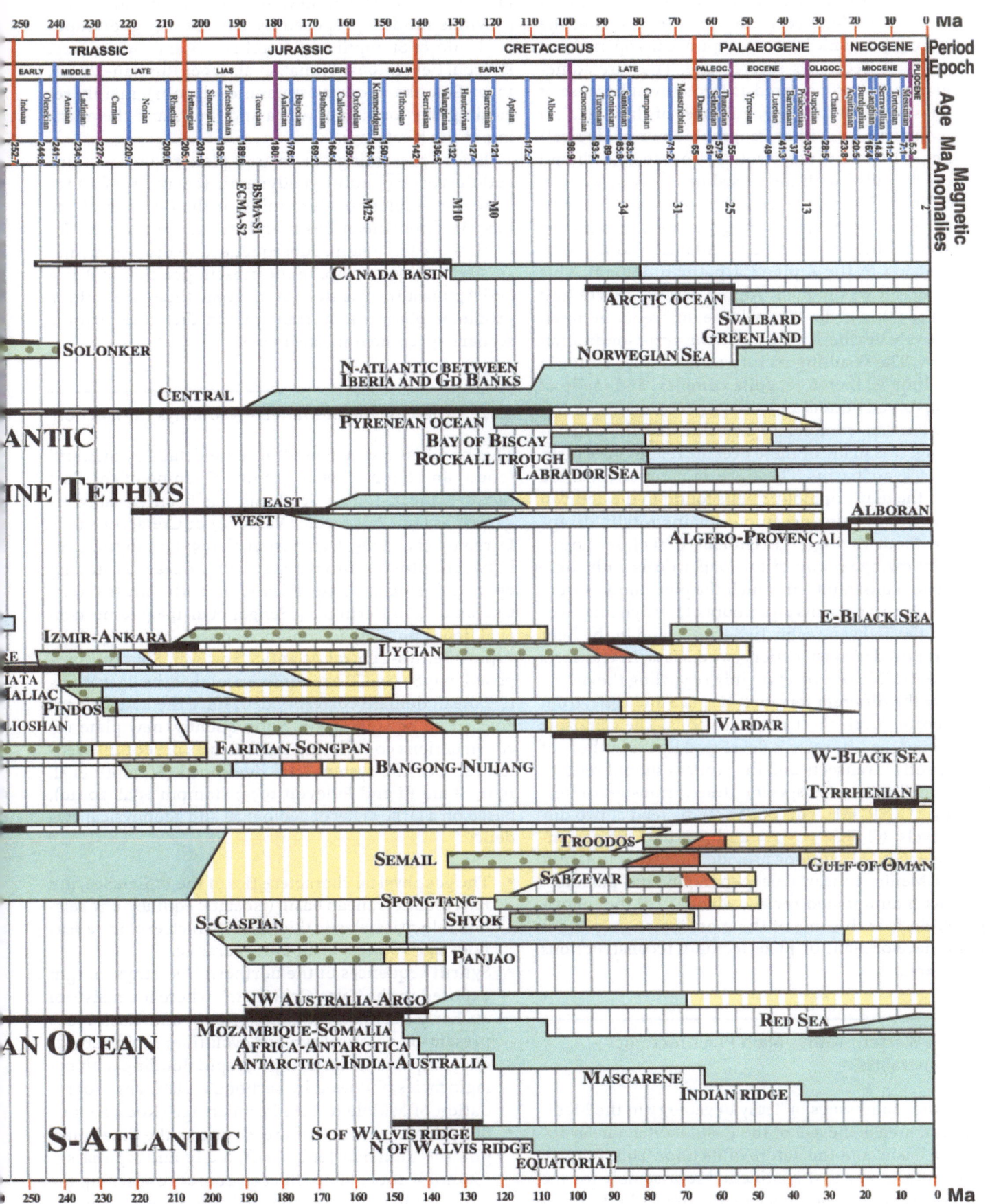

Besides the two large Paleotethyan and Neotethyan oceanic domains (one replacing the other during the Triassic) many oceanic back-arc type oceans opened just north of the Paleotethys suture zone. They are sometimes erroneously considered as Neotethyan because of their Triassic to Jurassic age, but most of these had no direct connection (neither geographic nor geological) with the peri-Gondwanan Neotethys ocean, and should therefore be called with their local names (e.g. Meliata, Maliac, Pindos, Vardar). During the break-up of Pangea, another relatively long, if not large, oceanic domain appeared, consisting of the Central Atlantic and its western extension in the Alpine Carpathian domain. This Jurassic ocean was named "Alpine Tethys" (Favre and Stampfli 1992) in order to mark the difference between this relatively northerly ocean and the peri-Gondwanan Neotethys. The resulting picture of the Tethys realm in Jurassic time is, therefore, quite complex, and made of numerous small oceans and a large peri-Gondwana Neotethys. Further complexity arose during the convergence stages, as many of these oceanic realms gave birth to new back-arc basins. These are, in most cases, at the origin of the many ophiolitic belts found in the Tethyan realm, whereas older oceanic domains totally disappeared without leaving large remnants of their sea floors.

The following section of this paper deals with some of the main constraints on which the plate models were built. Section 3.3 presents an account of the dynamic evolution of the Tethyan realm. In Sect. 3.4, and in the light of the proposed reconstructions, we discuss the transects of the TRANSMED Project mainly regarding the distribution of the different terranes in space and time, from the Paleozoic until the Late Cretaceous. The Cenozoic evolution of the transects is developed by the authors of the respective transects and the readers should refer to their contributions for specific descriptions and references (see CD-ROM). This paper can be read at two different levels: (i) Figs. 3.2–3.14 and 3.19 and the accompanying extended captions provide the reader unfamiliar with Mediterranean paleogeography and paleotectonics with an updated review, whereas (ii) the following text deals with some of the most interesting issues for the development of plate tectonic reconstructions of such area.

3.2 The Western Tethys Main Plate Tectonic Constraints

One of the main issues of Tethyan geology in the Mediterranean area is the age of the East Mediterranean-Ionian Sea basin, and the nature of its floor. This, in turn, influences the way that continental re-assembly for the Mesozoic is done. A new continental fit was built up recently (Stampfli, in press) in an attempt at reconciling plate tectonics with peri-Italian geophysical data (e.g.

Finetti et al. 2001). This new fit differs significantly from previous models proposed by ourselves (e.g. Stampfli et al. 2001c; Stampfli and Borel 2002) and others (e.g. Dercourt et al. 1993; Yilmaz et al. 1996; Russian-I.T.L.P. 1997; Golonka 2000; Wortmann et al. 2001), mainly regarding the position of Adria and Iberia and the opening of the Neotethys.

Before presenting our revised plate reconstructions and the corresponding geodynamic scheme, we first review the problem of the timing of the opening of the East Mediterranean basin and its relation to the Neotethys oceanic realm. Subsequently, we discuss the necessity to adopt a close to the present position of Adria and Apulia already since Late Triassic times. Both these problems play an important role in the plate tectonic scenario presented in this paper.

3.2.1 The East Mediterranean-Neotethys Connection

This problem was reviewed in some details by Stampfli (2000) and Stampfli et al. (2001c). Opening of the East Mediterranean-Ionian Sea basin has been variably regarded as Late Palaeozoic (Vai 1994) or Cretaceous (e.g. Dercourt et al. 1985; Dercourt et al. 1993) or anything in between. Most authors regard this ocean as having opened during the Late Triassic or Early Jurassic (e.g. Garfunkel and Derin 1984; Sengör et al. 1984; Robertson et al. 1996) possibly in conjunction with opening of the Alpine Tethys-Central Atlantic. However, Stampfli proposed in 1989 (IGCP 276 conference) that the East Mediterranean domain corresponded, since the Late Palaeozoic, to an oceanic basin. Subsequently, new plate reconstructions considering this basin as part of the Neotethyan oceanic system were developed (Stampfli et al. 1991; Stampfli and Pillevuit 1993; Stampfli et al. 2001c), based on a large array of geological and geophysical evidence:

- The geophysical characteristics of the Ionian Sea and East Mediterranean basin (isostatic equilibrium, seismic velocities, elastic thickness), which exclude a seafloor age younger than Early Jurassic.
- Synrift sequences of the northern Neotethys margin exposed in the Talea Ori (Crete), where it consists of an Early Permian sequence (König and Kuss 1980) presenting a typical syn-rift evolution: rapid subsidence, important clastic input, rapid flooding with pelagic facies, followed by carbonate platform progradation of Middle to Late Permian age (Kock 2003).
- Subsidence patterns of areas such as the Sinai margin, the Tunisian Jeffara rift, Sicily and Apulia s.s. (Stampfli 2000; Stampfli et al. 2001c) confirming the Late Permian onset of thermal subsidence along the margins of the East Mediterranean and Ionian Sea basins and the absence of younger thermal events.

- Late Permian Hallstatt-type pelagic limestones, similar to those found in Oman – where they sometimes rest directly on MORB (Pillevuit 1993; Niko et al. 1996) – have been reported from the Sosio complex in Sicily (Kozur 1995). These yielded a pelagic macro- and microfauna (Kozur 1990; Kozur 1991c; Kozur 1991d) that show affinities with both Oman and Timor. This implies a Middle Permian deep water connection of the westernmost East Mediterranean basin with the Neotethys (Stampfli et al. in press).
- Triassic MORB found in Cyprus in the Mamonia complex (Malpas et al. 1993) is derived from the East Mediterranean sea floor.
- Recent seismostratigraphy of the Ionian Sea (Stampfli et al. 2001; Finetti, in prep.) support the oceanic nature of the Ionian sea floor, as well as its Permo-Triassic age.

In conclusion, we propose a Middle to Late Permian onset of sea-floor spreading in the eastern Mediterranean basin, concomitant with the opening of the Neotethys (Stampfli et al. 2001c) and the northward drift of the Gondwana-derived Cimmerian continents (Sengör 1984) since the late Early Permian. This model implies also a Middle to Late Triassic closure of the Paleotethys in the Mediterranean domain of the Tethyan realm that was accompanied by the opening of back-arc basins in the active Eurasian margin (Meliata, Maliac and Pindos back-arc basins) (Stampfli 1996; Ziegler and Stampfli 2001; Stampfli et al. 2003).

3.2.2 The Apulia-Adria Problem

Paleomagnetic data show that the Apulian plate s.l. (Italy and the Adriatic Sea) underwent relatively little rotation with respect to Africa since the Triassic (e.g. Channell 1992, 1996). Yet, the continuity of convergence between the active subduction zone under Greece and the outer Dinarides (de Jonge et al. 1994; Wortel and Spakman 1992) shows that there is a possible plate boundary between Apulia s.l. and the autochthonous of Greece. We regard this as a recent feature that has no bearing on the fact that, in Mesozoic times, a Greater Apulia plate existed, which included all the autochthonous units of Greece, as well as the Bey-Daglari massif of SW Turkey (Poisson 1984). The apparent present intra-Apulian plate boundary accounts for the fact that the Hellenic orogenic/accretionary wedge is oblique with respect to former paleogeographic domains, still colliding with Apulia on an Albanides transect (Transect III, CD-ROM) and already subducting the East Mediterranean sea floor on a Greek transect (Transect VII, CD-ROM).

A first reassemblage of the western Tethyan microplates in a pre-break-up position led us to consider that the Apulian plate s.l. was most likely cut into two pieces, an Apulian plate s.s. to the south and an Adriatic plate s.s. to the north (Stampfli et al. 2001a). This was based on the concept that the location of the Paleotethys suture marks the boundary between these two plates (Stampfli and Mosar 1999). The CROP seismic lines through the Adriatic domain (Finetti, in prep.) clearly show that there is no major tectonic discontinuity cutting Italy into two units, at least since the Jurassic, and that our model had to be reassessed.

Our basic hypothesis is still that the Apulian s.s. part of Italy represents an African promontory that, from Middle Triassic to Recent times, underwent little displacement with respect to Africa. Such domain represents the easternmost Cimmerian element, detached from Gondwana in Middle Permian times during the opening of the East Mediterranean-Neotethys basin. We also assume that the Adriatic and Apulian microplates were welded together in a Late Variscan-Eocimmerian collision phase starting in the Middle Permian and ending in Carnian times (Stampfli et al. in press). With this the two amalgamated units formed what is referred to indifferently as Apulia s.l. or Adria s.l. and became part of the African plate. The need of cutting Apulia s.l. into two micro-plates stems from the fact that in a Triassic Pangean fit, as used in our former model (Stampfli and Mosar 1999; Stampfli and Borel 2002), which corresponds to the classic Pangea A fit revisited, there was very little room to insert this present form of the Apulian promontory in its proper place, knowing that it had reached its present location already in the Middle Triassic. Thus, we had to reconsider the fit of Iberia with Europe, as well as the dimension and position of the Alboran microplates. We constructed a much tighter fit of these elements with Europe, following similar proposals by Srivastava and Tapscott (1986) and Srivastava et al. (1990) and considering that the Variscan crust was systematically overthickened west of the Apulian promontory. Still, there was not enough room to accommodate the entire present length of Adria s.l. in its proper position (Figs. 3.8 and 3.9). This led to the conclusion that Adria s.l. was shorter in Triassic time that it is now (or already in the Late Jurassic) and that it was stretched by a few hundred kilometres during major rifting phases related to the Late Triassic to Middle Jurassic break-up of Pangea and the opening of the Alpine Tethys.

Based on subsidence analysis, we had already shown that the large Lombardian rift formed an aborted branch of the Central Atlantic rift and its extension into the Alpine region (Stampfli 2000). This rift certainly contributed to changing the geometry of the Adriatic plate in Late Triassic-Early Jurassic times. For Italy, the CROP seismic data provide evidence for pervasive Early to Middle Jurassic rifting, extending eastward up to Greece (Ionian rift system) and southward to the Pelagian domain, where it is accompanied by important volcanism

Table 3.1. Abbreviations of key localities for Figs. 3.2–3.14 and 3.20

Abbrev.	Locality	Abbrev.	Locality	Abbrev.	Locality
AA	Austroalpine	Ce	Cetic	Ha	Hadim
Ab	Alboran	Cg	Chagai arc	He	Helvetic rim basin
Ad	Adria s.str.	Ch	Channel	Hg	Huglu-Boyalitepe
Ae	Abadeh	cI	central Iberia	HK	Hindu-Kush
Af	N-Afghanistan, Band-e-Turkestan	Ci	Ciotat flysch	hK	high karst
Ag	Aladag-Bolkardag	Ck	Chehel Kureh ophiolite	Hr	Hronicum
Ah	Agh-Darband	CL	Campania-Lucania	Hy	Hydra
Aj	Ajat	Cm	Cadomia	Hz	Harz
Al	Alborz	cM	central Mongolia	IA	Izmir-Ankara ocean
Am	Armorica	Co	Codru	iA	intra-alpine terrane
An	Antalya, lower nappe	Cn	Carnic-Julian	Ib	Iberia, NW allochthon
Ao	Argoland	CP	Calabria-Peloritani	IC	Indochina
Ap	Apulia s.str.	cR	circum-Rhodope	Ig	Igal trough
Aq	Aquitaine	Cr	Carolina	Il	Ishim-Ishkeolmes (N Kazakhstan)
AP	Aspromonte, Peloritani	Cs	Chortis	Io	Ionian
Ar	Arna accretionary complex	Ct	Cantabria-Asturia	iP	intra-Pontides
As	Apuseni-south, ophiolites	Cu	Chuchi	Ir	Iranshar ophiolite
At	Attika	Cv	Canavese	Is	Istanbul
Au	Asterousia	Da	Dacides	Ja	Jadar
AT	Alpine Tethys	Db	Dent Blanche	Jf	Jeffara rift
Av	Arvi	DD	Dniepr-Donetz rift	Jo	Jolfa
Ay	Antalya, upper nappe	Dg	Denizgören ophiolite	Jv	Juvavic
Ba	Balkanides, external	DH	Dinarides-Hellenides	Ka	Kalnic
Bb	Band e Bayan	Di	Dizi accretionary complex	Kb	Karaburun
Bc	Biscay, Gascogne	Dm	Domar	Kd	Kopet-Dagh
Bd	Bey Daglari	DN	Dras-Naktol arc	Ke	Kotel flysch.
Be	Betic	Do	Dobrogea	Kf	Koufra basin
Bf	Baft ophiolite	Dr	Drina-Ivanjica	Kg	Karabogaz Gol
BH	Baer Basit-Hatay ophiolites	Ds	Drimos, Samothace ophiolites	Kh	Kohistan arc
Bh	Bihar	Du	Durmitor	Ki	Kirsehir
Bi	Ba'id margin	Dz	Dzirula	Kk	Karakaya forearc
Bj	Birjan ophiolite	eA	east Albanian ophiolites	Kl	Kabul block
Bk	Bozdag-Konya forearc	El	Elazig, Guleman ophiolites+arc	Km	Karakoram terrane
Bm	Burma	eP	east Pontides	KM	Kanty-Mansi
BM	Bela, Muslim-bagh ophiolites	Er	Eratosthen	Kn	Kunlun north
Bn	Bernina	Es	Esfandareh ophiolites	Ko	Korab
BN	Bangong-Nuijang ophiolites	Fa	Fatric	KQ	Kunlun-Qaidam
Bq	Bechar basin	Fc	Flamish cap	Kp	Karpinski
Br	Briançonnais	FM	Fanuj, Maskutan ophiolite	Kr	Kermanshah ophiolite
Bs	Bisitoun seamount	Fr	Farah basin	Ks	Kunlun south
BS	Bator-Szarvasko ophiolites	Ga	Gandese arc	KS	Kotel-Stranja rift
Bt	Batain	GB	Grand Banks	KT	Karakum-Turan
Bu	Bucovinian	gC	great Caucasus	Ku	Kura
Bü	Bükk	GC	Gascoyne-Cuvierland	Kü	Küre ocean
Bv	Budva	Gd	Geydag-Anamas-Akseki	KW	Khost, Waziristan ophiolites
BV	Bruno-Vistulian	Gi	Giessen	Ky	Kabylies
By	Beyshehir	Ge	Gemeric	Kz	Kazakhstan
Bz	Beykoz basin	GS	Gory-Sovie	La	Lagonegro
Ca	Calabria autochton	GT	Gavrovo-Tripolitza	lA	lower Austroalpine
cA	central Afghanistan, Hazarajat	Gt	Getic	Lb	Longobucco
cB	central Bosnia	Gü	Gümüshane-Kelkit	Le	Lesbos ophiolites
cD	central Dinarides ophiolites	hA	high-Atlas	Lg	Ligerian

Table 3.1. *Continued*

Abbrev.	Locality	Abbrev.	Locality	Abbrev.	Locality
Li	Ligurian	Ow	Owen basin	Sr	Severin ophiolites
Lm	Lamayuru seamount	Oz	Otztal-Silvretta	SS	Sanandaj-Sirjan
LM	Lysogory-Malopolska	**Pa**	**Panormides**	sT	south Tibet
Lo	Lombardian	Pb	Porcupine bank	St	Sitia
Ls	Lusitanian	Pc	Pamir central	Su	Sumeini margin
LT	Lut-Tabas-Yazd	Pd	Pindos rift/ocean	Sv	Svanetia rift
Lu	Lut	Pe	Penninic ocean	Sw	Seward
Ly	Lycian ophiolitic complex	Pi	Piemontais ocean	Sx	Saxo-Thuringian
Lz	Lizard ophiolitic complex	Pj	Panjao, Waser ocean	Sy	Seychelles
mA	middle Atlas	Pk	Paikon island arc	Sz	Sabzevar ophiolite
Ma	Mani	Pl	Pelagonian	**Ta**	**Taurus s.l.**
Mb	Magnitogorsk back-arc	Pm	Palmyra rift	Tb	Tabas
Mc	Maliac rift/ocean	Pn	Pienniny rift	TB	Tirolic-Bavaric
Md	Mozdak	Pp	Paphlagonian ocean	tC	Transcaucasus
MD	Moldanubian	Pr	Pamir, northern	Td	Taoudeni basin
Me	Meliata rift/ocean	PS	Pamir, southern & Shaksgam	TD	Transdanubian
Mf	Misfah seamount	Px	Paxi	Tf	Tindouf basin
Mg	Meguma	Py	Pyrenean rift	Th	Thrace basin
Mh	Mugodzhar ocean	**Qa**	**Qamar**	Ti	Timor
Mi	Mirdita autochton	Qi	Qilian	Tk	Tuarkyr
Mk	Mangyshlak rift	Qn	Qinling north	Tm	Tarim block
Ml	Meglenitsa ophiolite	Qs	Qinling south	Tn	Tarim north
Mm	Mamonia accretionary complex	**Rb**	**Rockall-Hatton bank**	To	Talea Ori
Mn	Menderes	Re	Réunion hot-spot	Tp	Troodos ophiolite
Mo	Moesia	Rf	Rif, external	TR	Taimyr
MP	Mersin, Pozanti ophiolites	Rg	Reggan basin	Tr	Turan
Mq	Moursouk basin	Rh	Rhodope	Ts	Tarim south
Mr	Mrzlevodice forearc	RH	Rheno-Hercynian ocean	Tt	Tatric
MR	Masirah, Ra's Madrekah ophiolites	Ri	Rif, internal	Tu	Tuscan
Ms	Moroccan Meseta	Rs	Rhadames basin	Tv	Tavas+Tavas seamount
MS	Margna-Sella margin	Ru	Rustaq seamount	Ty	Tyros forearc
Mt	Monte Amiata forearc	Rw	Ruwaydah seamount	Tz	Tizia
Mu	Maizuru arc	**Sa**	**Salum**	uJ	upper Juvavic
Mz	Munzur dag-Keban	sA	south Alpine	UM	Umbria-Marches
NA	**Nan-Uttaradit-Ailoshan ocean**	Sb	Sibumasu	Uy	Ust-Yurt
nC	north Caspian	sB	sub-Betic rim basin	**Va**	**Valais trough**
NCA	North Calcareous Alps	Sc	Scythian platform	VC	Valerianov-Chatkal (s-Kazakhstan)
Nh	Naga Hill ophiolite	sC	south Caspian basin	Ve	Veporic
Ni	Nilüfer seamount	Sd	Srednogorie rift-arc	Vo	Vourinos (Pindos)-Mirdita ophiolites
Nj	Njurol massif	Se	Sesia (western Austroalpine)	Vr	Vardar ocean
Nk	Nakhlak forearc	Sh	Shemshak foreland basin	**Wa**	**Wartovsky massif**
Nr	Neyriz seamount	Si	Sicanian	wC	western-Crete (Phyl-Qrtz) accretionary complex
Nn	Nain ophiolite	Sj	Strandja		
Ns	Niesen flysch	Sk	Sakarya	WC	West Carpathian
nT	north Tibet	sK	south-Karawanken forearc	wK	western Klamath
Nt	Nish-Troyan trough	Sl	Slavonia	wS	western Sumatra
Ny	Neyriz ophiolite	Sm	Silicicum	Xi	**Xigaze island-arc**
OA	**Oaxacan-Arequipa**	SM	Serbo-Macedonian	**Ya**	**Yazd**
Ok	Okhotsk-Chukotka ocean	sM	southern Mongolia	Yu	Yucatan
OM	Ossa-Morena	Sn	Sevan ophiolites	zH	**zone Houillère**
Or	Ordenes ophiolites	So	Solonker arc	Zl	Zlatibar ophiolites
Ot	Othrys-Evia-Argolis ophiolites	sP	south Portuguese	Zo	Zonguldak
				Zt	Band e Ziarat ophiolite

(Finetti 1985; Argnani and Torelli 2001). Paleofaults have been observed in many parts of the Ionian zone of Greece and Albania (Baudin and Lachkar 1990; Karakitsios 1991; Dodona et al. 1994; Karakitsios and Dermitzakis 1997) and also in the more autochthonous sequences found along the Adriatic coast and in the Adriatic Sea, as seen on the seismic data (Finetti, in prep.) and onshore. The reason for this widespread Jurassic rifting is that for a successful Pangean break-up, the Atlantic rift system had to join another plate limit located in the Neotethyan domain, either its mid-ocean ridge or its northern active margin. Other possible avenues to break through the Alpine-Mediterranean lithosphere had been tried and many resulted in aborted rifts. This pervasive Jurassic rifting finally gave birth to the opening of the Alpine Tethys which is flanked by large passive margins and rim basins (e.g. Subbrianconnais, Helvetic-Dauphinois, Lombardian and Subbetic basins), and a narrow oceanic strip dominated by mantle denudation (Stampfli and Marchant 1997). Opening of the Alpine Tethys involved rift propagation through an already thinned and thermally not yet stabilized lithosphere, and the evolving rift remained below sea level, with little isostatic rebound of its shoulders.

The fact that the Alpine Tethys ocean floor is dominated by denuded continental mantle of mainly Permian age (Rampone and Piccardo 2000) suggests that this basin opened very slowly and that large amount of extension was distributed elsewhere in Greater Apulia. The Adriatic plate presumably attained its present length and geometry by Middle to Late Jurassic times (Fig. 3.10).

3.3 The Geodynamic Evolution of Greater Apulia and Surrounding Regions

The reconstructions shown on Figs. 3.2–3.14 are based on a tight Permian pre Pangea break-up fit and on the magnetic anomalies from the world oceans. Europe is fixed and paleolatitudes were derived from Baltica paleomagnetic data (Torsvik et al. 2001). Plate tectonic concepts were systematically applied to our palinspatic models of the western Tethys, moving away from pure continental drift models that are not constrained by plate boundary dynamics, in an effort to arrive at a model that is self-constrained. Most other constraints and databases used in our reconstructions can be found in Stampfli et al. (2001a, 2001c), and Stampfli and Borel (2002).

In this section we review the major steps in the geodynamic evolution of the periApulian domain, starting with the Paleotethys ocean, then moving to the Cimmerian event, and finally the Alpine cycle proper. The abbreviations of key localities used in Figs. 3.2–3.14 and 3.20 are given in Table 3.1.

3.3.1 Paleotethys Evolution (Figs. 3.2–3.6)

The opening of Paleotethys is related to the Early Paleozoic drifting away from Gondwana of a ribbon-shaped array of microcontinents, grouped under the label of Hun superterrane (von Raumer et al. 2002). Among these terranes we find the Adria s.s. part of Italy, which together with other European Hunic terranes was accreted to Laurussia in Devonian times (Stampfli et al. 2002b). Following this accretion, Paleotethys started subducting northward, creating the Variscan cordillera system. Paleotethys was a circumGondwana ocean whose history is well constrained on an Iranian transect (Alborz range, North Iran; Stampfli 1978; Stampfli et al. 1991; Stampfli et al. 2001c) representing the southern Gondwanan margin of the eastern part of such ocean.

Westward, along the northern African margin, opening of the western part of Paleotethys and the detachment of the European Hunic terranes from Gondwana is less well constrained. In the High Atlas of Morocco (Destombes 1971), the Ordovician is unconformably covered by Silurian strata, the top of which is locally conglomeratic. The overlying Emsian-Eifelian sequence is locally very condensed and represented by open marine carbonates. This starvation event reflects the onset of important thermal subsidence, which can be related to the detachment of the Meseta and neighbouring terranes from Africa.

The northern margin of Paleotethys is preserved in the middle part of the European Hunic terranes in the Carnic Alps (Schönlaub and Histon 1999), in Tuscany and Sardinia, and in the Alboran fragments (cf. Stampfli 1996) where it is also characterized by a Late Ordovician-Early Silurian clastic – and often volcanic – syn-rift sequence (Silurian flood basalts are known in Sardinia and the Rif; Piqué 1989). Rift-related thermal uplift, erosion and block tilting occurred in Silurian time and is often wrongly related to a compressive rather than an extensional Taconic event. Open-marine conditions, represented by graptolites facies, were established during the Silurian. During the Early Devonian a general flooding marked the onset of widespread thermal subsidence that can be related to the progressive opening of Paleotethys. Along its northern margin, the Visean usually marks the onset of widespread flysch deposition (Culm), often accompanied by volcanic activity that reflects a change from a passive to an active margin setting.

Potential Paleotethyan accretionary sequences are located along the southern part of the Variscan orogen, Such sequences are everywhere metamorphosed and intruded by Late Carboniferous granites (Vavassis et al. 2000). These sequences, which were generally overprinted by Eocimmerian and Alpine deformation, are mainly found in the Chios-Karaburun region and in Crete (Stampfli et al. 2003), but certainly extend in most

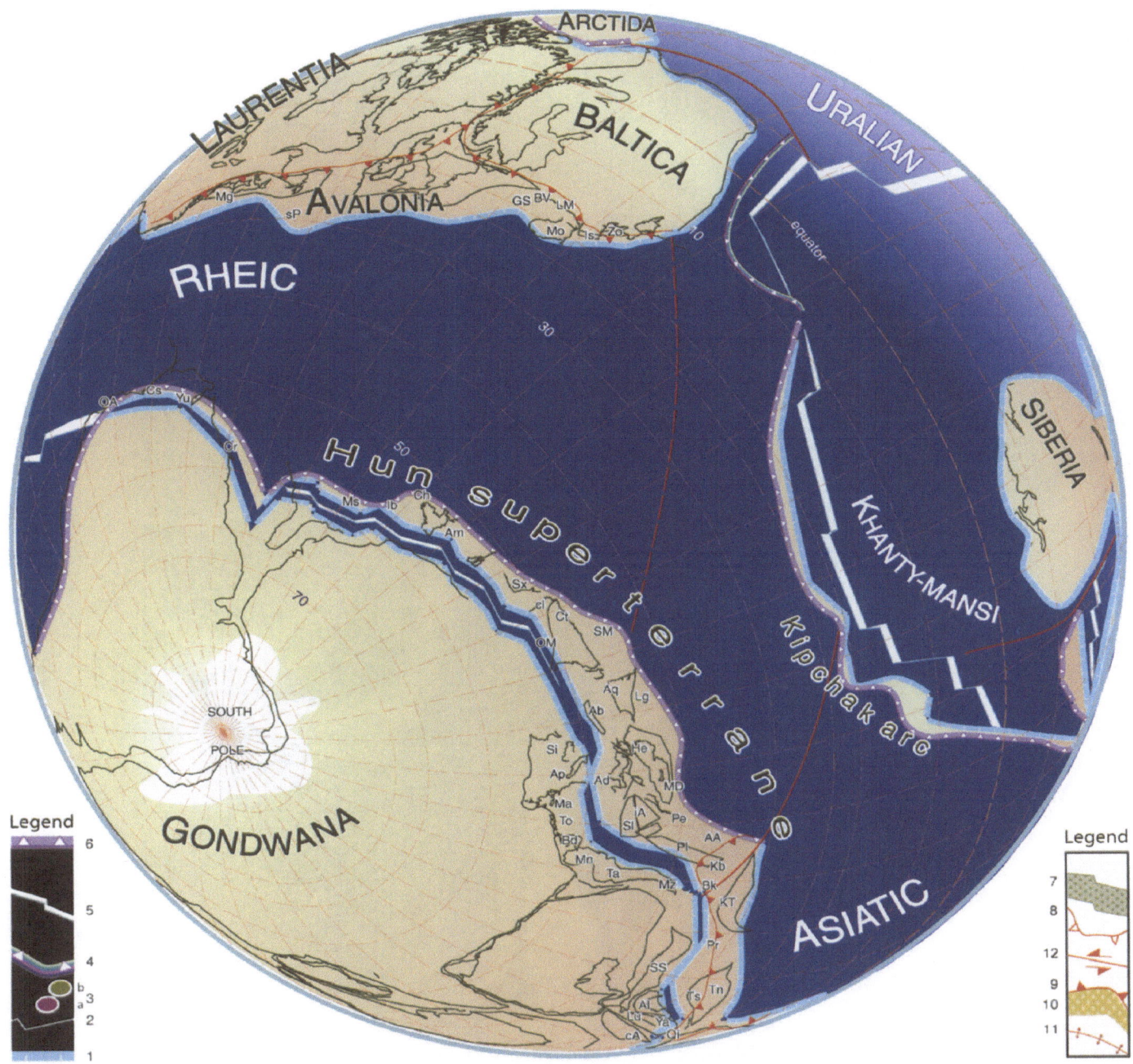

Fig. 3.2. 420 Ma (Late Silurian) paleotectonic-paleogeographic reconstruction. Legend for Figs. 3.2–3.13 (modified after Stampfli and Borel 2002): *1* passive margin; *2* magnetic anomalies or synthetic anomalies; *3* seamount (*a* active, *b* inactive); *4* intraoceanic subduction/arc complex; *5* spreading ridges; *6* subduction zone; *7* rifts; *8* sutures; *9* active thrusts; *10* foreland basins; *11* flexural bulge; *12* strike-slip faults. By this time the peri-Gondwanan Prototethys ocean was replaced by the Rheic ocean in the west and by the Asiatic ocean in the east. The opening of the Rheic detached the Avalonia terrane from Gondwana. Baltica underwent major rotation during early Paleozoic times, to finally collide with Laurentia through the closure of the Iapetus ocean and the resulting Caledonian orogeny. Avalonia just collided with Laurentia. The eastern tail of Avalonia left some Gondwana-derived terranes stranded on the SW side of Baltica, from Turkey to Poland, following the closure of the Tornquist oceanic domain. The subduction of young oceanic lithosphere under northern Gondwana generated a mid-Ordovician orogenic phase, followed soon by the creation of a new peri-Gondwana cordillera due to southward subduction of the Rheic ocean. This, in turn, generated the opening of the Paleotethys and drifting of the Hun superterrane away from Gondwana

areas of Turkey comprised between the Pontides and Taurides domains (e.g. in the Konya region; Kozur et al. 1998; Stampfli et al., in press), and should also be present beneath Apulia (e.g. volcaniclastic Permian of the well Gargano), to reappear westward in southern Spain (Stampfli et al. 2002b). The youngest fore-arc-type se-

quences related to the northward subduction of Paleotethys are found in Crete, Turkey, Iran and Afghanistan, where they extend into the Carnian (Stampfli et al., in press) and were affected by the Eocimmerian tectonic event that marked the final closure of Paleotethys in the western Tethyan region.

Fig. 3.3. 380 Ma (Eifelian-Givetian). The Hun superterrane split into two major terranes; the European and the Asiatic Hunic terranes. The former collided with terranes derived from Laurussia (Hanseatic terrane), as shown by a widespread HP-LT metamorphic phase along the leading edge of the terrane, and also by exchange of Gondwana and Laurussian fauna at that time. The Hanseatic terrane detached from Laurussia through the opening of the Rheno-Hercynian ocean, which is not a back-arc, but opened through strong slab-pull forces generated by the southward subduction of the Rheic ocean. On its eastern side, Baltica developed another passive margin through the opening of the Mughodzar ocean. A link between the Rheno-Hercynian and Mughodzar oceans is found in Turkey, the Paphlagonian ocean. In the Asiatic ocean, amalgamation of intra-oceanic arcs created an embryonic Khazakstan plate. Such arcs locally collided with the east side of the European Hunic terranes, whereas the west side of these (e.g. Yucatan, Chortis) collided with southern Laurentia

3.3.2   Cimmerian Events and Triassic Marginal Oceans (Figs. 3.6–3.9)

The apparent lack of major tectonic events during the Late Permian and Triassic in SW Europe, or in the Appalachian domain, follows the final welding of Gondwana and Laurasia, which together formed the Permo-Triassic Pangea. In earlier papers we discussed the diachronous closure of the large Paleotethyan ocean and focused on the likely Permo-Triassic development of a system of back-arc oceans or basins along the southern Eurasian margin (Stampfli 1996; Ziegler and Stampfli 2001). East of the paleo-Apulian promontory, a domain of Eocimmerian (Middle to Late Triassic) deformation occurs just south of a relatively "undeformed" Variscan domain which is represented by Late Carboniferous-Early Permian arc and clastic sedimentation of Verrucano type,

Fig. 3.4. 340 Ma (Visean). At this time, the amalgamated European Hunic and Hanseatic terranes were colliding with Laurussia. Subsequently, subduction prograded to the southern margin of the Hunic terranes and northward subduction of Paleotethys created the Variscan cordillera. Westward collapse of the latter was induced by the oblique subduction of the Paleotethys mid-oceanic ridge which, in turn, generated imbrication and lateral displacement of the European Hunic terranes as well as major plutonism in the Variscan domain. Northward subduction of Paleotethys was also active under the Asiatic Hunic terranes. The closure of peri-Khazakstan back-arc basins contributed to the enlargement of this continental block; closure of the Khanti-Mansi and Uralian oceans will eventually lead to the formation of Laurasia. The Visean Sea (*dashed blue line*) is now reaching far inside the northern margin of Gondwana, still affected by episodic glaciations

mainly affected by extension (Cassinis et al. 1995). In contrast, Cimmerian deformation was accompanied by the deposition of Triassic flysch, the development of mélanges, and by the emplacement of collisional type volcanics (Castellarin et al. 1988; Pe-Piper 1998; Stampfli et al. 2003) and intrusives (Reischmann 1998) marking the closing of Paleotethys. The final closure of some of the European marginal oceans (e.g. Küre, Svanetia) only happened during the Jurassic and represents the Cim-

merian deformation cycle, which characterizes the Balkan, Black Sea, and Caucasus regions.

As shown in the plate tectonic reconstructions of Fig. 3.6, the Middle Permian margin of SE Europe was of a transform-type and little subduction occurred along it due to locking of the Gondwana-Laurasia suture. However, roll-back of the Paleotethys slab induced "back-arc" rifting along the entire Paleotethys margin starting from the Late Permian. East of the paleo-Apulian promontory,

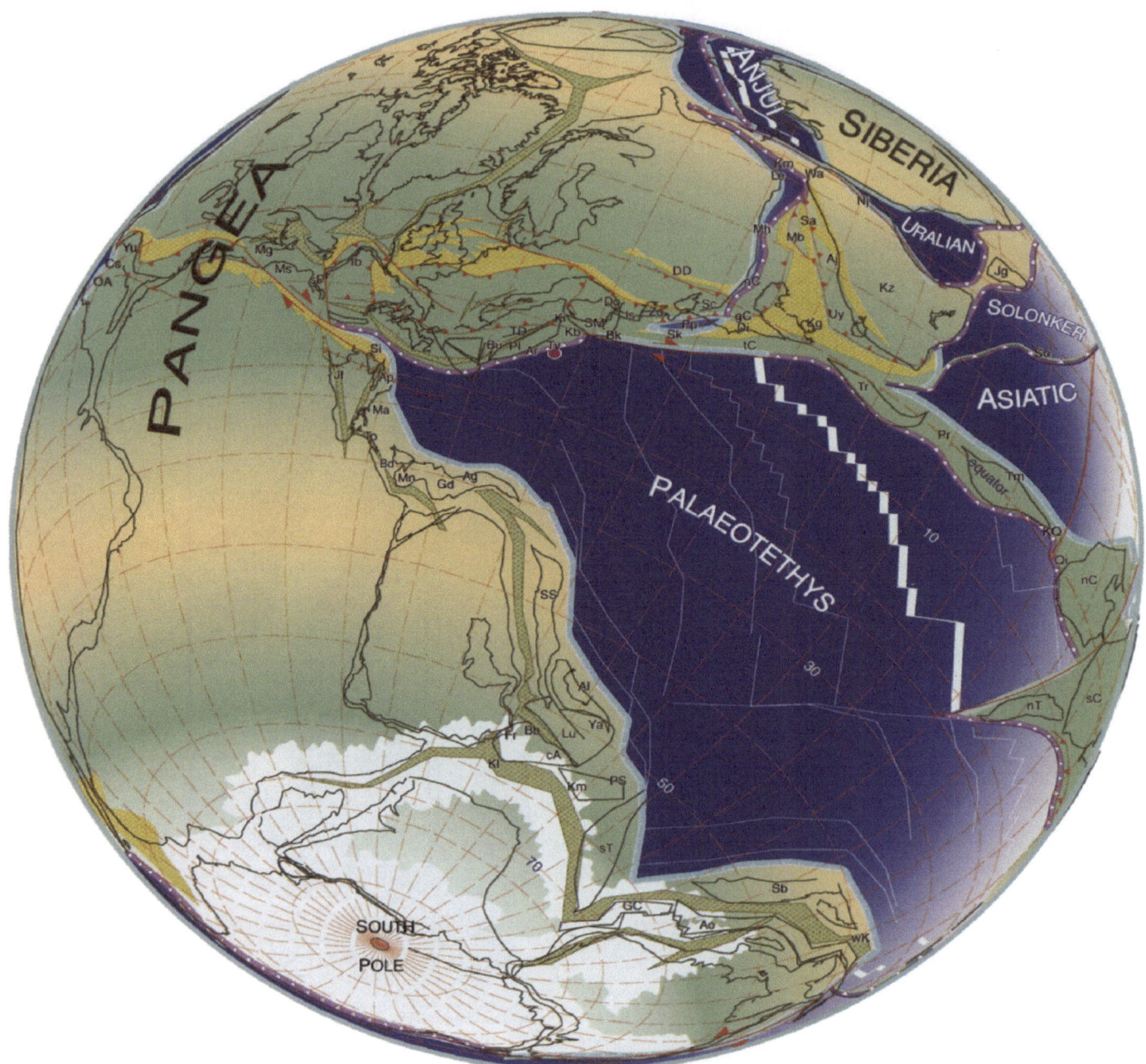

Fig. 3.5. 300 Ma (Kazimovian). Westward collapse of the Variscan cordillera was stopped by the collision of Gondwana with Laurentia, which resulted in the Alleghenian orogeny and the formation of Pangea. In southern Europe, subduction of Paleotethys continued; however, the coupling of Gondwana and Laurussia prevented further narrowing of the oceanic space between them. This was compensated by two processes: roll-back of the Paleotethyan slab and extension along the northern border of Gondwana, which eventually lead to the opening of Neotethys. The Neotethyan rift zone around India and Australia was already quite active, and its rift shoulders were covered by ice caps, which were invading the rift zone. The Paleotethys mid-ocean ridge was now rapidly approaching the active margin; this will result in ridge failure and opening of Permian intra-oceanic back-arc basins, whose remnants are found from Iran to China. The Khazakstan and Siberian blocks were soon to collide with Laurussia by consumption of the intervening oceanic basins, to be replaced by mountain belts such as the Urals

this back-arc rifting progressed to sea-floor spreading and opening of a series of marginal oceanic basins. Due to accelerated roll-back of the Paleotethys, sea-floor spreading in the Hallstatt-Meliata basin terminated during the Early Ladinian at the expense of the development of a new spreading axis in the Maliac basin, and again in the Early Carnian, the spreading axis changed from the Maliac to the Pindos basin. Therefore, the Late Carboni-

ferous Pelagonia arc terrane got stranded between the Maliac and Pindos oceans and never collided with Greater Apulia; the boundary between the two plates being of a transform type (Figs. 3.6 and 3.7). South of the Pindos, the small Variscan ribbon-shaped Sitia terrane and its Triassic fore-arc collided with Greater Apulia during the Late Carnian (Stampfli et al. 2003); the same applies to the Tavas region in SW Turkey (Kozur et al. 1998).

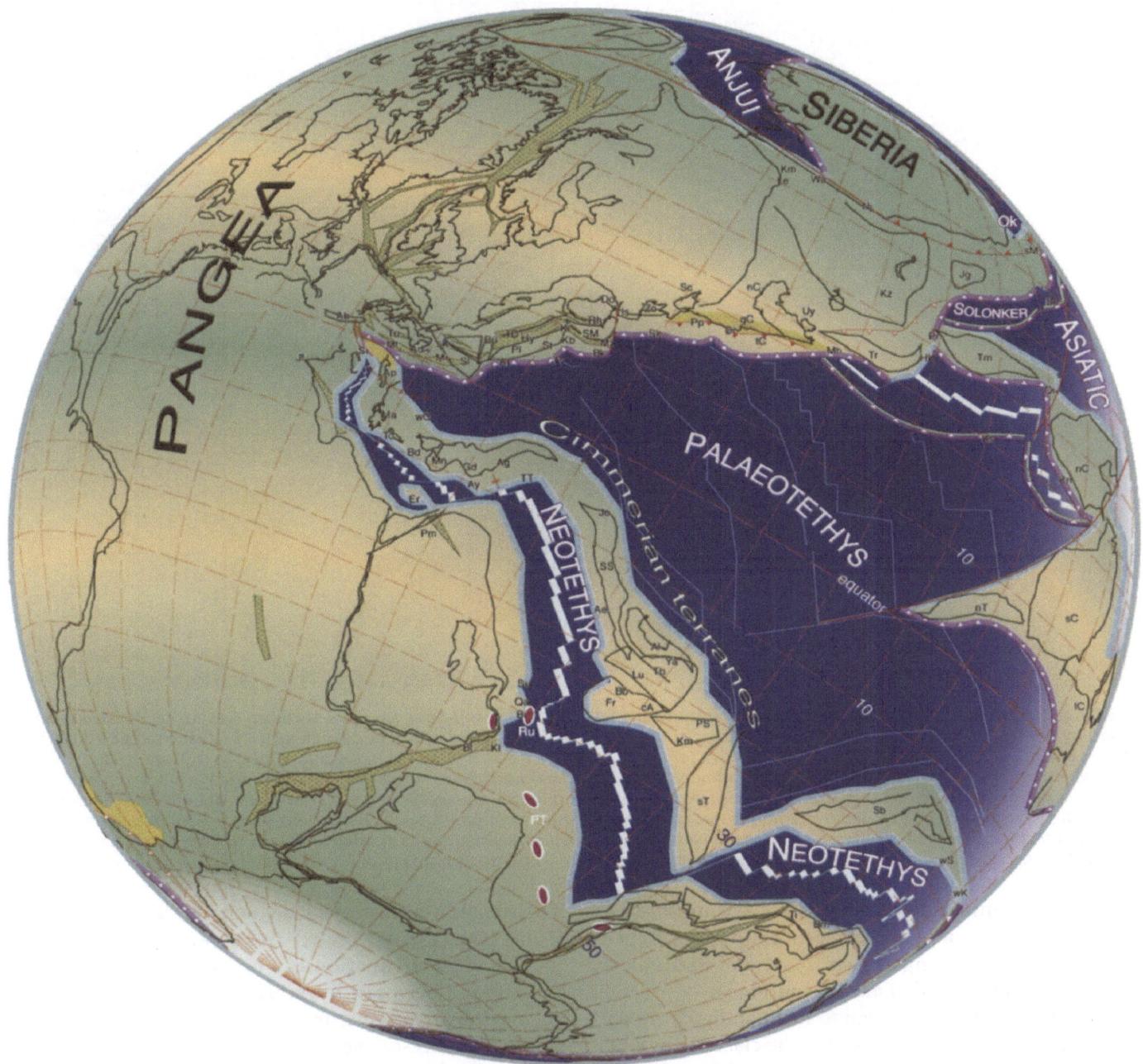

Fig. 3.6. 260 Ma (Middle-Late Permian boundary). Neotethys sea-floor spreading was now active from Sicily to Timor (as shown by similar ammonite and conodont faunas found in both places, and in Oman). Opening of the Neotethyan ocean detached the ribbon-like Cimmerian terranes from Gondwana, and invasion of warmer water around the latter, as well as its drifting away from the pole, brought an end to the Gondwanan glaciation. The Asiatic Hunic terranes started to disaggregate and to collide with Laurasia; this will lead to the rotation of the Chinese blocks and their collision in Triassic times. Back-arc extension was quite active along the Paleotethys northern margin and was characterized by a general collapse of the Variscan cordillera from Italy to Iran. The back-arc rift zones are locally invaded by the sea, as was the rift that developed between Greenland and Europe. Laurasia was forming a single block, enlarging considerably the Pangean supercontinent. The other hemisphere of the planet consisted entirely of the Panthalassa ocean

Along the Cimmerian orogen, development of carbonate platforms resumed in Norian to Liassic times (Fig. 3.8), marking the end of this orogenic cycle. This Middle-Late Triassic tectonic pulse is well documented in the Dolomites where it was also accompanied by the emplacement of salt diapirs (Castellarin et al. 1996) and preceded by important volcanism, first in the Permian, then in the Middle Triassic (Pietra Verde event). This shows that northern Italy was strongly affected by processes related to the opening of Triassic marginal seas and closing of Paleotethys.

Apart from these Hellenic Paleotethyan remnants, continental to marine upper Carboniferous to lower Permian sequences have been identified at the following locations along the arc of the Eurasian active margin (Figs. 3.5 and 3.6):

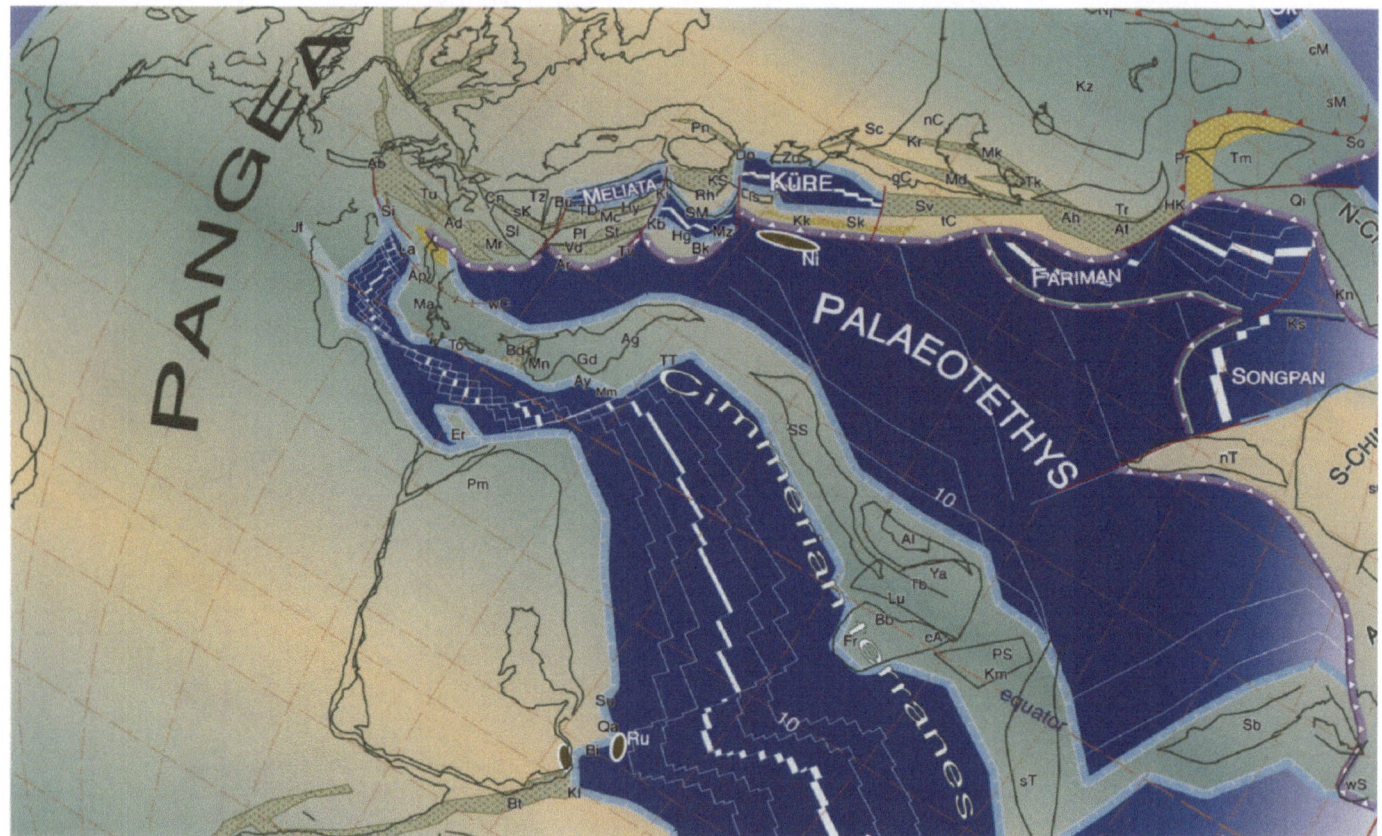

Fig. 3.7. 240 Ma (Anisian). As Paleotethys subduction came to a final stage, slab roll-back along its northern margin accelerated and was marked by the opening of oceanic back-arc basins such as Meliata and Küre. There was still enough space available to open successive back-arc basins south of Meliata (Maliac, then Pindos) before the final closure of Paleotethys in Carnian times. To the east, back-arc rifting was very active also in the Caucasus (Svanetia) and north Iran (Agh-Darband), in an attempt to link with the Fariman-Songpan intra-oceanic back-arc system. To the west, extension was affecting Adria and Iberia, a possible link with the north Atlantic rift system existed through the Pyrenean rift. The Nilüfer seamount was colliding with the Karakaya fore-arc basin, and both will soon collide with the Cimmerian terrane. The East Mediterranean part of Neotethys was still spreading; on its northern margin Greater Apulia represents at this time the westernmost part of the Cimmerian terranes, which were colliding (Sibumasu) with the Anamia block to the east

- In the Carboniferous of the Tuscan Apennines (e.g. Gattiglio et al. 1989, and references therein; Engelbrecht et al. 1989; Engelbrecht 1997) and Permian of Monte Amiata (Pandeli and Pasini 1990).
- In the deep-water Kungurian to Roadian flysch (Kozur and Mostler 1992; Kozur 1999) of the clastic Trogkofel beds (Ramovs 1968) found just south of the Periadriatic line (Mrzlevodice fore-arc).
- In the Permo-Carboniferous deep-marine clastic sequences of the Carnic and Karawanken Alps.
- In the deep-water Permian of Sicily (Sicanian basin) and its Early Permian pelagic fauna (Catalano et al. 1988, 1995) which presents clear-cut Paleotethyan affinities (Kozur 1990).

These series are attributed to a system of late Variscan fore-arc-type basins that was located southward and adjacent to a cordillera characterized by Late Carboniferous to Early Permian subduction-related plutonism (Stampfli 1996; Vavassis et al. 2000). These fore-arc basins were slightly deformed or uplifted during Middle Permian times and unconformably covered by

continental Verrucano s.l. sequences, deposited in rift basins (Cassinis et al. 1995). In Tuscany (Aldinucci et al. 2001), a clear distinction can be made between rift-related upper Permian-lower Triassic deposits, containing marine incursions, unconformably covering the Carboniferous fore-arc sequences, and younger Verrucano s.s. clastics of Middle to Late Triassic age (Cirilli et al. 2002). The older deposits correspond to the rifting phase responsible for the opening of the Meliata-Maliac back-arc system to the east; the continental Verrucano s.s. deposits of Late Triassic age are related instead to the Eocimmerian tectonic inversion events.

3.3.3 The Jurassic Oceans: Alpine Tethys, Central Atlantic and Vardar (Figs. 3.8–3.11)

The results of field work on the Canary Islands and in Morocco (Favre et al. 1991; Favre and Stampfli 1992; Steiner et al. 1998) indicate that the onset of sea-floor spreading in the northern part of the Central Atlantic occurred in the Toarcian. Similar subsidence patterns between this

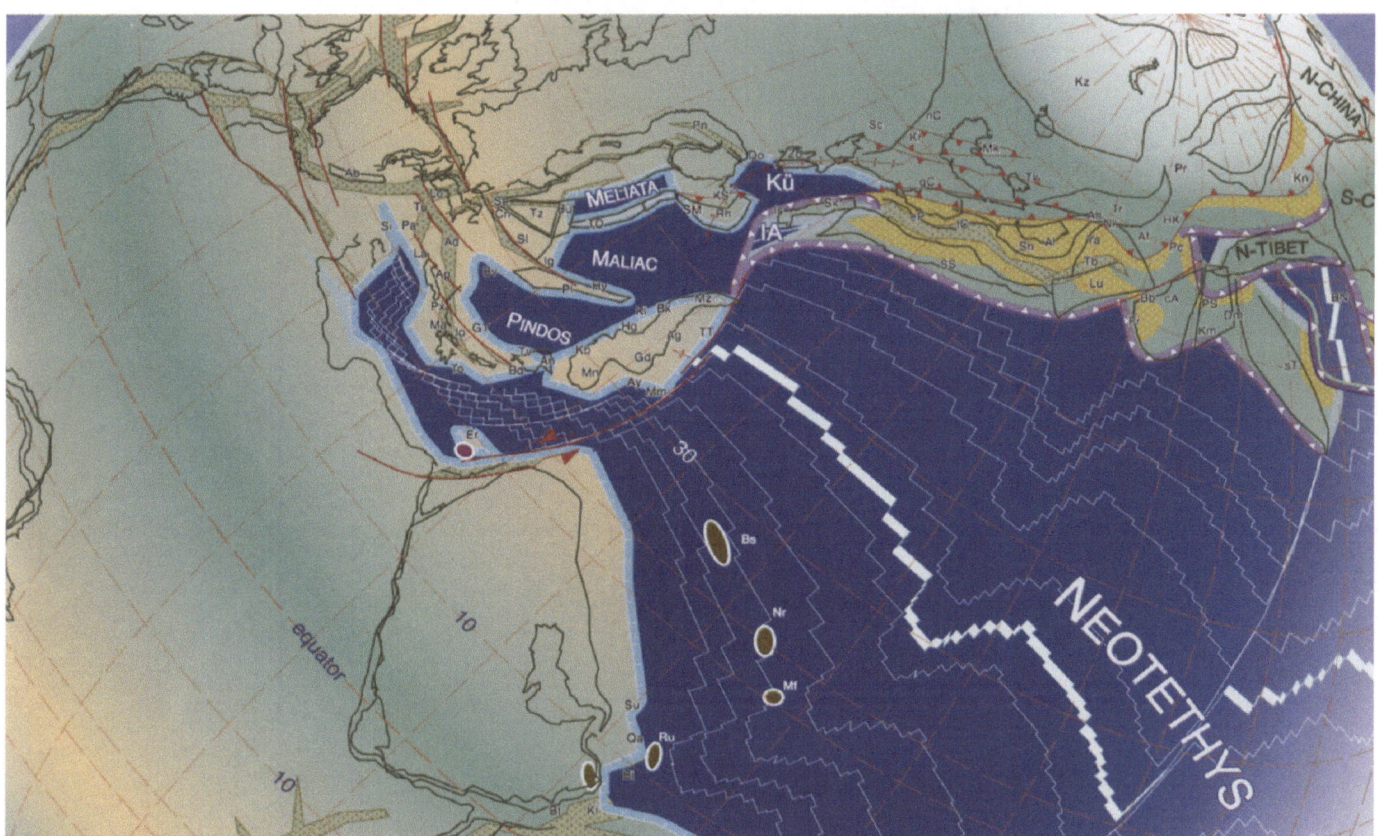

Fig. 3.8. 200 Ma (Sinemurian). At this time Paleotethys was completely closed except for some small remnants around the north-Tibet promontory. In Iran, the closing of Paleotethys generated the development of a large molassic basin (Shemshak) and subduction progradation to the northern side of Neotethys, marked by the onset of subduction-related volcanism in the Sanandaj-Sirjan and Lut blocks. In turn, this subduction created the possibility for Gondwana to separate from Laurasia, as space was provided for Gondwana to move eastward, despite the fact that Neotethys was still spreading. Thus, the Central Atlantic rift widened but had difficulty finding a way to link with a plate limit to the east. This activated many rift zones in the Alpine-Mediterranean area, even along the Levant transform margin. Most of these rifts will eventually abort, only the Alpine Tethys rift will open in Bajocian times, south of Iberia, south of France, and along the Alpine region up to the Carpathians. The closure of Paleotethys south of the Küre basin generated the southward subduction of the latter. Slab roll-back, both in Küre and the Neotethys, allowed the opening of the Izmir-Ankara (IA) ocean. This opening will prograde eastward up to the south Caspian region in middle Jurassic times. The reason for Pangea break-up is also to be searched in the Panthalassa domain, because East- and West-Gondwana started also to separate in Early Jurassic times, with the development of a major rift between Africa and India

region and the Lombardian basin (Stampfli 2000) led us to propose a direct connection between these areas (Fig. 3.9). The Lombardian basin aborted in Middle Jurassic times (Bertotti et al. 1993) as it could not link up with the oceanic Meliata-Maliac-Pindos domains whose already cold lithosphere was rheologically considerably stronger than the surrounding continental areas. Therefore, the Alpine Tethys rift opened to the north of the Meliata basin, separating Adria and the Austro-Carpathian domain from Europe. Thermal subsidence of areas flanking the Alpine Tethys commenced in the Aalenian in the west (Briançonnais margin: Stampfli and Marchant 1997; Stampfli et al. 1998; Stampfli et al. 2002) and in the Bajocian further to the east (Helvetic and Austroalpine margin: Froitzheim and Manatschal 1996; Bill et al. 1997). Alpine Tethys spreading was considerably delayed with respect to the Central Atlantic, very slow spreading gave birth to a limited amount of oceanic crust, the oceanic area being dominated by continental

mantle denudation. A larger transform Maghrebide ocean linked the central Atlantic and the Alpine Tethys, and was also characterized by delays in thermal subsidence (e.g Rif area, Favre 1995; Stampfli 2000).

Within the Alpine domain s.s. there is a fundamental difference between the Austroalpine-Carpathian and Western Alps systems. Austroalpine-Carpathian evolution was rooted in the dynamics of the Triassic back-arc basins located to the south (Meliata-Maliac domain). These back-arc basins were shortened in conjunction with the opening of the Central Atlantic and rotation of Africa with respect to Europe. Subsequent slab roll-back of the Maliac-Meliata sea floor induced the opening of the Vardar supra-subduction ocean, which, by Late Jurassic times, had completely replaced the pre-existing oceanic basins (Fig. 3.10). Continued rotation of Africa provoked ridge failure in the Vardar and large scale Late Jurassic ophiolitic obduction onto the Dinaride-Hellenide passive margin of the Pelagonian terrane (e.g. Laubscher and

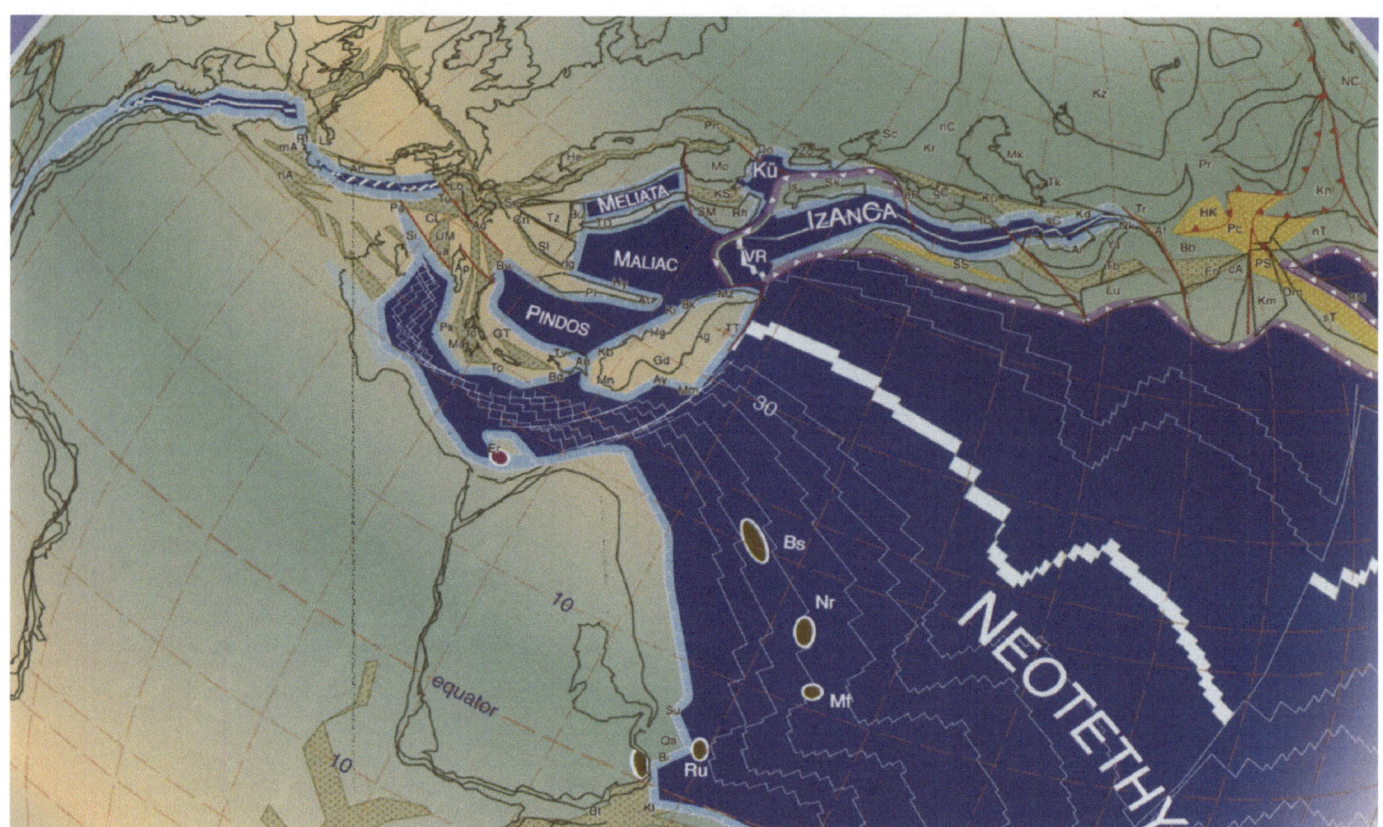

Fig. 3.9. 180 Ma (Toarcian-Aalenian). Spreading is now active in the Central Atlantic and the segment of the Alpine Tethys located south of Iberia. From there, two main rift zones were spreading, one branch went southward toward Greece (Ionian rift system), the other went eastward to the Carpathians domain (Alpine Tethys). The latter will successfully open in Bajocian times as it was able to connect with a plate limit through the Dobrogean transform and from there to the active subduction zone around the Vardar ocean (*VR*). Opening of the Vardar corresponds to subduction progradation from the Küre domain toward the Maliac domain, but can also be regarded, together with the Izmir-Ankara ocean, as back-arc extension of the large Neotethys. However, roll-back of the Maliac sea floor, was generating westward spreading of the Vardar basin in a scenario of intra-oceanic subduction. Closure of the Küre basin was entering a collisional stage in the Black Sea domain and around the Rhodope promontory. The Izmir-Ankara ocean was extending eastward to the south Caspian domain (*IzAnCa*), following more or less the Paleotethys suture zone, which was extensionally reactivated up to central Afghanistan (Panjao ocean, Fig. 3.10). The rift between Africa and India was still active, with new branches developing towards central Africa. Gondwana was soon to be separated in two large landmasses

Bernoulli 1977; Dercourt et al. 1986). Roll back of the oceanic Maliac-Meliata slab was a centrifugal phenomenon that controlled successive collision of the Vardar arc with all passive margins surrounding the Meliata-Maliac basin. Following obduction on the western Dinaric margin in Late Jurassic times, the northeastern part of the Vardar arc-trench system collided with the northern passive Meliata margin – corresponding to the Northern Calcareous Alps (NCA), the Western Carpathian domain and the Rhodope – from Late Jurassic to Early Cretaceous times. Closure of the Balkan rift system between Moesia and the Rhodope (Fig. 3.11) controlled the development of the Balkan orogen, accompanied by large-scale Early Cretaceous northward nappe emplacement and metamorphism (Georgiev et al. 2001). Remnants of the Vardar arc are found in northern Greece and Bulgaria as tectonic klippen (Bonev and Stampfli 2003). This circum Vardar orogenic event commenced in the Late Jurassic and was sealed by Albian to Cenomanian molasse-type sediments.

Along the NCA margin, elements of the Austro-Alpine micro-continent were scraped off and incorporated into the accretionary wedge, to form the different internal units of the Austro-Carpathian orogen (Kozur 1991b; Plasienka 1996; Faupl and Wagreich 1999). This event was accompanied by Early Cretaceous HP-LT metamorphism (e.g. Thöni and Jagoutz 1992). Subsequently, the enlarged accretionary wedge began to incorporate the entire eastern segment of the Alpine Tethys (Figs. 3.11 and 3.12) (Penninic-Vahic ocean), to finally collide with its northern passive margin (Helvetic domain s.l., Magura rim basin) thus forming the present Eastern Alps and Carpathian orogen (Wortel and Spakman 1993). This process involved continuous slab roll-back that had commenced during the Carnian in the Küre domain and continued into the Neogene period in the Eastern Carpathians.

A relatively large remnant of the Vardar ocean subducted northward under Moesia during the Late Cretaceous, as evidenced by the large Srednogorie volcanic arc of the Balkans and the Late Cretaceous opening of the Black Sea (Nikishin et al. 2003), representing the third generation of back-arc opening in that region.

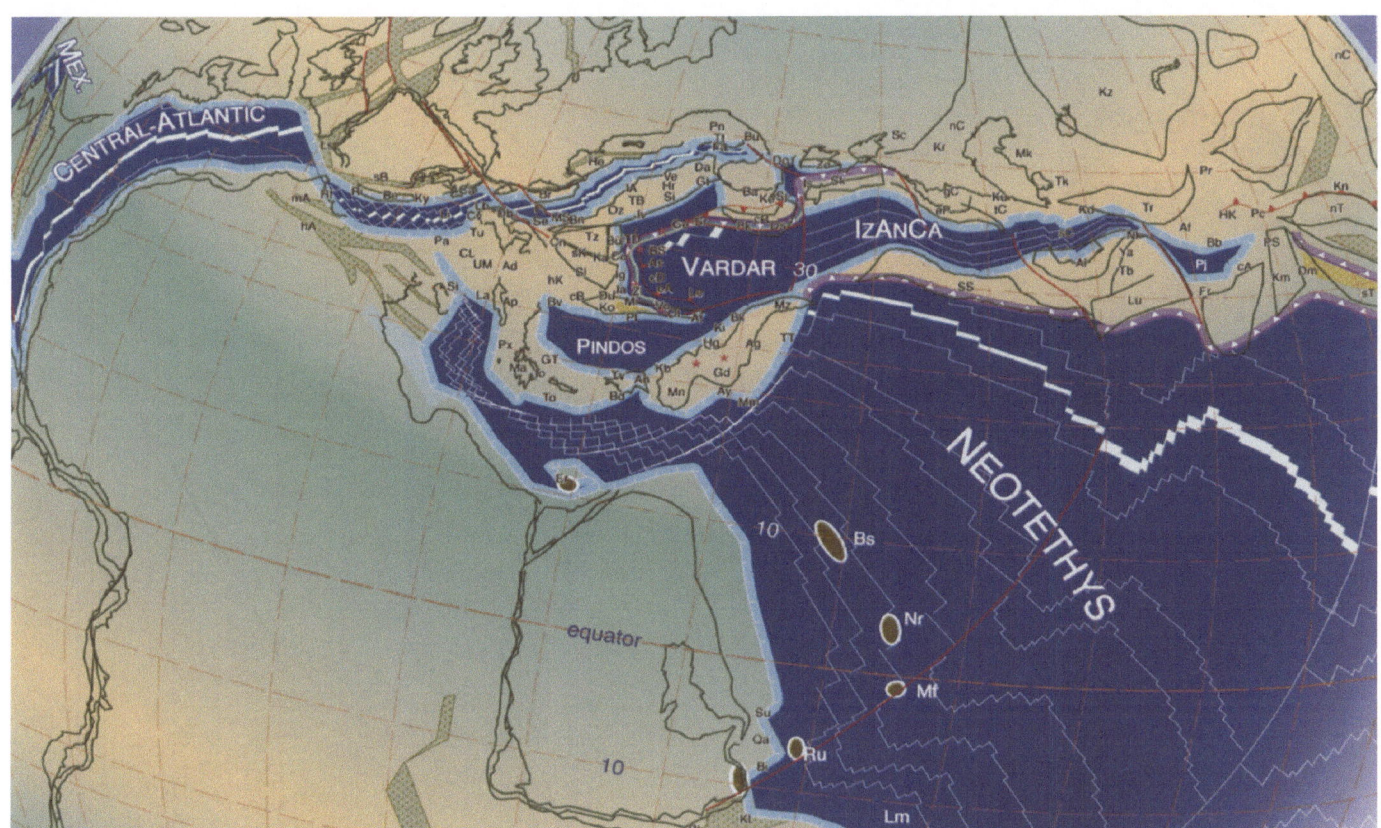

Fig. 3.10. 155 Ma (Oxfordian; magnetic anomaly M25). Spreading in the Alpine Tethyan ocean now reached the Carpathian domain. In the process, Moesia was detached from Europe by only of a few hundreds of km, the point of rotation of Gondwana being located in Moesia. Northward extension of the Central Atlantic was very active between Iberia and Newfoundland, and the northern limit of the Iberian plate was affected by pull-apart rifting extending eastward into the Briançonnais domain, through the Pyrenees and Provence. The Küre ocean was nearly closed, collision of its arc-trench system with the Rhodope was causing the first phases of the Balkan orogeny, accompanied by inversion of former rift zones. Roll-back of the Maliac slab allowed rapid westward expansion of the Vardar back-arc basin, its arc trench system was soon to collide with the Pelagonian and Dinaric landmass. This east-west shortening in the Maliac-Vardar domain was due also to the anti-clockwise rotation of Gondwana with respect to Europe. We regard this event as creating a change of spreading direction in the western Neotethyan domain, at the origin of a mid-ocean ridge failure. Differential movement between Africa and India was reactivating a Neotethys former transform, separating the future Indian plate from Africa

3.3.4 The Cretaceous Oceans: North Atlantic and the Pyrenean Domain (Figs. 3.10–3.14)

Starting in Early Cretaceous time (Figs. 3.10 and 3.11) the Pangean break-up had more and more difficulty to link up eastward with another plate limit due to intra-oceanic subduction in the Neotethys and shortening affecting the Vardar region. As a result, the Pyrenean and Biscay oceans opened, with a rifting phase starting in the Oxfordian, and the onset of spreading in the Aptian for the Portuguese-Galician ocean and Pyrenean area, and in the Albian for the Gulf of Biscay (Stampfli et al. 2002). Then, by the end of the Santonian the break-up between North America and Greenland took place, and in Campanian time the Biscay ocean aborted. Closing of the Pyrenean domain already took place during the opening of the Gulf of Biscay due to the accelerated rotation of the Iberian plate together with Africa. It is uncertain how wide the Pyrenean ocean was, and whether it was limited to mantle denudation as indicated by the

lherzolites at Lherz (Fabries et al. 1998); a minimum extension of 60 to 80 km can be calculated from restored cross-sections (Vergés and Garcia-Senz 2001). Sea-floor spreading was recently dated as late Early Cretaceous sea in the eastern Alps, thus confirming that the Iberian-Briançonnais northern plate boundary reached that region (Froitzheim, pers. comm.). The rotation of Iberia also placed the Briançonnais peninsula in front of the Helvetic margin, creating a repetition of the European margin in the western Alps domain. The space between these two margin segments is generally referred to as the Valais ocean, a domain that actually formed part of the Piemont (Alpine Tethys) ocean, which was trapped by the eastward displacement of the Iberian-Briançonnais block during the opening of the North Atlantic. The Middle Jurassic age of the Valais sea floor was recently confirmed by Liati et al. (2003).

At least from the Aptian (Mo magnetic anomaly) to the Maastrichtian (Stampfli and Borel 2002) and most likely up to the Thanetian (anomaly C25), Iberia rotated with Africa and Apulia-Adria, without appreciable N-S

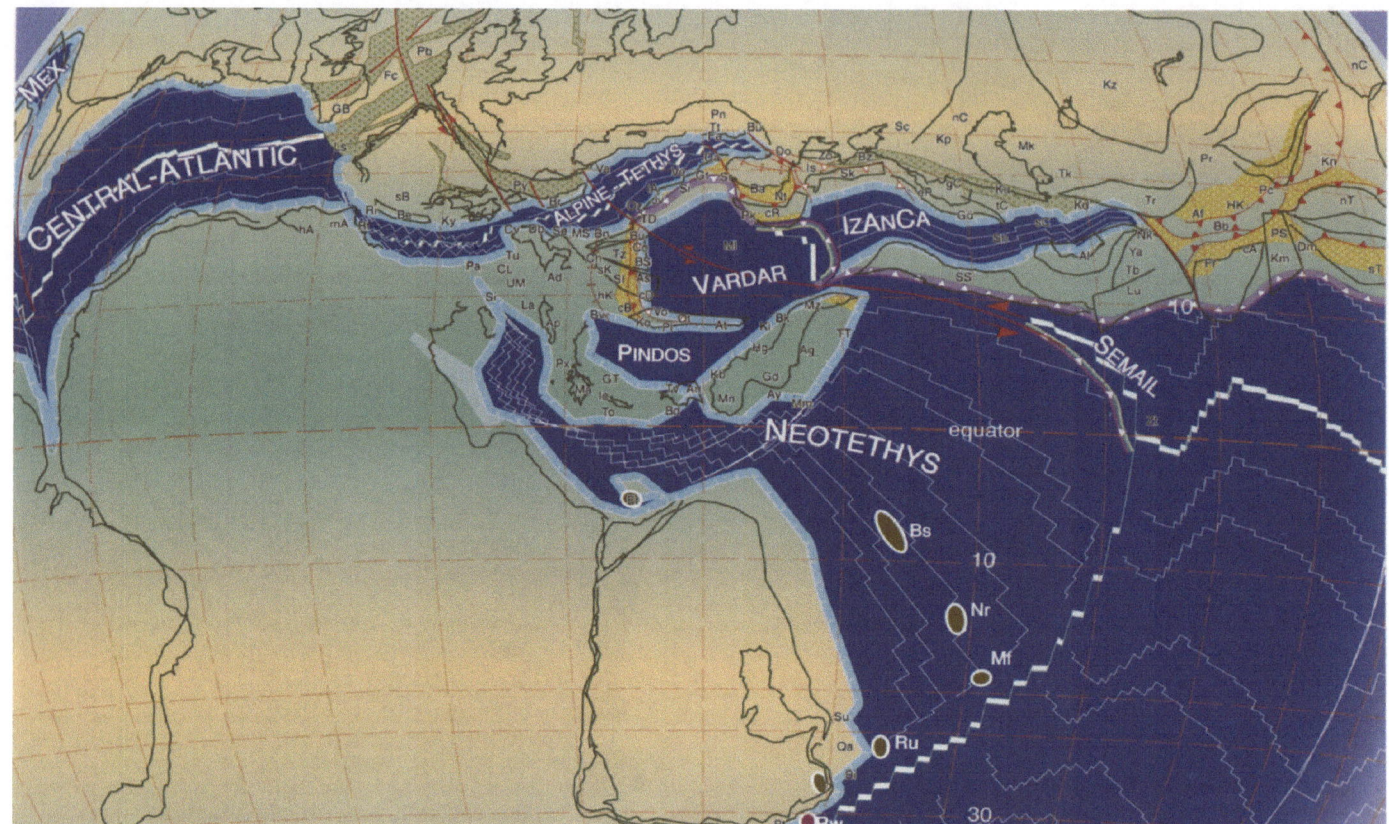

Fig. 3.11. 131 Ma (Valanginian-Hauterivian; magnetic anomaly M10). Accelerating anti-clockwise rotation of Gondwana was responsible for the obduction of part of the Vardar mid-ocean ridge system onto the Pelagonia, Dinaride and Tizia blocks. Collision of the Vardar arc-trench system with the northern margin of Meliata was taking place at this time, detaching the future North Calcareous Alps domain from its basement (internal Austroalpine). Collision of the Vardar arc-trench system continued also in the Balkan, where parts of the Rhodope cover and basement were transported northward and thrust onto the Nish-Troyan trough. The major changes affecting the Neotethyan domain brought to an end the opening of the IzAnCa (south Caspian) back-arc basin system, and closed the Panjao basin in Afghanistan. The Izmir-Ankara slab started retreating eastward, allowing the opening of a new supra-subduction spreading centre. In the western Neotethys, ridge failure generated a new intra-oceanic subduction zone along the former spreading centre; this new oceanic domain will eventually obduct onto Arabia (e.g. Semail ophiolites). Hot and buoyant sea-floor was subducting under the Iranian blocks generating a cordillera stage, and marine deposits are only found along the South-Caspian margin. Spreading was also taking place along the plate limit between the now separated West- and East-Gondwana blocks. The Iberian plate was nearly totally detached from Laurasia; spreading was still active in the Alpine Tethys, but soon Iberia and Africa will form a single plate

shortening. However, the southern margin of the Piemont ocean (Southern Alps-Austroalpine domain of the western Alps) was affected by tectonic movements since the Coniacian-Santonian, as evidenced by the onset of flysch deposition in the Piemont area (Gets and Dranse flysch; Caron et al. 1989) and in the Lombardian basin (Bernoulli and Winkler 1990). We relate such tectonism to large-scale sinistral strike-slip movements which affected the boundary between Adria and the Alpine domain, due to large scale E-W shortening taking place in the Vardar area during Albian to Campanian times (Figs. 3.10–3.13). These very large-scale lateral movements are well known also to the east (Trümpy 1988; Trümpy 1992) and finally placed Adria-Tizia behind the Austroalpine accretionary prism. In the process, the Piemont oceanic lithosphere was progressively detached from the northern margin of Adria and subducted beneath it. In conjunction with its westward escape the Austroalpine prism probably collided locally with the northern Helvetic margin already in Late Cretaceous times (e.g. area of the Tauern window) and

with Calabria (Austroalpine Longobucco unit) and the eastern Briançonnais peninsula (Tasna-Falknis) in Late Cretaceous-Paleocene times. Frontal pieces of the western Adria-Austroalpine margin were dragged into the subduction zone, as evidenced by the Late Cretaceous-Early Paleocene HP/LT metamorphism recorded in the Sesia domain (Oberhänsli et al. 1985; Rubatto 1998) and Calabria (Colonna and Piccarreta 1975).

During the Late Cretaceous oceanic basins opened in Turkey (Lycian) and further east (Semail-Spongtang) (Figs. 3.13 and 3.14). These are supra-subduction intra-oceanic back-arc type oceans, which developed at the expense of preexisting oceans, remnants of which totally disappeared or where partly recycled in accretionary or fore-arc units found along the suture zones (e.g. the famous Ankara mélanges: Gansser 1974; Transect VIII, this publ.). This is also the case in Oman where remnants of Neotethys are only found as exotic blocks at the sole of the Cretaceous Semail ophiolite (Pillevuit et al. 1997), a situation which can be extended eastward to the Hima-

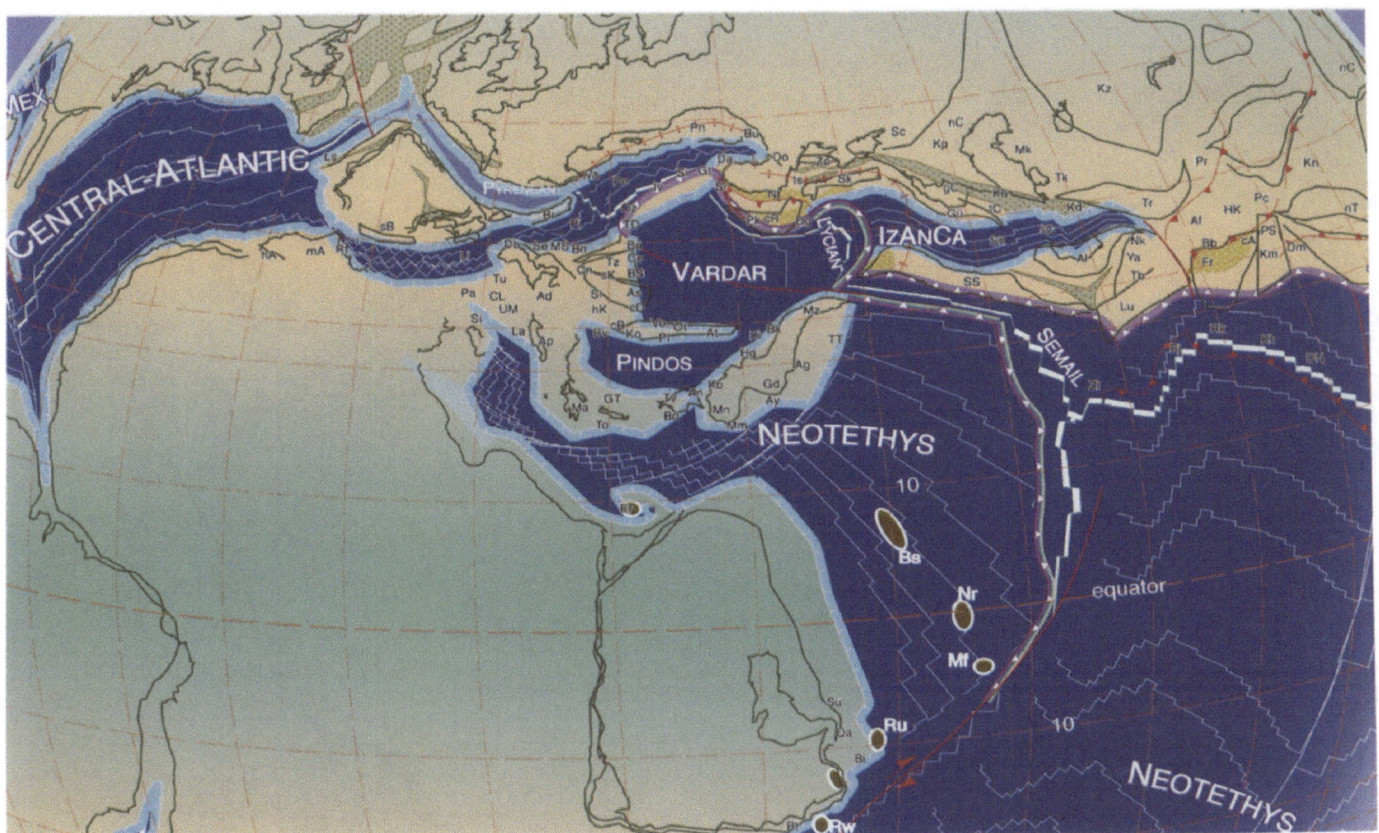

Fig. 3.12. 121 Ma (Barremian-Aptian; magnetic anomaly Mo). At this time spreading in the Central Atlantic extended northward, between Iberia and Newfoundland, and the African plate limit was north of Iberia, in the Pyrenees. The southward movement of Africa-Iberia with respect to Europe, generated a new spreading centre in the Alpine Tethys, separating its west and east segment. In the latter, subduction progradation brought the exotic Austroalpine terranes onto the Alpine Tethys, whereas its west segment was being passively transported with the African plate. Orogenic processes were soon to come to an end in the Balkan (sealed by Albian molasses), whereas the Lycian intra-oceanic system extended eastward. In this process a flexural basin developed along the northern margin of the Izmir-Ankara ocean, with emplacement of mélanges on the Pontides domain. Thereafter, the Lycian ocean obducted southward on the Anatolide-Taurus block, the obduction is sealed by Maastrichtian shallow-water deposits. The Semail intra-oceanic back-arc system was expanding southward following slab retreat of the Neotethys in that direction. Spreading between the African and Indian plates stopped and was replaced by a major dextral transform system. The remnant Neotethyan spreading ridge, south of the Afghan-south Tibet cordillera, will soon fail as it is too buoyant to be subducted. Its eastern segment will give birth to a supra-subduction back-arc spreading system, whereas its western segment will obduct onto the Lut block. Thereafter, subduction of negatively buoyant Neotethyan sea floor will allow the Indian plate to start its fast motion northward. The presence of a spreading ridge attached to the Indian plate until this Mo anomaly reconstruction is necessary, as India was going southward with respect to Africa

laya (e.g. Bassoulet et al. 1981; Robertson and Degnan 1993) and westward to Cyprus (Transect VIII, this publ.) (e.g. Blome and Irwin 1985; Robertson and Xenophontos 1993).

3.4 The TRANSMED Transects in Space and Time

In this section we review the TRANSMED transects in terms of their location in space and time within the framework of the paleotectonic reconstructions presented above. In support of the following discussion, we show the different – mainly continental – segments of the transects (except the Alpine ones) on some key reconstructions (Figs. 3.15–3.18), in order to analyze how the final arrangement of the transects is the result of multiple episodes of dispersion and juxtaposition of such segments. Such exercise gives an appreciation of

the complex geological history underlying the Mediterranean region and provides examples of how 2-D evolutionary models often fail to portray adequately tectonic displacements at high angle with the plane of the cross sections.

3.4.1 Transects I-II-III West

The southern segments of Transects I and II belonged to Gondwana-Africa from the Early Palaeozoic until now, whereas all other segments were detached from Gondwana, accreted to Laurussia during the Variscan cycle, and juxtaposed through major strike-slip displacement inside the Variscan cordillera (segment IIIa–b and IIIc–d). After a first Pangean juxtaposition (Fig. 3.16) some segments were displaced again during the opening and closing of the Alpine Tethys and Pyrenean rift system. Seg-

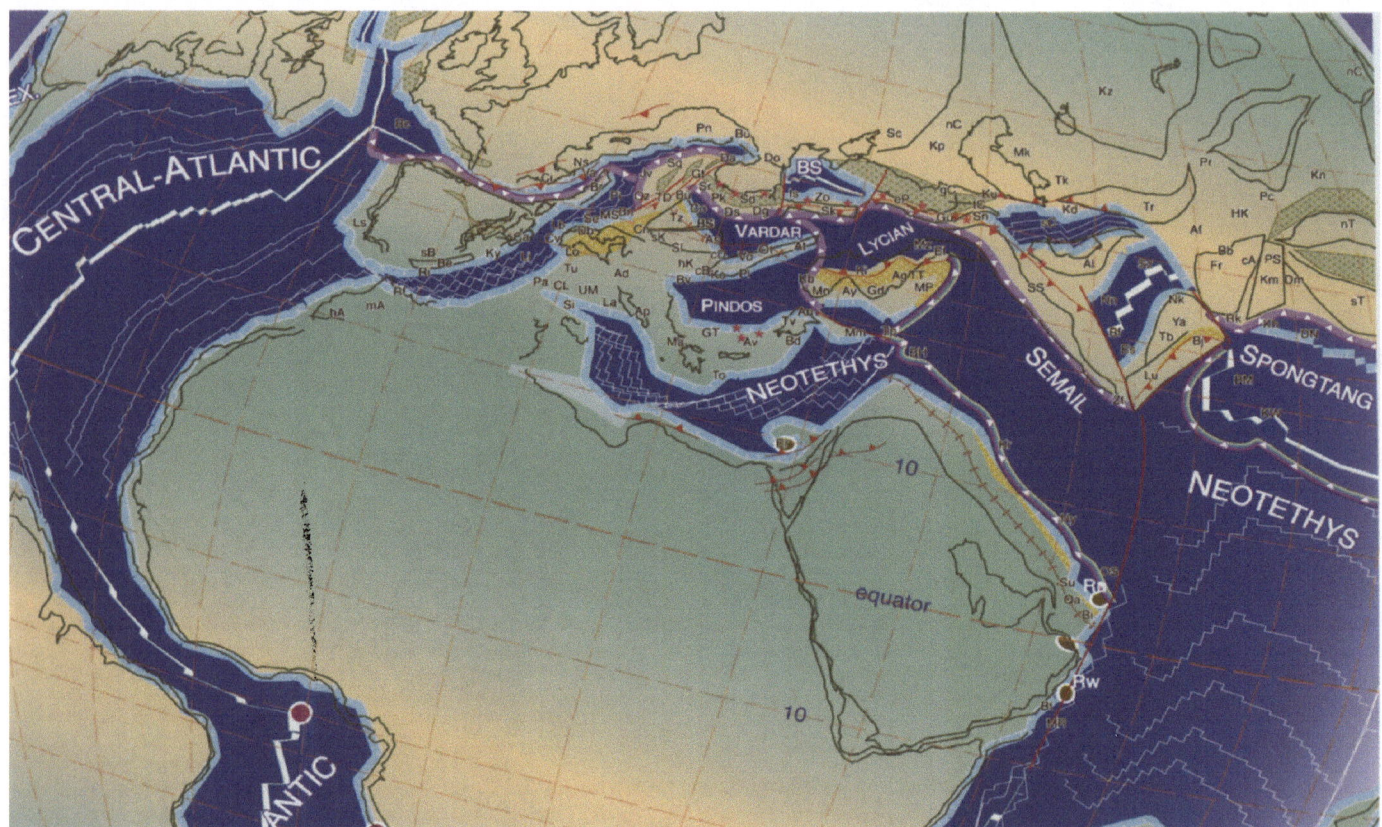

Fig. 3.13. 84 Ma (Santonian). The absence of magnetic anomalies between the Mo and C34 anomalies reduces the constraints on the paleogeographic evolution during the intervening 37 Ma. The main event was the accelerated rotation of Africa, which narrowed down significantly the Neotethyan domain south of the Semail ocean. Transform motion between the latter and the remnant Neotethys north of India changed from dextral to sinistral. Collapse of the old Neotethyan lithosphere along the transform generated the opening of an intra-oceanic basin east of Arabia. Before that, roll-back of the Neotethyan sea floor south of the Lut block had detached the latter – together with the Sanandaj-Sirjan block – from Eurasia. Eastward movement of these two Iranian blocks was accompanied by the expansion of the Lycian back-arc ocean, which totally replaced the Izmir-Ankara ocean. Then, east-west shortening, due to the rotation of Africa, triggered the southward obduction of the Lycian ocean onto the Anatolian-Tauric plate, whereas the Semail ocean partially obducted along its southern transform margin. The Anatolian-Tauric plate was nearly totally covered by ophiolitic-type mélange. The northern limit of the African plate in the Pyrenees became a zone of convergence, partly extending into the Atlantic. The latter was now spreading between Greenland and North America, whereas the South Atlantic was fully open. Large scale east-west shortening in the Alpine-Vardar region was bringing Adria behind the Austroalpine prism, whereas the northward subducting Vardar remnant ocean generated an active margin setting in the Balkan, accompanied by the opening of the Black Sea back-arc basin

ments belonging to the Alboran-Corsica-Sardinian domain (Ib, IIc and IIIc–d) were displaced only in Tertiary times. Segments of Transect III are highly dispersed in view of its U shape geometry, and its eastern part is treated together with Transects VII and VIII.

Geodynamic Evolution

As discussed above, southward subduction of the Alpine Tethys ocean was related to the closure of the Meliata-Maliac and Vardar domains, and was inherited in the western Alps from the northward vergence of the Austro-alpine accretionary wedge. This northward vergence is quite unique in the entire Alpine and Tethyan domain, in which most orogens are south-vergent. Moreover, a change from west- to east-dipping of the Alpine Tethys subduction occurs at the connecting point between the Alps and the northern Apennines (Fig. 3.19). In that re-

spect it should be noted that the Penninic Austroalpine prism is older (Late Cretaceous-Eocene) than the Apenninic one (Oligocene-Pliocene). Actually the latter was activated after the former had collided with the European-Iberian margins. Things are further complicated by the fact that the Apenninic prism remobilized parts of the Penninic prism as exotic elements (e.g. Bracco ophiolitic ridge: Elter et al. 1966; Hoogenduijn Strating 1991).

When large scale constraints are taken into consideration, the ophiolites of the Apenninic prism (Transect III) must be considered as being mainly derived from the former Alpine prism, which had collided in Eocene times with the Iberian plate (Corsica and Calabria; Fig. 3.19). These oceanic elements were reworked when northward subduction of the remnant oceanic domain of the Alpine Tethys (Ligurian basin) beneath Iberia commenced. This subduction corresponds to north-south shortening between Europe and Africa (Transects I and II), and postdates anomaly C25 (Thanetian,

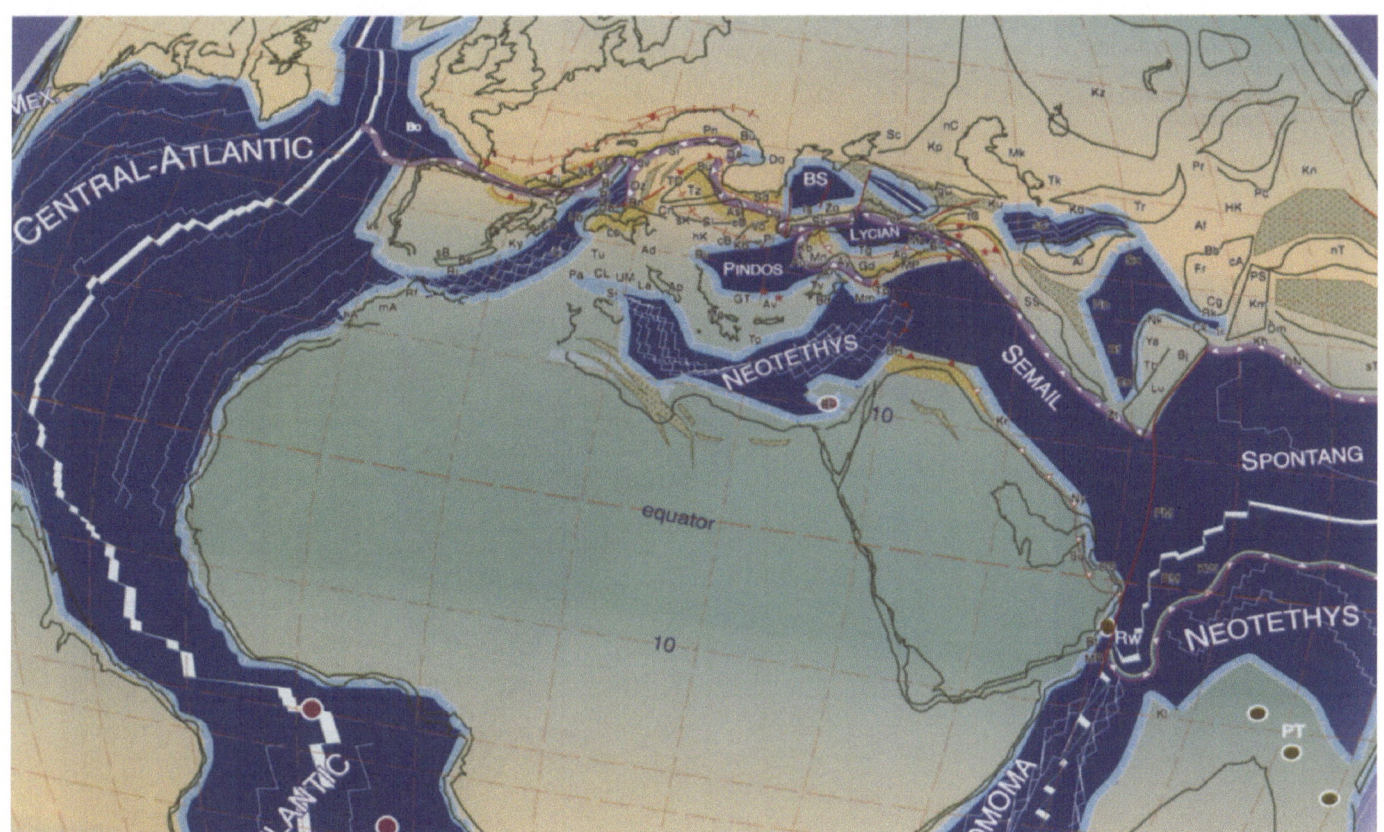

Fig. 3.14. 70 Ma (Maastrichtian). Continued counterclockwise rotation of Africa was responsible for the obduction of the Semail ocean onto Arabia (from Oman to Syria). The two intra-oceanic subduction zones, which had developed around the Neotethys sea floor north of India are now joined in a single back-arc system. Such back-arc system will obduct onto the northern margin of India in latest Cretaceous-Paleocene time. Around Arabia – as well as in the Himalayas – these obductions completely obliterated the Neotethyan ocean, which in this time frame is represented only by a few exotic blocks and by Permo-Triassic pelagic sediments found at the sole of the Cretaceous ophiolites. East-west shortening was still very active in the Alpine and Vardar domains. The latter is now totally closed, whereas roll-back of the remnant Lycian slab allows the opening of the east Black Sea back-arc basin. At this time, the Anatolian-Tauric plate was a free moving entity, pulled westward by roll-back in the Pindos ocean, generating a collision between the "greater Apulia" eastern promontory (Bey Daglari domain) and the Taurus margin, now forming the Antalya (Pamphylian) suture zone, mixing Tauric and Pindos elements. In the Alps, the northern sinistral Adriatic plate boundary was extending westward into the Piedmont ocean, which soon will start subducting southward beneath Adria. This was also allowing the Austroalpine prism to collapse westward and collide with the Briançonnais peninsula. Slight north-south shortening was taking place along the northern boundary of the Iberian plate, whereas N-S shortening between Africa and Iberia will occur only in the Eocene

c. 57 Ma). In fact, Atlantic magnetic anomalies (e.g. Torsvik et al. 2001) show that there was no shortening or differential movement of Europe and Africa between anomaly C31 (Maastrichtian, c. 70 Ma) and anomaly C25, and Iberia and Africa. This implies that the Paleocene was a period of no major tectonic movements, as already pointed out by Trumpy (1980) for the Alps.

Deep seismic data clearly show that the southward subducted Piemont ocean partially underplated the Adriatic indenter (Finetti et al. 2001). Subsequently, the tectonically underplated material moved together with Adria northward, whereas the not yet subducted Ligurian part of the Alpine Tethys started to subduct northward under the Iberian plate, generating HP/LT rocks (Puga et al. 1995). In the Middle Eocene reconstruction (Fig. 3.19), subduction of the Alpine Tethys is well underway. During the Late Oligocene important roll-back of the Alpine Tethys subduction system led to the opening of the Algero-Provençal back-arc basin (Transects II

and III) (Fig. 3.19). Our Late Oligocene reconstruction suggests that the Apenninic accretionary system had overridden the margin of the Tuscan-Campano-Lucanian block. Oligocene volcanism of Sardinia may, therefore, be regarded as subduction related (Monaghan 2001), although the onset of oceanic slab detachment could also be the cause of this volcanism. Subsequently, the Alpine accretionary prism (Ligurian units) stranded in the border of Corsica and Calabria started to collapse backward, following the build up of the new Apenninic accretionary prism. Accordingly, the Ligurian units are now found in an higher structural position that the one they occupied at the end of the Alpine collision.

The Lucano-Campanian promontory acting as an indenter, separated the developing northern and southern Apenninic accretionary prisms. Both prisms advanced into preexisting depressions, the Lombardian rift in the north and the Ionian Sea oceanic corridor in the south. Accretionary processes are still active in these areas.

Fig. 3.15.
Distribution of the segments
of the TRANSMED transects
in space and time (Visean).
See text for comments

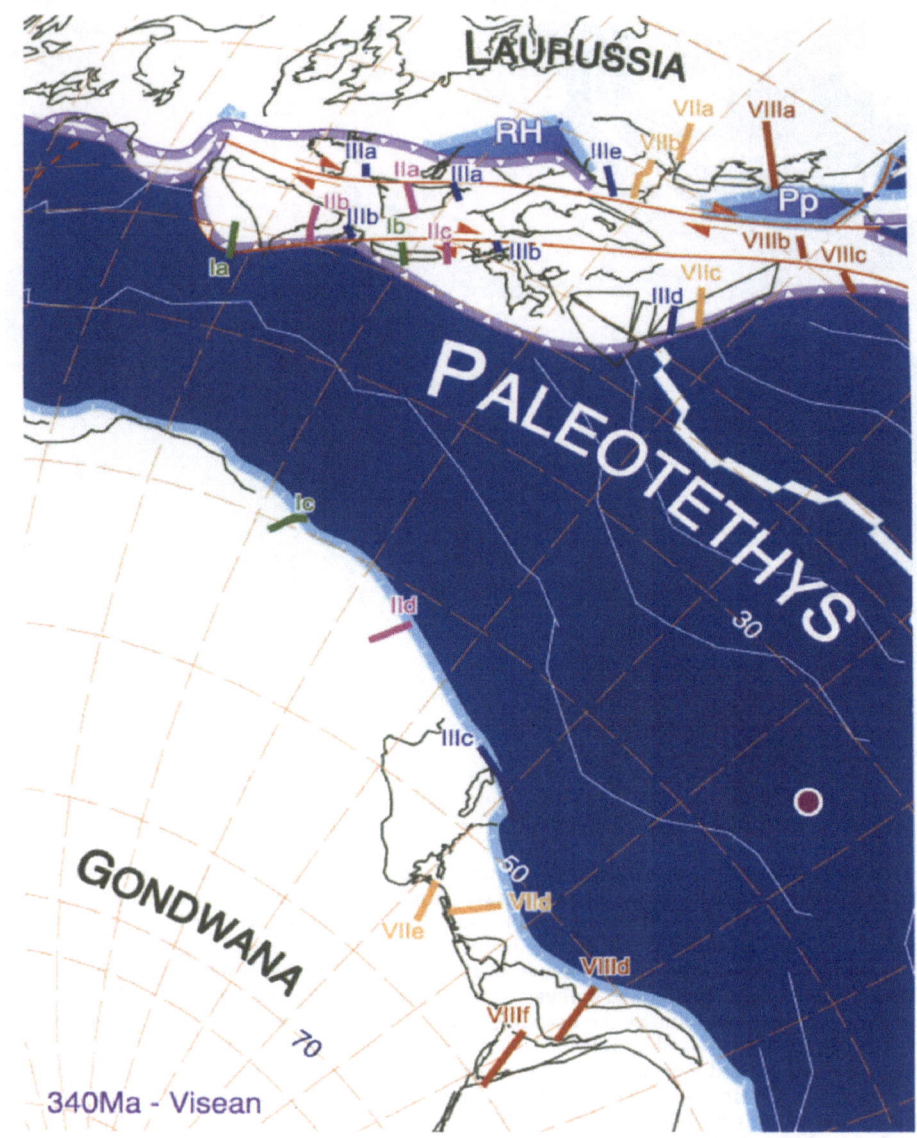

Fig. 3.16.
Distribution of the segments
of the TRANSMED transects
in space and time (Anisian).
See text for comments

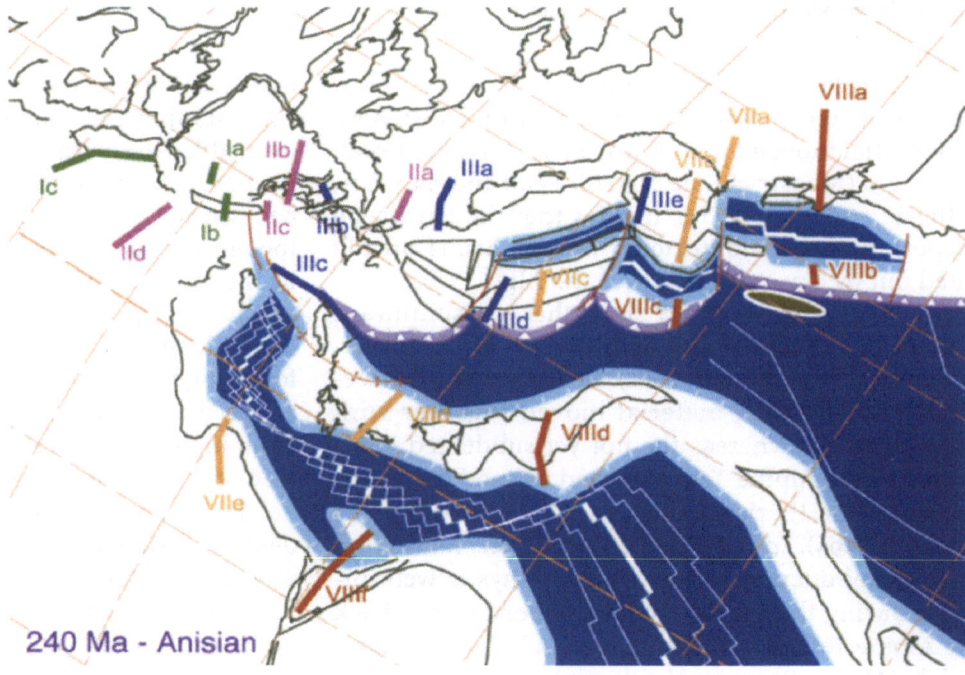

Fig. 3.17.
Distribution of the segments of the TRANSMED transects in space and time (Oxfordian). See text for comments

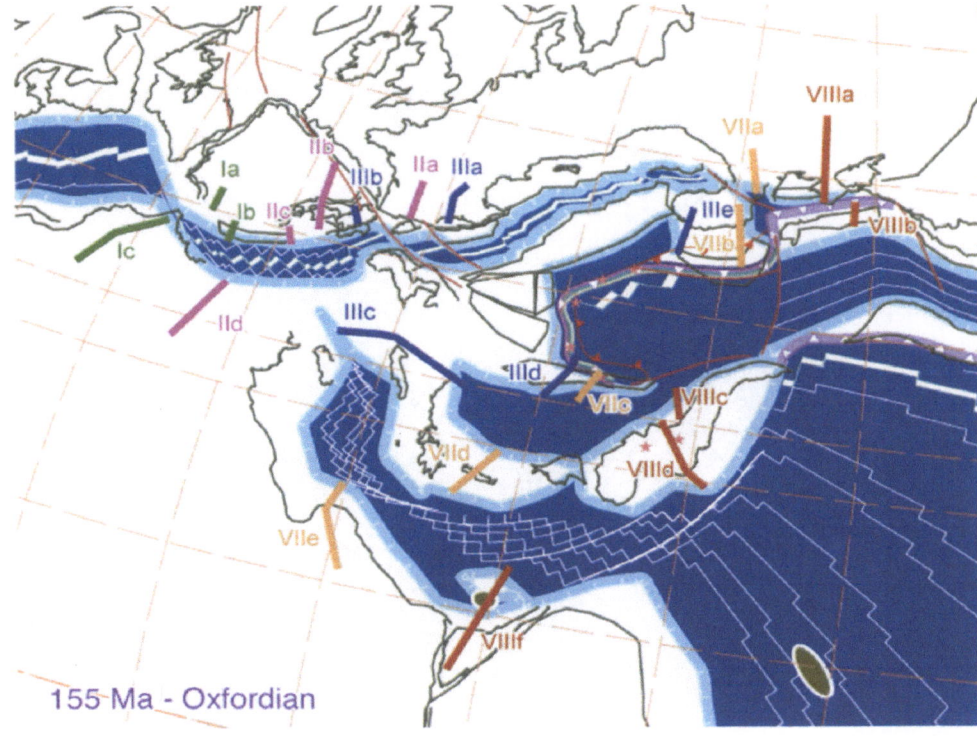

Fig. 3.18.
Distribution of the segments of the TRANSMED transects in space and time (Maastrichtian). See text for comments

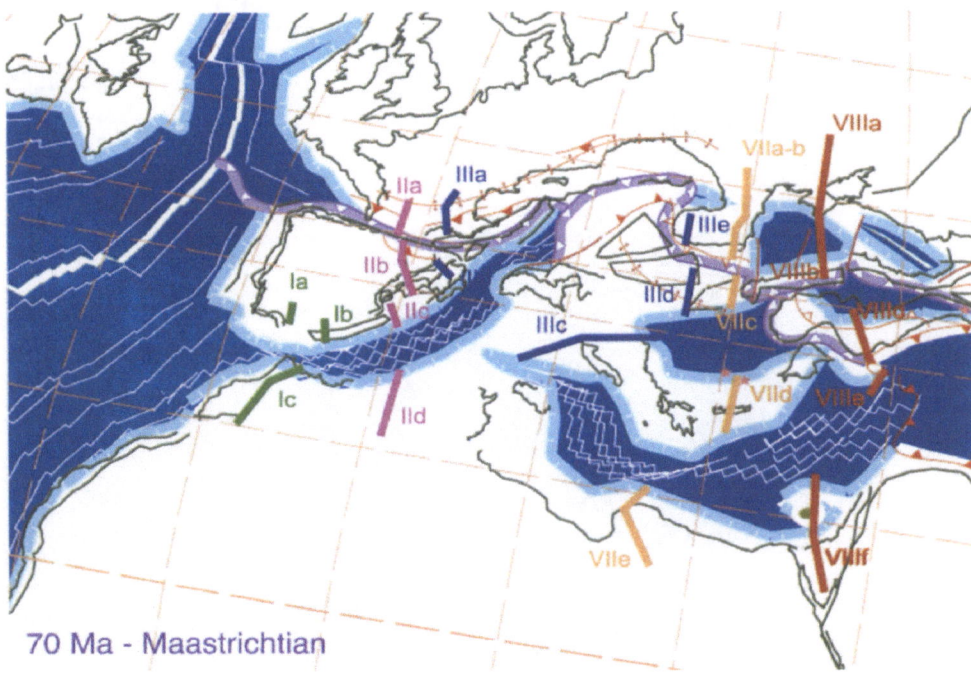

Northward subduction of the western Alpine Tethys – extending from Italy (Liguria) to Morocco (Maghrebian Tethys) – beneath Iberia probably commenced during the Late Paleocene (Finetti et al. 2001), just after the C25 magnetic anomaly. Roll-back of the west Alpine Tethys slab was accompanied by the development of the Apenninic-Maghrebian orogenic wedge (Transects I and II), and the Early Miocene detachment of the Corsica-Sardinian and Kabylian block led to the opening of the Algero-Provençal basin (Roca 2001). As roll back was proceeding, this orogenic wedge finally reached the Ionian basin (Neotethys wes-

ternmost tip: Catalano et al. 2001; Stampfli et al. 2001c). During the Late Miocene and Pliocene opening of the Tyrrhenian back-arc basin, the Calabrian block was detached from Sardinia (Mantovani et al. 1994; Gueguen et al. 1998; Argnani and Savelli 2001; Bonardi et al. 2001). To the west, slab roll-back proceeded along the Maghrebian oceanic corridor (Gutscher et al. 2002) giving rise to the collision of the Kabylian accretionary prism with North Africa in Langhian times (Frizon de Lamotte et al. 2000) and eventually to the development of the Betic and Rif orogenic wedges and the opening of the Alboran marginal basin (Transect I).

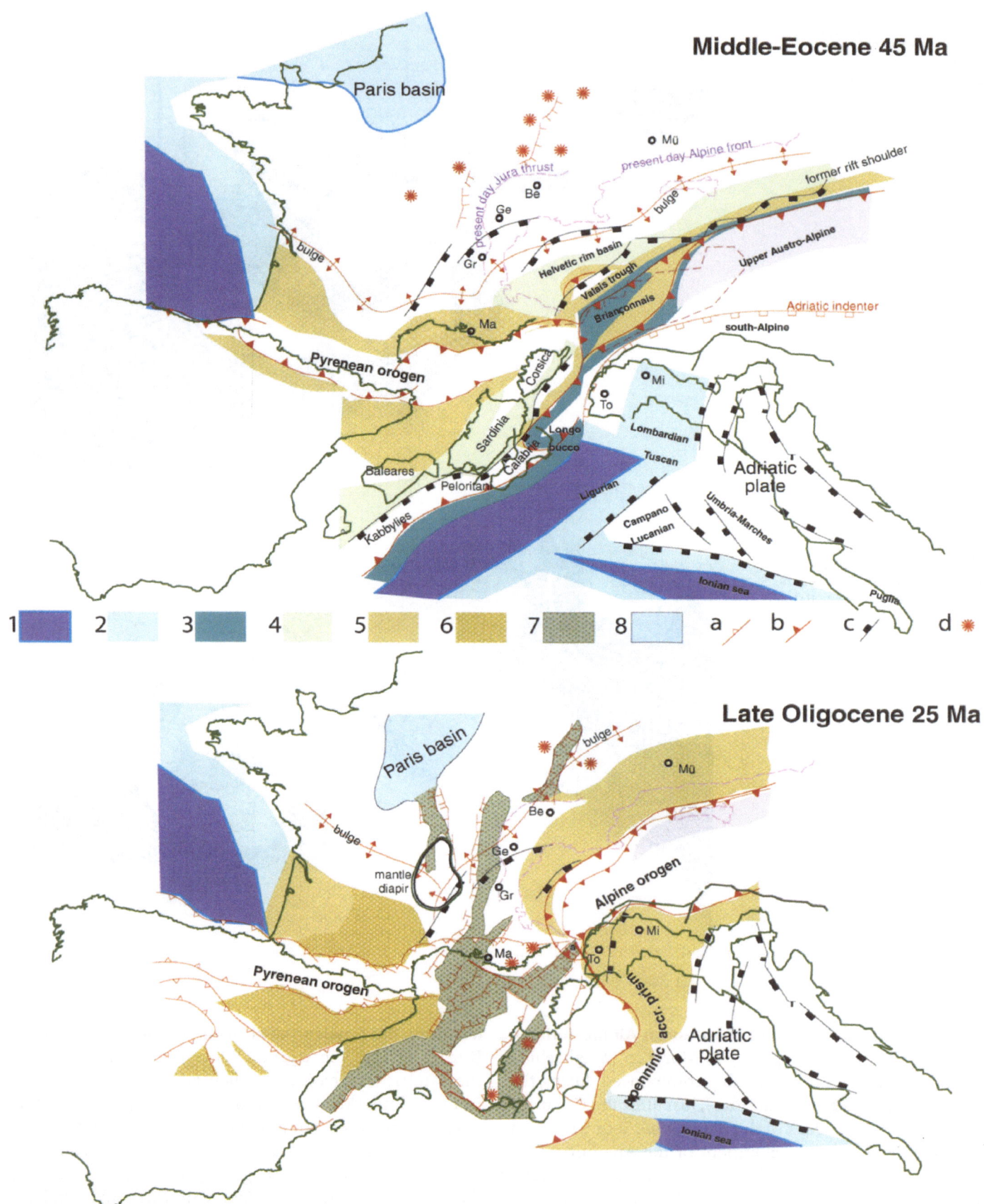

Fig. 3.19. Western Tethys reconstructions for the Eocene and Oligocene, modified from Stampfli et al. (2002). *1* Ocean; *2* passive margin; *3* active margin; *4* nummulitic platform; *5* marine foreland basin; *6* continental foreland basin; *7* rifts; *8* epicontinental basin; *a* inactive thrusts; *b* active thrusts; *c* Jurassic normal faults; *d* active volcanic centres. *Be,* Berne; *Ge,* Geneva; *Gr,* Grenoble; *Ma,* Marseilles, *Mi,* Milan; *Mü,* Munich; *To,* Turin

Crustal shortening in the western and central Alps continued whilst in the western Mediterranean area extensional opening of back-arc basins progressed. By Late Eocene times the Briançonnais terrane was incorporated into the Western Alps orogenic wedge (Stampfli et al. 2002) which had collided with the European passive margin from the Western Alps to the Eastern Carpathians. During the Late Eocene and the Oligocene the distal parts of this margin were subducted as shown by pervasive metamorphism, with some HP relics (Gebauer et al. 1992). During the Miocene, imbrication of the External Massifs of the Alps (e.g. Mont-Blanc) commenced, culminating in Late Miocene-Pliocene time in folding of the external Jura mountain range, the frontal folds of which were thrust onto the Oligo-Miocene Bresse graben.

3.4.2 Transects IV-V-VI

As lateral displacement between the southern and northern margins of the Alpine domain was quite large, the concept that Adria was the conjugate margin of the Helvetic shelf of the western Alps domain is erroneous. Adria was located further to the west, southeast of Spain, whereas the lower Austroalpine domain formed the conjugate margin of the Briançonnais and Helvetic segment of the northern Alpine Tethys margin. Following large scale Late Jurassic to Cretaceous left-lateral movements, Adria reached its present position adjacent to the western Alps only in Tertiary times (Figs. 3.12–3.14 and 3.19).

Geodynamic Evolution

Our plate reconstructions show that the Alpine Tethys must be considered as an extension of the Central Atlantic ocean into the Tethyan realm, rather than as a branch of the large and older Neotethys ocean (Stampfli 2000). In this respect, the onset of the Alpine cycle ought to be placed into the Carnian, a period corresponding to the final closure of Paleotethys in the Mediterranean and Middle East domains and to the beginning of rifting in the Central Atlantic-Alpine domain.

After several phases of rifting, as described above, and the development of mature passive margins during the Cretaceous, the Alpine region was dominated by the convergence of the African plate and Europe. The tectonic evolution of the western Alps commenced with the development of an accretionary prism and the southward subduction of the Alpine Tethys, most likely in latest Cretaceous times. The following tectonic units were successively incorporated into the accretionary prism:

- the Adriatic-Austroalpine back-stop, comprising an aborted Early Jurassic rifted basin (Lombardian basin);

- the oceanic accretionary prism of the Piemont ocean (west Alpine part of the Alpine Tethys), including crustal elements of the toe of the southern passive margin (lower Austroalpine elements);
- sedimentary and crustal material of the Briançonnais terrane derived from the Iberian plate;
- sedimentary and crustal material of the Valais domain, and the toe of the European (Helvetic s.l.) passive margin;
- sedimentary and crustal material of the European continental margin and rim basin (Helvetic s.s. domain).

From the Late Cretaceous to the Middle Eocene, the Helvetic margin was subjected to compression causing its partial inversion (Ziegler 1990; Ziegler et al. 1995). This was followed by mid-Eocene to Miocene flexural subsidence of the European lower plate under the load of the advancing tectonic wedge (Burkhard and Sommaruga 1998; Stampfli et al. 1998).

Following collision of the accretionary prism with the European margin the lithospheric slab was detached (or delaminated) in the Early Oligocene (e.g. Stampfli and Marchant 1995). However, convergence between Adria and Europe continued, resulting in the subduction of continental material and the development of the present Alpine orogenic wedge (Marchant and Stampfli 1997) which includes as pro-wedge external units (*i*) the external Crystalline Massifs and their cover, (*ii*) the Molasse foreland basin, and (*iii*) the Jura Mountains. Continued convergence resulted also in the retro-wedge thrust belt of the southern Alps that encroaches on the Po plain foreland basin, which is superimposed on the Jurassic Lombardian rim basin.

Considering that the Alpine Chain extends from Nice to the Carpathians, there is a fundamental difference between its western (Transects IV and V) and Austro-alpine-Carpathian sectors (Transect VI). Although in both areas an orogenic wedge was ultimately emplaced on the European margin, the Jurassic and Cretaceous evolution of the western-central Alpine domain differs considerably from that of the Austroalpine-Carpathian system (Faupl and Wagreich 1999). The latter was first involved in the collision of the Vardar domain with the Triassic Meliata-Maliac margin (Bernoulli 1981; Kozur 1991a; Haas et al. 1995), causing imbrication of the upper Austroalpine nappes, represented mainly by the Northern Calcareous Alps. This implies that the Austro-alpine wedge contains two sutures, namely the Meliata and the Penninic sutures, both of which are associated with HP/LT rocks (Thöni 1999).

As seen above, the drift patterns of Iberia and Africa were similar during most of the Cretaceous and until the Paleocene (M0 to C25 anomaly). Therefore, the northern limit of the African plate has to be placed during the Cretaceous between Iberia and Europe. Separation of the Iberia-Corsican-Briançonnais micro-continent from southern France through opening of the Pyrenean rift system resulted in its juxtaposition with the Alpine domain (Frisch 1979; Stampfli 1993; Stampfli et al. 1998; Stampfli et al. 2002). With the insertion of the Briançonnais domain into the West Alpine domain, the northern margin of the Alpine Tethys was repeated. Following the eastward drift of Iberia, the eastern tip of the Briançonnais collided with the western part of the Austroalpine-Carpathian prism during the Late Cretaceous-Paleocene (Fig. 3.14) (angular unconformity and Palaeocene Wildflysch of the Falknis nappe, eastern Briançonnais; Alleman 2002). The Valais domain, underlain by Piemont ocean floor trapped between the Helvetic passive margin and the Briançonnais terrane, links up to the east with the north Penninic domain, both of which were affected by Oligocene HP/LT metamorphism (Goffé and Oberhänsli 1992; Bousquet et al. 1998). However, closing of the Valais domain already commenced during Late Cretaceous times, witnessed by the deposition of the Maastrichtian Niesen flysch followed in the Middle Eocene by the Meilleret flysch. These Pyrenean tectonic inversion phases can be followed from the western Alps (Transects IV and V) via Provence (Transect III) to the Pyrenees (Transect II) (Stampfli et al. 1998, 2002).

This east-west shortening was followed by latest Cretaceous onset of subduction of the Piemont part of the remnant Alpine Tethys, that was accompanied by HP-LT metamorphism of some elements of its southern passive margin (e.g. Sesia Massif, Rubatto 1998). This subduction was located along a major sinistral transcurrent plate boundary (a paleo-Insubric line), accommodating eastward translation of the Adria-Tizia plate to a position south of its present location. Eastward movement of Adria-Tizia was triggered by subduction of the remnant Vardar ocean beneath Moesia and eastward slab retreat in the Izmir-Ankara ocean (Figs. 3.11–3.14).

This implies that on all three TRANSMED Alpine transects, the present Adria continent was located in a pre-rift Early Jurassic fit more than a thousand kilometres to the west and formed the southern margin of the West-Central Alpine Tethys. The lower Austroalpine units formed the original southern margin of the eastern part of the Alpine Tethys (e.g. Dent-Blanche, Margna-Sella, Bernina). The Insubric line which separated these units from Adria, acted during the Cretaceous as a major left-lateral plate boundary, which was reactivated as a right-lateral fault during post-collisional phases of the Alpine orogeny (Laubscher 1983). On the other hand, the Briançonnais terrane and Iberia, which formed the Jurassic northern margin of the western Alpine Tethys, were displaced eastward during the opening of the North-Atlantic and Bay of Biscay.

3.4.3 Transects III East, VII and VIII

The different segments of these transects underwent large lateral displacements both in an E-W and a N-S

direction. The southernmost African segments of Transects VII and VIII are still separated from Europe by the East Mediterranean-Neotethys ocean. Large scale imbrication of terranes located between Europe and Africa caused the duplication of major suture zones (Fig. 3.20). This is particularly the case in Turkey, with segments presently juxtaposed along Transect VIII being derived from often remote and unrelated areas. Major strike-slip movements during the Variscan orogenic cycle juxtaposed segments. Subsequently, they were dispersed again during the collapse of the Variscan cordillera and the opening of Triassic back-arc basins. Some segments were re-assembled by the Cimmerian terrane collage (segment VIIId–e). Some segments came only in evidence during the Late Cretaceous-Early Tertiary, such as the Cyprus area (VIIIf), which was derived from the Semail ocean to the east. Other southern continental segments were mainly displaced to the NE with respect to Europe in conjunction with the rotation of Africa.

Geodynamic Evolution

East-west shortening in the Vardar region started in mid-Cretaceous time, after the Balkanic orogenic event (Georgiev et al. 2001). Thereafter, the subduction polarity changed and the remnant Vardar ocean was subducted northward beneath the Balkanic orogen (Transects III and VII) and further east beneath the western Pontides (Transect VIII), where it was accompanied by HP-LT metamorphism dated between 100 and 60 Ma (Okay et al. 1991; Okay and Tansel 1992). Roll-back of the Vardar slab controlled the development of the Late Cretaceous Srednogorie-Pontides volcanic arc and of the Black Sea back-arc basin (Figs. 3.13 and 3.14). Closure of the Vardar ocean was diachronous, and was completed first in the Dinaric region (Transect III) during the Maastrichtian-Palaeocene (Pamic 2002), then, in the Rhodope-Hellenic region (Transect VII) in the Palaeocene-Eocene (Yanev and Bardintzeff 1997). An oceanic basin remained open longer along the Pontides (Transect VIII) (Okay and Tüyzüs 1999). There, the Late Cretaceous opening of the Lycian back-arc ocean followed the eastward slab retreat of the Izmir-Ankara ocean. Subduction-related processes lasted until the final Eocene closure of the remnant Lycian ocean along the entire Pontides segment (Koçyigit 1991; Okay and Sahintürk 1997; Kaymakçi et al. 2000). Roll-back of the Lycian slab triggered the opening of the eastern Black Sea in Late Cretaceous-Paleocene times (Robinson 1997; Nikishin et al. 2003).

Along Transect VIII north-south-oriented Tertiary shortening involved obduction/subduction of a young Cretaceous ocean (Lycian domain), whereas along Transect VII this time interval saw the closure of the Triassic Pindos basin (Richter et al. 1993b; Degnan and

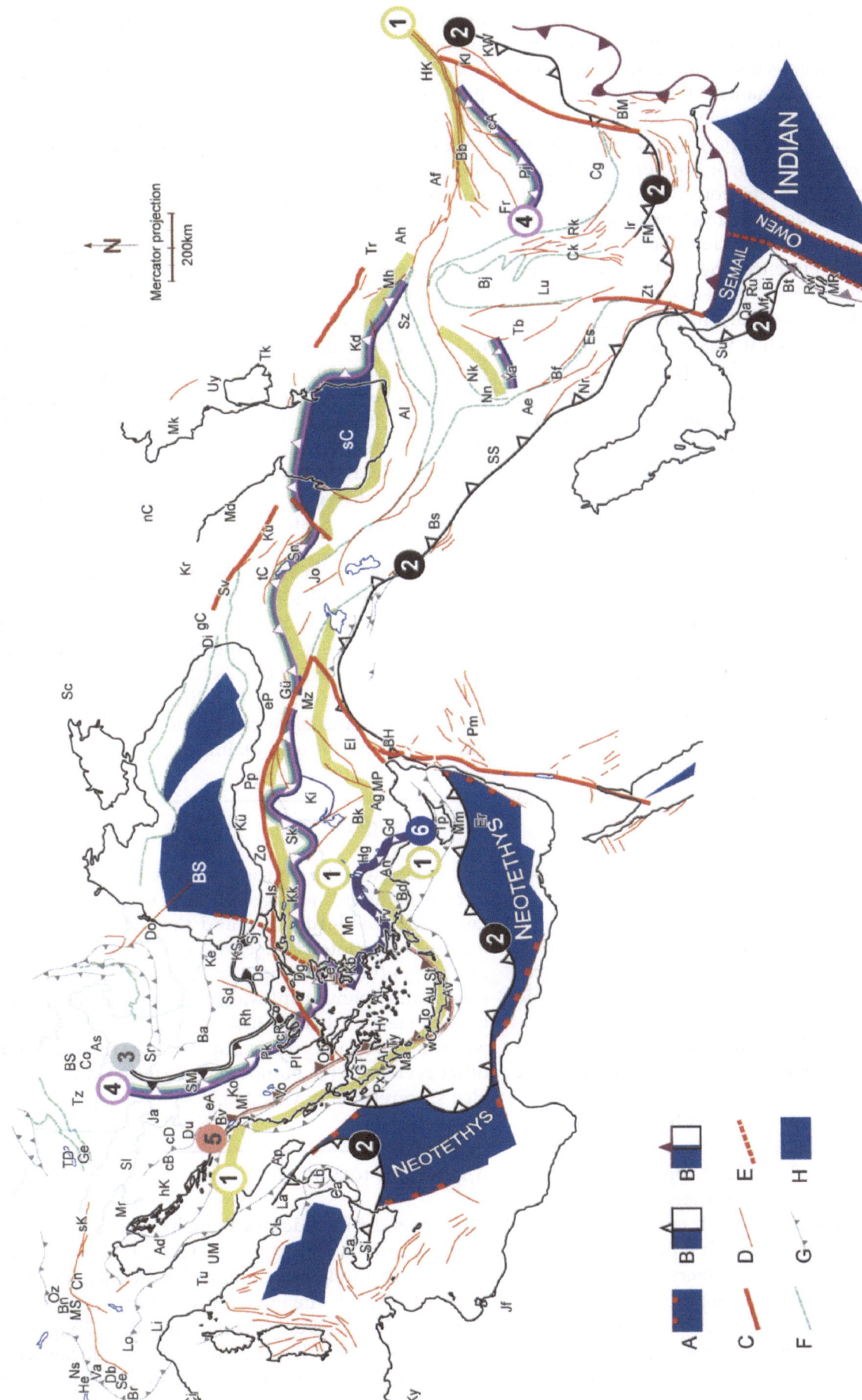

Fig. 3.20. Location of major sutures in the western Tethys area, modified from Stampfli et al. (in press): *1* Paleotethys; *2* Neotethys (in the eastern part it includes the sutures of Neotethys-derived intra-oceanic back-arc oceans, e.g. Semail); *3* Balkanic suture; *4* Vardar, Izmir-Ankara (including Lycian), Sevan, south-Caspian, Panjao; *5* front of the Pindos nappe; *6* Pamphylian (Antalya-Cycladic). *A* passive margins of Neotethys; *B* subduction zones; *C* major active faults; *D* minor faults; *E* former transform faults; *F* block limits; *G* major thrusts; *H* oceanic basement. For abbreviations of key localities, see Table 3.1

Robertson 1998). However, subduction of the eastern part of the Pindos under the Tauride plate started already in the Late Cretaceous due to important east-west shortening movements. Remnants of this event are found in the Late Cretaceous metamorphic sequences of the Cyclades (Bröcker and Enders 1999) and the Asteroussia nappe of Crete (Seidel et al. 1976; Bonneau 1984); it is also recorded in the Late Cretaceous Pindos first flysch (Neumann and Zacher 1996; Wagreich 1996). Obduction of the Lycian nappe onto the Anatolian-Tauric plate (Robertson 2002 and references therein) should also be viewed in the frame of Late Cretaceous east-west shortening and final emplacement of this plate to the south of the Pontides, from which it was separated by more than 2 000 km in Permo-Triassic times.

Eastward subduction of the Pindos basin implies that elements located to the north (Pelagonia) and south of it (Greater Apulia, i.e. Bey Daglari and lower Antalya domain) were imbricated on both sides of the Tauric-Anatolian plate, creating a duplication of older features such as the Paleotethys suture zone (Fig. 3.20). This is certainly one of the main difficulties inherent to the geology of Turkey. Elements related to this Late Cretaceous juxtaposition correspond (*i*) south of the Tauric-Anatolian plate, to the Pamphylian (Antalya) suture (Monod 1977; Gutnic et al. 1979), which extends westward under the Lycian nappes and joins the Cycladic domain; (*ii*) north of the Tauric-Anatolian plate, to the Tavsenli and Kütahya-Bolkardag zones (e.g. Okay and Tüyzüs 1999), and (*iii*) eastward, to the Kirsehir massif (Transect VIII), accreted to the Pontides margin around 95 Ma as witnessed by the presence of syn-collisional and younger (c. 72 Ma) post-collisional granites (Göncüoglu et al. 1997).

3.5 Conclusions

Through this review of the TRANSMED transects we tried to emphasize the importance of lateral displacements and duplications of geological units, a large-scale complication upon which smaller scale tectonic imbrication are superimposed. This makes the geology of the Tethyan area so fascinating and at the same time difficult to grasp in its entirety.

Palinspastic models are too often built by retro-deforming cross sections ignoring the exotic nature of some of their segments. Such a 2-D approach to geodynamics is counterproductive and breeds many erroneous concepts through which tectonic complexity is simplistically explained by a complex paleogeography. Clearly, the construction of lithospheric scale cross sections remains a very desirable approach, provided that stratigraphic and tectonic problems have been satisfactorily solved, and represents a useful step in the development of plate tectonic models. Modern investigation techniques allow to decipher the geodynamic context

of each tectonostratigraphic unit. However, this is not enough if one considers that not all of the units now occurring along a given cross section were located from the beginning of time along its trace. In other words, the possibility of major terrane motions at high angle with respect to the trace of the cross section should always be considered. For example, using plate tectonic models well constrained in space and time, it became evident that, for reconstructing a Triassic N-S cross section through the Alpine Mediterranean domain, one has to use segments IIa, IIIf, VIIe and VIIIg, and complete it with missing segments not found in any of the transects. A three dimensional approach is therefore a basic requirement for meaningful paleotectonic-paleogeographic reconstructions. This implies access to large-scale plate tectonic models that are constrained in space and time. Of course, key information comes first from the development of the transects themselves, which are rooted in fieldwork done over decades and in more recent geophysical data. This means that an iterative approach must be systematically applied to palinspastic models.

It was the aim of this project to create the necessary conditions for multidisciplinary research and international collaboration because, obviously, terranes know no political borders. We hope the pooling of knowledge and experience during the TRANSMED Project through international and multidisciplinary collaboration has contributed to the advancement of the understanding of our fascinating Tethyan realm.

Acknowledgements

W. Cavazza was the central person to put a large team of circum-Mediterranean geologists at work on this project, and under his leadership and through our numerous meetings and interchanges, it became possible to build up this synthesis; his input in the final stage of elaboration of this manuscript was highly appreciated. A thorough review of this manuscript was done by P. Ziegler, whose suggestions greatly improved the final document, other reviewers' (O. Monod and J. Mosar) comments are also sincerely acknowledged.

Appendix 3 (CD-ROM)

This appendix includes the complete set of twenty-three paleogeographic-paleotectonic reconstructions elaborated by G. Stampfli and G. Borel for the TRANSMED Atlas. Of these, only the most representative (fourteen) are shown in the printed portion of the atlas (Chapt. 3). See Chapt. 3 for legend, list of abbreviations of localities, and for a general discussion of the paleogeographic-paleotectonic reconstructions.

References

Preface, Chapters 1, 2 and 3

Acosta J, Munoz A, Herranz P, Palomo C, Ballesteros M, Vaquero M, Uchupi E (2002) Geodynamics of the Emile Baudot Escarpment and the Balearic Promontory, western Mediterranean. Mar Petrol Geol 18:349–369

Aksu AE, Hiscott RN, Yasar D, Isler FI, Marsh S (2002a) Seismic stratigraphy of Late Quaternary deposits from the southwestern Black Sea shelf: evidence for non catastrophic variation in sea level during the last 10.000 yr. Mar Geol 190:6–94

Aksu AE et al. (2002b) Persistent Holocene outflow from the Black Sea to the Eastern Mediterranean contradicts Noah's flood hypothesis. GSA Today 12:4–10

Aldinucci M, Cirilli S, Costantini A, Decandia FA et al. (2001) Stratigraphic-structural evolution of the Late Paleozoic-Triassic sequences of southern Tuscany (northern Apennines). Dipartimento di Scienze della Terra, Siena University, 72 pp

Alleman F (2002) Erläuterungen zur Geologischen Karte des Fürstentums Liechtenstein. Länggas Druck AG-Bern, 129 pp

Alonso B, Maldonado A (1992) Pliocene-Quaternary margin growth patterns in a complex tectonic setting: northeastern Alboran Sea. In: Maldonado A (ed) The Alboran Sea. Geo-Mar Lett 12:137–143

Alvarez W, Cocozza T, Wezel FC (1974) Fragmentation of the Alpine orogenic belt by microplate dispersal. Nature 248:309–314

Amato A, Alessandrini B, Cimini GB (1993) Teleseismic tomography of Italy. In: Iyer HM, Hirahara K (eds) Seismic tomography: theory and applications, Prentice-Hall, London, pp 361–396

Amato A, Margheriti L, Azzara R, Basili A, Chiarabba C, Ciaccio MG, Cimini GB, Di Bona M, Frepoli A, Lucente FP, Nostro C, Selvaggi G (1998) Passive seismology and deep structure in central Italy. Pure Appl Geophys 151:479–493

Argnani A, Savelli C (2001) Magmatic signature of episodic back-arc rifting in the southern Tyrrhenian Sea In: Ziegler PA, Cavazza W, Robertson AHF, Crasquin-Soleau S (eds) Peritethyan rift/wrench basins and passive margins. Mém Museum National Hist Nat, vol 186, pp 735–754

Argnani A, Torelli L (2001) The Pelagian shelf and its graben system. In: Ziegler PA, Cavazza W, Robertson AHF, Crasquin-Soleau S (eds) Peritethyan rift/wrench basins and passive margins. Mém Museum National Hist Nat vol 186, pp 529–544

Ballard RD, Coleman D, Rosenberg G (2000) Further evidence of abrupt Holocene drowning of Black Sea shelf. Mar Geol 170: 253–261 (see also: www.nationalgeographic.com/blacksea)

Bally AW, Burbi L, Cooper LC, Ghelardoni R (1988) Balanced sections and seismic reflection profiles across the Central Apennines. Mem Soc Geol Ital 35:257–310

Bassoulet JP, Colchen M, Marcoux J, Mascle G (1981) Les masses calcaires du flysch Triassico-Jurassique de Lamayuru (zone de la suture de l'Indus, Himalaya du Ladakh): klippes sédimentaires et éléments de plate-forme remaniés. Riv Ital Paleont Strat 86:825–844

Baudin F, Lachkar G (1990) Geochimie organique et palynologie du Lias superieur en zone ionienne (Grece) exemple d'une sedimentation anoxique conservee dans un paleo-marge en distension. Bull Soc Geol France 8(6):123–132

Becker TW, Faccenna C, O'Connell RJ, Giardini D (1999) The development of slabs in the upper mantle: Insights from numerical and laboratory experiments. J Geophys Res 104(B7):15207–15226

Beghein C, Resovsky JS, Trampert J (2002) P and S tomography using normal-mode and surface waves data with a neighbourhood algorithm. Geophys J Int 149:646–658

Belousov VV, Volvovsky BS, Arkhipov IV, Buryanov VB, Evsyukov YD, Goncharov VP, Gordienko VV, Ismagilov DF, Kislov GK, Kogan LI, Kondyurin AV, Kozlov VN, Lebedev LI, Lokholatnikov VM, Malovitsky YP, Moskalenko VN, Neprochnov YP, Ostisty BK, Rusakov OM, Shimkus KM, Shlezinger AE, Sochelnicov VV, Sollogub VB, Solovyev VD, Starostenko VI, Starovoitov AF, Terechov AA, Volvovsky IS, Shigunov AS, Zolotarev VG (1988) Structure and evolution of the Earth's crust and upper mantle of the Black Sea. Bollettino di Geofisica Teorica e Applicata 30: 109–196

Bernier P, Berné S, Rabineau M, Thollet G (2004) Last Glacial maximum shoreline position in the Gulf of Lions determined by lithified coastal sediments. Sedimentology (in press)

Bernoulli D (1981) Ancient continental margins of the Tethyan Ocean. In: Bally AW, Watts AB, Grow JA, Manspeizer W, Bernoulli D, Schreiber C, Hunt JM (eds) Geology of passive continental margins; history, structure and sedimentologic record (with special emphasis on the Atlantic margin). Am Ass Petrol Geol, Education Course Note Series 19/5, pp 1–36

Bernoulli D, Winkler W (1990) Heavy-mineral assemblages from Upper Cretaceous South Alpine and Austro-alpine flysch sequences (N-Italy and S-Switzerland): source terranes and paleotectonic implications. Eclogae geol Helv 83:287–310

Bertotti G, Picotti V, Bernoulli D, Castellarin A (1993) From rifting to drifting: tectonic evolution of the South-Alpine upper-crust from the Triassic to the Early Cretaceous. Sed Geol 86:53–76

Biju-Duval B, Dercourt J, Le Pichon X (1977) From the Tethys ocean to the Mediterranean seas: a plate tectonic model of the evolution of the western Alpine system. In: Biju-Duval B, Montadert L (eds) Structural history of the Mediterranean basins. Ed. Technip, Paris

Bijwaard H, Spakman W (1999) Fast kinematic ray tracing of first and later arriving global seismic phases. Geophys J Int 139: 359–369

Bijwaard H, Spakman W (2000) Nonlinear global P-wave tomography by iterated linearised inversion. Geophys J Int 141:71–82

Bijwaard H, Spakman W, Engdahl ER (1998) Closing the gap between regional and global travel time tomography. J Geophys Res 103L:30,055–30,078

Bill M, Bussy F, Cosca M, Masson H, Hunziker JC (1997) High precision U-Pb and ^{40}Ar/^{39}Ar dating of an Alpine ophiolite (Gets nappe, French Alps). Eclogae geol Helv 90:43–54

Blanco MJ, Spakman W (1993) The P-velocity structure of the mantle below the Iberian Peninsula: evidence for subducted lithosphere below southern Spain. Tectonophysics 221:13–34

Blome CD, Irwin WP (1985) Equivalent radiolarian ages from ophiolitic terranes of Cyprus and Oman. Geology 13:401–404

Boccaletti M, Manetti P, Peccerillo A (1974) Hypothesis on the plate tectonic evolution of the Carpatho-Balkan areas. Earth Planet Sci Lett 23:193–198

Bonardi G, Cavazza W, Perrone V, Rossi S (2001) Calabria-Peloritani terrane and northern Ionian Sea. In: Vai GB, Martini IP (eds) Anatomy of an orogen: the Apennines and adjacent Mediterranean basins. Kluwer Academic Publishers, Dordrecht, pp 287–306

Bonev NG, Stampfli GM (2003) Mesozoic units in SE Rhodope (Bulgaria): new structural and geodynamic implications for the Early Jurassic to Mid-Cretaceous evolution of the Vardar basin. EGS-AGU-EUG Joint Assembly, Abstract 344

Bonneau M (1984) Correlation of the Hellenide nappes in Southeast Aegean and their tectonic reconstruction In: Dixon JE, Robertson AHF (eds) The Geological Evolution of the Eastern Mediterranean. Spec Publ Geol Soc London, vol 17, pp 517–527

Bos AG, Spakman W, Nyst MCJ (2003) Surface deformation and tectonic setting of Taiwan inferred from a GPS velocity field. J Geophys Res 108(B10), art no 2458

Bousquet R, Oberhansli R, Goffe B, Jolivet L, Vidal O (1998) High-pressure-low-temperature metamorphism and deformation in the Bundnerschiefer of the Engadine Window; implications for the regional evolution of the eastern Central Alps. J Met Geol 16:657–674

Bracene R, Frizon de Lamotte D (2002) The origin of intraplate deformation in the Atlas system of western and central Algeria: from Jurassic rifting to Cenozoic-Quaternary inversion. Tectonophysics 357:207–226

Bröcker M, Enders M (1999) U-Pb zircon geochronology of unusual eclogite-facies rocks from Syros and Tinos (Cyclades, Greece). Geol Mag 136:111–118

Buforn E, Sanz de Galdeano C, Udías A (1995) Seismotectonics of the Ibero-Maghrebian region. Tectonophysics 248:247–261

Burkhard M, Sommaruga A (1998) Evolution of the Swiss Molasse basin: structural relations with the Alps and the Jura belt. In: Mascle A, Puigdefabregas A, Luterbacher HP, Fernandez M (eds) Cenozoic basins of western Europe. Spec Publ Geol Soc London vol 134, pp 279–298

Burrus J (1989) Review of geodynamic models for extensional basins; the paradox of stretching in the Gulf of Lions (northwest Mediterranean). Bull Soc Géol France 8:377–393

Butler RWH, Lickorish WH, Grasso M, Pedley HM, Ramberti L (1995) Tectonics and sequence stratigraphy in Messinian basins, Sicily: constraints on the initiation and termination of the Mediterranean salinity crisis. Geol Soc Am Bull 107:425–439

Butler RWH, McClelland E, Jones RE (1999) Calibrating the duration and timing of the Messinian salinity crisis in the Mediterranean: linked tectonostratigraphic signals in thrust-top basins of Sicily. J Geol Soc London 156:827–835

Caby R, Hammor D, Delor C (2001) Metamorphic evolution, partial melting and Miocene exhumation of lower crust in the Edough metamorphic core complex, west Mediterranean orogen, eastern Algeria. Tectonophysics 342:239–273

Calvert A, Sandvol E, Seber D, Barazangi M, Roecker S, Mourabit T, Vidal F, Alguacil G, Jabour N, (2000) Geodynamic evolution of the lithosphere and upper mantle beneath the Alboran region of the western Mediterranean: constraints from travel time tomography. J Geophys Res 105:10871–10898

Carminati E, Wortel MJR, Meijer PT, Sabadini R (1998a) The two-stage opening of the western-central Mediterranean basins: a forward modeling test to a new evolutionary model. Earth Planet Sci Lett 160:667–679

Carminati E, Wortel MJR, Spakman W, Sabadini R (1998b) The role of slab detachment processes in the opening of the western-central Mediterranean basins: some geological and geophysical evidence. Earth Planet Sci Lett 160:651–665

Carminati E, Doglioni C, Scrocca D (2003) Apennines subduction-related subsidence of Venice (Italy). Geophys Res Lett 30 (13): art no 1717

Caron C, Homewood P, Wildi W (1989) The original Swiss Flysch: A reappraisal of the type deposits in the Swiss Prealps. Earth-Sci Rev 26:1–45

Casero P, Roure F (1994) Neogene deformations at the Sicilian-North African plate boundary. In: Roure F (ed) Peri-Tethyan platforms. Ed. Technip, Paris, pp 27–50

Cassinis G, Toutin-Morin N, Virgili C (1995) A general outline of the Permian continental basins in southwestern Europe. In: Scholle PA, Paryt TM, Ukmer-Scholle DS (eds) The Permian of northern Pangea. Springer Verlag, Heidelberg, pp 137–157

Castellarin A, Lucchini F, Rossi PL, Simboli G (1988) The Middle Triassic magmatic-tectonic arc development in the Southern Alps. Tectonophysics 146:79–89

Castellarin A, Selli L, Picotti V, Cantelli L (1996) La tettonica medio Triassica e il diapirismo nelle Dolomiti occidentali. Geologia delle Dolomiti, pp 25–46

Catalano R, Di Stefano P, Kozur H (1988) New results in the Permian and Triassic stratigraphy of western Sicily with special reference to the section at Torrente San Calogero, SW of Pietra di Salome (Sosio valley). Soc Geol Ital, Atti del 74° Congresso Nazionale, pp 119–125

Catalano R, Di Stefano P, Vitale FP (1995) Structural trends and paleogeography of the central and western Sicily belts: new insight. Terra Nova 7:189–199

Catalano R, Franchino A, Merlini S, Sulli A (2000) A crustal section from the Eastern Algerian basin to the Ionian ocean (Central Mediterranean). Mem Soc Geol It 55:71–85

Catalano R, Doglioni C, Merlini S (2001) On the Mesozoic Ionian Basin. Geophys J Intern 144:49–64

Cavazza W, DeCelles PG (1998) Upper Messinian siliciclastic rocks in southeastern Calabria (southern Italy): paleotectonic and eustatic implications for the evolution of the central Mediterranean region. Tectonophysics 298:223–241

Cavazza W, Wezel FC (2003) The Mediterranean region – a geological primer. Episodes 26(3):160–168

Channell JET (1992) Paleomagnetic data from Umbria (Italy): implications for the rotation of Adria and Mesozoic apparent polar wander paths. Tectonophysics 216:365–378

Channell JET (1996) Paleomagnetism and paleogeography of Adria. In: Morris A, Tarling DH (eds) Paleomagnetism and tectonics of the Mediterranean region. Geol Soc London Spec Publ 105, pp 119–132

Channell JET, D'Argenio B, Horvath F (1979) Adria, the African promontory in Mesozoic Mediterranean paleogeography. Earth Sci Rev 15:213–292

Choukroune P, Ecors Team (1989) The Ecors Pyrenean deep seismic profile reflection data and the overall structure of an orogenic belt. Tectonics 8:23–39

Christensen UR (1995) The influence of trench migration on slab penetration into the lower mantle. Earth Planet Sc Lett 140(1–4):27–39

Christensen UR (2001) Geodynamic models of deep subduction. Phys Earth Planet Inter 127(1–4):25–34

Cimini GB, De Gori P (1997) Upper mantle velocity structure beneath Italy from direct and secondary P-wave teleseismic tomography. Annali di Geofisica XL:175–194

Cimini GB, De Gori P (2001) Nonlinear P-wave tomography of subducted lithosphere beneath central-southern Apennines (Italy). Geophys Res Lett 28(23):4387–4390

Cirilli S, Decandia FA, Lazzarotto A, Pandeli E, Rettori R, Spina A (2002) Stratigraphy and depositional environment of the Mt Argentario sandstones Fm. (southern Tuscany, Italy). Boll Soc Geol It spec vol 1, pp 489–498

Cita MB, Colombo L (1979) Sedimentation in the latest Messinian at Capo Rossello (Sicily). Sedimentology 26:497–522

Cita MB, McKenzie JA (1986) The terminal Miocene event. In: Hsu KJ (ed) Mesozoic and Cenozoic oceans. Geodynamic series, vol 15, Am Geophys Union, pp 123–140

Cizkova H, van Hunen J, van den Berg AP, Vlaar NJ (2002) The influence of rheological weakening and yield stress on the interaction of slabs with the 670 km discontinuity. Earth Planet Sc Lett 199 (3–4):447–457

Clauzon G, Suc J-P, Popescu S-M, Marunteanu M, Rubino J-L, Marinescu F, Jipa D (2004) Influence of the Mediterranean sea-level changes over the Dacic Basin (Eastern Paratethys) in the late Neogene. Basin Res, in press

Cloetingh S, van der Beek PA, van Rees D, Roep TB, Biermann C, Stephenson RA (1992) Flexural interaction and the dynamics of Neogene extensional basin formation in the Alboran-Betic region. In: Maldonado A (ed) The Alboran Sea. Geo-Mar Lett 12:66–75

Cloetingh S, van Wees JD, van der Beek PA, Spadini G (1995) Role of pre-rift rheology in kinematics of extensional basin formation : constraints from thermomechanical models of Mediterranean and intracratonic basins. Mar Petrol Geol 12:793–807

Colonna V, Piccarreta G (1975) Metamorfismo di alta pressione/bassa temperatura nei micascisti di Zangarona-Ievoli-Monte Dondolo (Sila Piccola, Calabria). Boll Soc Geol Ital 94:17–25

Comas MC, Garcia-Duenas V, Maldonado A, Megias AG (1990) The Alboran Basin: tectonic regime and evolution of the Northern Alboran Sea. In: Global events and Neogene evolution of the Mediterranean. Abs., IX Congr Reg Comm Med Neogene Strat, 107–108

Comas MC, Platt JP, Soto JI, Watts AB (1999) The origin and tectonic history of the Alboran Basin : insights from Leg 161 results. In: Zahn R, Comas MC, Klaus A (eds) ODP Proceedings, leg 161, sites 974–979, pp 555–580

Compagnoni R (2003) HP metamorphic belts of the western Alps. Episodes 26:200–204

Corinth Natural Laboratory, www.corinth-rift-lab.org/index_fr.html

Coulon C, Megartsi M, Fourcade S, Maury RC, Bellon H, Louni-Hacini A, Cotten J, Coutelle A, Hermitte D (2002) Post-collisional transition from calc-alkaline to alkaline volcanism during the Neogene in Oranie (Algeria): magmatic expression of a slab breakoff. Lithos 62:87–110

Dal Piaz GV, Bistacchi A, Massironi M (2003) Geological outline of the Alps. Episodes 26:175–180

Davies JH, Von Blanckenburg F (1995) Slab breakoff – a model of lithosphere detachment and its test in the magmatism and deformation of collisional orogens. Earth Planet Sc Lett 129:85–102

DeCelles P, Cavazza W (1995) Upper Messinian fanglomerates in eastern Calabria (southern Italy): response to microplate migration and Mediterranean sea-level changes. Geology 23:775–778

Decima A, Wezel FC (1973) Late Miocene evaporites of the central Sicilian basin. In: Ryan WBF, Hsu KJ (eds) Initial reports of the Deep Sea Drilling Project, Leg 13, Washinghton DC, US Printing Office 1234–1240

Degnan PJ, Robertson AHF (1998) Mesozoic-early Tertiary passive margin evolution of the Pindos ocean (NW Peloponnese, Greece). Sedim Geol 117:33–70

de Jong MP, Wortel MJR, Spakman W (1993) From tectonic reconstruction to upper mantle model: an application to the Alpine-Mediterranean region. Tectonophysics 223:53–65

de Jong MR, Wortel MJR, Spakman W (1994) Regional scale tectonic evolution and the seismic velocity structure of the lithosphere and upper mantle: the Mediterranean region. J Geophys Res 99:12091–12108

Dercourt J, Zonenshain LP, Ricou LC, Kazmin VG, Le Pichon X, Knipper AL, Grandjacquet C, Sborshchikov IM, Boulin J, Sorokhtin O, Geyssant J, Lepvrier C, Biju-Duval B, Sibuet JC, Savostin L, Westphal M, Lauer JP (1985) Présentation des 9 cartes paléogéographiques au 1/20 000 000 s'étendant de l'Atlantique au Pamir pour la période du Lias à l'Actuel. Bull Soc géol France 8:637–652

Dercourt J, Zonenshain LP, Ricou L-E, Kazmin VG, Le Pichon X, Knipper AL, Grandjacquet C, Sbortshikov IM, Geyssant J, Lepvrier C, Perchersky DH, Boulin J, Sibuet J-C, Savostin LA, Sorokhin O, Westphal M, Bazhenov ML, Lauer JP, Biju-Duval B (1986) Geological evolution of the Tethys from the Atlantic to the Pamirs since the Lias. Tectonophysics 123: 241–315

Dercourt J, Ricou LE, Vrielinck B (1993) Atlas Tethys, paleoenvironmental maps. Gauthier-Villars, Paris

Dercourt J, Gaetani M, Vrielynk B, Barrier E, Biju-Duval B, Brunet MF, Cadet JP, Crasquin S, Sandulescu M (eds) (2000) Peri-Tethys palaeogeographic atlas. Paris

Destombes J (1971) L'Ordovicien du Maroc. Essai de synthèse stratigraphique. Mém Bureau Rech Géol Min Orléans 73: 237–263

De Voogd B, Nicolich R, Olivet JL, Fanucci F, Burrus J, the ECORS-CROP Profile Group (1991) First deep seismic reflection transect from the Gulf of Lions to Sardinia (ECORS-CROP profiles in western Mediterranean). In: Continental lithosphere: Deep seismic reflections. Geodyn Ser, Am Geophys Union, Washington DC, vol 22, pp 265–274

De Voogd B, Truffert C, Chamot-Rooke N, Huchon P, Lallemant S, Le Pichon X (1992) Two-ship deep seismic soundings in the basins of the Eastern Mediterranean Sea (Pasiphae Cruise). Geophysical Journal International 109:536–552

Devoti R et al. (2002) Geophysical interpretation of geodetic deformations in the central Mediterranean area. In: Stein S, Freymueller JT (eds) Plate boundary zones. Geodyn Ser, Am Geophys Union, Washington DC, vol 30, pp 57–65

Dewey JF (1988) Extensional collapse of orogens. Tectonics 7: 1123–1139

Dewey JF, Helman ML, Turco E, Hutton DHW, Knott SD (1989) Kinematics of the Western Mediterranean. In: Coward MP, Dietrich D, Park RG (eds) Alpine Tectonics, Geol Soc Spec Publ 45, pp 421–443

Dèzes P, Ziegler PA (2002). Moho depth map of Western and Central Europe. WWW address: http://www.unibas.ch/eucor-urgent

Dixon JE, Robertson AHF (eds) (1984) The geological evolution of the Eastern Mediterranean. Geol Soc London Spec Publ 17

Dodona E, Farinacci A, Kanani J, Nicosia U, Tonielli R (1994) Mid Jurassic events in the Ionian Zone (SW Albania). Paleopelagos 4:73–85

Doglioni C, Mongelli F, Pieri P (1994) The Puglia Uplift (SE Italy); an anomaly in the foreland of the Apenninic subduction due to buckling of a thick continental lithosphere. Tectonics 13: 1309–1321

Doglioni C, Guegen E, Sabat F, Fernandez M (1997) The Western Mediterranean extensional basins and the Alpine orogen. Terra Nova 9:109–112

Doglioni C, Fernandez M, Gueguen E, Sabat F (1999a) On the interference between the early Apennines-Maghrebides backarc extension and the Alps-Betics orogens in the Neogene geodynamics of the Western Mediterranean. Boll Soc Geol Ital 118:75–89

Doglioni C, Harabaglia P, Merlini S, Mongelli F, Peccerillo A, Piromallo C (1999b) Orogens and slabs vs. their direction of subduction. Earth Sc Rev 45(3–4):167–208

Droz L, Kergoat R, Cochonat P, Berné S (2001) Recent sedimentary events in the western Gulf of Lions (Western Mediterranean). Mar Geol 176:23–37

Duggen S, Hoernle K, van den Bogaard P, Rupke L, Morgan JP (2003) Deep roots of the Messinian salinity crisis. Nature 422(6932): 602–606

Durand B, Jolivet L, Horvath F, Séranne M (eds) (1999) The Mediterranean basins: Tertiary extension within the Alpine orogen. Geol Soc London Spec Publ 156

Ellouz N, Roca E (1994) Palinspastic reconstructions of the Carpathians and adjacent areas since the Cretaceous: a quantitative approach. In: Roure F (ed) Peri-Tethyan platforms . Ed. Technip, Paris, pp 51–78

Elter P, Sturani C, Weidmann M (1966) Sur la prolongation du domaine ligure de l'Apennin dans le Montferrat et les Alpes et sur l'origine de la nappe de la Simme s.l. des Préalpes romandes et chablaisiennes. Archives des Sciences, Genève 19:279–377

Elter P, Grasso M, Parotto M, Vezzani L (2003) Structural setting of the Apennine-Maghrebian thrust belt. Episodes 26:205–211

Engdahl ER, van der Hilst R, Buland R (1998) Global teleseismic earthquake relocation with improved travel times and procedures for depth determination. Bull Seismolog Soc Am 88(3): 722–743

Engelbrecht H (1997) From Upper Paleozoic extensional basin fill to late Alpine low-grade metamorphic core complex: preliminary note on the sedimentary and tectonic development of the Monticiano-Roccastrada-Zone (MRZ, southern Tuscany, Italy). Z Dt Geol Gesell 148:523–546

Engelbrecht H, Klemm DD, Pasini M (1989) Preliminary notes on the tectonics and lithotypes of the "verrucano s.l." in the Monticiano area (Southern Tuscany, Italy) and the finding of Fusulinids within the M.te Quoio FM. (Verrucano group). Riv Ital Paleont Strat 94:361–382

Fabries J, Lorand JP, Bodinier JL (1998) Petrogenetic evolution of orogenic lherzolite massifs in the central and western Pyrenees. Tectonophysics 292:145–167

Faccenna C, Mattei M, Funiciello R, Jolivet L (1997) Styles of back-arc extension in the Central Mediterranean. Terra Nova 9:126–130

Faccenna C, Becker TW, Lucente FP, Jolivet L, Rossetti F (2001a) History of subduction and back-arc extension in the Central Mediterranean. Geophys J Int 145(3):809–820

Faccenna C, Funiciello F, Giardini D, Lucente P (2001b) Episodic back-arc extension during restricted mantle convection in the Central Mediterranean. Earth Planet Sc Lett 187(1–2):105–116

Faccenna C, Jolivet L, Piromallo C, Morelli A (2003) Subduction and the depth of convection in the Mediterranean mantle. J Geophys Res 108(B2):2099, doi:10.1029/2001JB0011690

Faupl P, Wagreich M (1999) Late Jurassic to Eocene palaeogeography and geodynamic evolution of the Eastern Alps. Mit Öster Geol Gesell 92:79–94

Favre P (1995) Analyse quantitative du rifting et de la relaxation thermique de la partie occidentale de la marge transformante nord-africaine: le Rif externe (Maroc). Geodinamica Acta 8:59–81

Favre P, Stampfli GM (1992) From rifting to passive margin: the example of the Red Sea, Central Atlantic and Alpine Tethys. Tectonophysics 215:69–97

Favre P, Stampfli G, Wildi W (1991) Jurassic sedimentary record and tectonic evolution of the northwestern corner of Africa. Paleogeogr Paleoecol Paleoclim 87:53–73

Finetti I (1982) Structure, stratigraphy and evolution of the central Mediterranean Sea. Boll Geof Teor Appl 24:247–312

Finetti I (1985) Structure and evolution of the Central Mediterranean (Pelagian and Ionian Seas). In: Stanley DJ, Wezel FC (eds) Geological evolution of the Mediterranean basin. Springer Verlag, Heidelberg, pp 215–230

Finetti IR (in press) CROP deep seismic exploration of the Mediterranean region. Elsevier, Amsterdam

Finetti I, Bricch, G, Del Ben A, Pipan M, Xuan Z (1988) Geophysical study of the Black Sea. Bolletino Geofisica Teorica e Applicata 30:197–324

Finetti IR, Boccaletti M, Bonini M, Del Ben A, Geletti R, Pipan M, Sani F (2001) Crustal section based on CROP seismic data across the North Tyrrhenian-Northern Apennines-Adriatic sea. Tectonophysics 343:135–163

Frasheri A, Nishani P, Bushati S, Hyseni A (1996) Relationship between tectonic zones of the Albanides, based on results of geophysical studies. In: Ziegler PA, Horvàth F (eds) Structure and prospects of Alpine basins and forelands. Mém Mus Natl Hist Nat 170:485–511

Frepoli A, Amato A (2000) Spatial variation in stresses in peninsular Italy and Sicily from background seismicity. Tectonophysics 317:109–124

Frey M, Desmons J, Neubauer F (1999) Metamorphic Maps of the Alps. CNRS (Paris) Swiss N.S.F. (Berne), BMfWFV and FWF (Vienna)

Frisch W (1979) Tectonic progradation and plate tectonic evolution of the Alps. Tectonophysics 60:121–139

Frizon de Lamotte D, Saint-Bezar B, Bracene R, Mercier E (2000) The two main steps of the Atlas building and geodynamics of the western Mediterranean. Tectonics 19:740–761

Froitzheim N, Manatschal G (1996) Kinematics of Jurassic rifting, mantle exhumation, and passive-margin formation in the Austroalpine and Penninic nappes (Eastern Switzerland). Geol Soc Am Bull 108:1120–1133

Gansser A (1974) The ophiolitic melange, a world-wide problem on Tethyan examples. Eclogae Geol Helv 67:479–507

Garfunkel Z, Derin B (1984) Permian-early Mesozoic tectonism and continental margin formation in Israel and its implications for the history of the Eastern Mediterranean. In: Dixon JE, Robertson AHF (eds) The geological evolution of the Eastern Mediterranean. Geol Soc Spec Publ 17:187–201

Gattiglio M, Meccheri M, Tongiorgi M (1989) Stratigraphic correlation forms of the Tuscan Paleozoic basement. In: Sassi FP, Zanferrari A (eds) Pre-Variscan and Variscan events in the Alpine-Mediterranean belts, Stratigraphic correlation forms. Rend Soc Geol Ital 12:245–257

Gebauer D, Gruenenfelder M, Tilton G, Trommsdorff V, Schmid S (1992) The geodynamic evolution of garnet-peridotites, garnet-pyroxenites and eclogites of Alpe Arami and Cima di Gagnone (Central Alps) from early Proterozoic to Oligocene. Schweizerische Min Petr Mitt 72:107–111

Gelabert B, Sabat F, Rodriguez-Perea A (1992) A structural outline of the Serra de Tramuntana of Mallorca (Balearic islands). Tectonophysics 203:167–183

Gelabert B, Sabat F, Rodriguez-Perea A (2002) A new proposal for the late Cenozoic geodynamic evolution of the western Mediterranean. Terra Nova 14:93–100

Gelati R, Rogledi S, Rossi ME (1987) Significance of the Messinian unconformity-bounded sequences in the Apenninic margin of the Padan foreland basin, northern Italy. Mem Soc Geol Ital 39:319–323

Georgiev G, Dabovski C, Stanisheva-Vassileva G (2001) East Srednogorie-Balkan rift zone. In: Ziegler PA, Cavazza W, Robertson AHF, Crasquin-Soleau S (eds) Peritethyan rift/wrench basins and passive margins. Mém Museum National Hist Nat vol 186, pp 259–294

Giardini D (ed) (1999). The Global Seismic Hazard Assessment Program (GSHAP) 1992–1999. Annali di Geofisica 42(6):957–976

Giese P, Reutter KJ, Jacobshagen V, Nicolich R (1982) Explosion seismic crustal studies in the Alpine-Mediterranean region and their implications to tectonic processes. In: Berckhemer H, Hsu K (eds) Alpine-Mediterranean Geodynamics. Am Geophys Union Geodyn Series vol 7, pp 39–73

Goes S, Govers R, Vacher P (2000) Shallow mantle temperatures under Europe from P and S wave tomography. J Geophys Res 105:11153–11169

Goffé B, Oberhänsli R (1992) Ferro- and magnesiocarpholite in the "Bündnerschiefer" of the eastern Central Alps (Grisons and Engadine window). Eur J Mineral 4:835–838

Golonka J (2000) Cambrian-Neogene plate tectonic maps. Wydawnictwo Uniwersytetu Jagielloskiego, 125 pp

Göncüoglu MC, Köksal S, Floyd PA (1997) Post-collisional A-type magmatism in the central Anatolian Crystalline Complex: petrology of the Idis Gagi Intrusives (Avanos, Turkey). Turkish J Earth Sci 2:195–203

Göncüoglu MC, Truhan N, Sentürk K, Özcan A, Uysal S, Kenan Yalinz M (2000). A geotraverse across northwestern Turkey: tectonic units of the Central Sakaraya region and their tectonic evolution. In: Bozkurt E, Winchester JA, Piper JD (eds) Tectonics and Magmatism in Turkey and the Surrounding Area. Geol Soc London Spec Publ, vol 173, pp 139–161

Gorini C, Le Marrec A, Mauffret A (1993) Contribution to the structural and sedimentary history of the Gulf of Lions, (Western Mediterranean), from the ECORS profiles, industrial seismic profiles and well data. Bull Soc Géol France 164:353–363

Gorini C, Lofi J, Duvail C, des Reis AT, Berné S, Guennoc P, Lestrat P, Mauffret A (2003) The Late Messinian – 5.6/5.32 M.a. – salinity crisis and Late Miocene tectonic interactions and consequences on the physiography and post-rift evolution of the Gulf of Lions margins. EGS-AGU-EUG Joint Convention, Nice, April 6–11, Abs

Griffiths RW, Hackney RI, Van der Hilst RD (1995) A laboratory investigation of effects of trench migration on the descent of subducted slabs. Earth Planet Sc Lett 133:1–17

Gueguen E, Doglioni C, Fernandez M (1998) On the post-25 Ma geodynamic evolution of the western Mediterranean. Tectonophysics 298:259–269

Gutnic M, Monod O, Poisson A, Dumont JF (1979) Géologie des Taurides occidentales (Turquie). Mém Soc Géol France 137/LVIII

Gutscher MA, Malod J, Rehault JP, Contrucci I, Klingelhoefer F, Mendes-Victor L, Spakman W (2002) Evidence for active subduction beneath Gibraltar. Geology 30:1071–1074

Haas J, Kovacs S, Krystyn L, Lein R (1995) Significance of Late Permian-Triassic facies zones in terrane reconstructions in the Alpine-North Pannonian domain. Tectonophysics 242:19–40

Hall R, Spakman W (2002) Subducted slabs beneath the eastern Indonesia-Tonga region: insights from tomography, Earth Planet Sc Lett 201(2):321–336

Hinz K (1972) Results of seismic refraction investigations (Project Anna) in Western Mediterranean, south and north of the island of Mallorca. Bull. Centre Rech. Pau-SNPA 6, 2:405–426

Hsü KJ (1972) Origin of saline giants: a critical review after the discovery of the Mediterranean evaporite. Earth Sci Rev, 8:371–396

Hsü KJ, Montadert L (1978) History of the Mediterranean salinity crisis. In: Hsü KJ, Montadert L (eds) Init Rep Deep Sea Drilling Proj, vol 42, I, US Gov Print Off, Washington DC, pp 1053–1078

Hoogenduijn Strating EH (1991) The evolution of the Piemonte-Ligurian ocean, a structural study of ophiolite complexes in Liguria (NW Italy). Geologica Ultraiectina vol 74

Jiménez-Munt I, Sabadini R, Gardi A, Bianco G (2003) Active deformation in the Mediterranean from Gibraltar to Anatolia inferred from numerical modeling and geodetic and seismological data. J Geophys Res 108 doi :10 :1029/2001IB001544

Jolivet L (2001) A comparison of geodetic and finite strain pattern in the Aegean, geodynamic implications. Earth Planet Sci Lett 187:95–104

Jolivet L, Brun JP, Gautier P, Lallemant S, Patriat M (1994) 3D-kinematics of extension in the Aegean region from the early Miocene to the Present: insights from the ductile crust. Bull Soc Géol France 165:195–209

Jolivet L, Faccenna C (2000) Mediterranean extension and the Africa-Eurasia collision. Tectonics 19(6):1095–1106

Kahle H-G, Cocard M, Peter Y, Geiger A, Reilinger R, Barka A, Veis G (2000) GPS-derived strain rate field within the boundary zones of the Eurasian, African and Arabian plates. J Geophys Res 105(23):353–370

Karakitsios V (1991) Study of Jurassic paleofaults in the Ionian Zone of Epirus. 5th Congr Geol Soc Greece, Proceedings, pp 307–318

Karakitsios V, Dermitzakis MD (1997) Lacunes sedimentaires, discordances et phenomenes paleokarstiques (pliensbachien-tithonique) dans la zone ionienne (Epire, Grece nord-occidentale). In: Anonymous (ed) A la memoire du Professeur G. M. Paraskevopoulos. Laboratoire de Geologie de l'Universite 37, pp 847–864

Kastens K, Mascle J (eds) (1990) Proceedings Ocean Drilling Program. Scientific Results, vol 107, College Station, Texas

Kaymakçi N, White SH, Van Dijk PM (2000) Paleostress inversion in a multiphase deformed area: kinematic and structural evolution of the Çankiri Basin (central Turkey). In: Bozkurt E, Winchester JA, Piper JDA (eds) Tectonics and magmatism in Turkey and surrounding area. Geol Soc London Spec Publ 173, pp 295–323

Kennett BLN, Engdahl ER, Buland R (1995) Constraints on seismic velocities in the earth from travel-times. Geophys J Int 122(1):108–124

Kissling E (1993) Deep structure of the Alps – what do we really know. Phys Earth Planet Inter 79(1–2):87–112

Kissling E, Spakman W (1996) Interpretation of tomographic images of uppermost mantle structure: Examples from the western and central Alps. J Geodyn 21:97–111

Kock S (2003) Nouvelles données stratigraphiques, métamorphiques et géochimiques sur la série autochtone des Talea Ori (Citère centrale). Masters Thesis, University of Lausanne

Koçyigit A (1991) An example of an accretionary forearc basin from Central Anatolia and its implications for the history of subduction of Neo-Tethys in Turkey. Geol Soc Am Bull 103:22–36

König H, Kuss S (1980) Neue Daten zur Biostratigraphie des permotriadischen Autochtonous der Insel Kreta (Griechenland). Neues Jahr Geol Paläont, Abhandlungen 9:525–540

Kopf A, Behrmann JH (2003) Quantitative approach to the extrusion of active mud volcanoes on the Mediterranean Ridge. EGS-AGU-EUG Joint Convention, Nice, April 6–11, Abs

Kozur H (1990) Deep-water permian in Sicily and its possible connection with the Himalaya-Tibet region. 5th Himalaya-Tibet-Karakorum Workshop

Kozur H (1991a) The geological evolution at the western end of the Cimmerian ocean in the Western Carpathians and Eastern Alps. Zbl Geol Paläont 1:99–121

Kozur H (1991b) The evolution of the Meliata-Hallstatt ocean and its significance for the early evolution of the Eastern Alps and western Carpathians. In: Channell JET, Winterer EL, Jansa LF (eds) Paleogeography and paleoceanography of Tethys. Palaeogeogr Palaeoclim Palaeoecol 87:109–135

Kozur H (1991c) Permian deep-water ostracods from Sicily (Italy) Part 1: Taxonomy. Geol Paläont Mitt Innsbruck, Sonderband 3: 1–24

Kozur H (1991d) Permian deep-water ostracods from Sicily (Italy) Part2: Biofacial evaluation and remarks to the Silurian to Triassic paleopsychrospheric ostracods. Geol Paläont Mitt Innsbruck, Sonderband 3:25–38

Kozur H (1995) First evidence of Middle Permian Ammonitico Rosso and further new stratigraphic results in the Permian and Triassic of the Sosio Valley area, Western Sicily. 1st Croatian Geological Congress, pp 307–310

Kozur H (1999) Permian development in the western Tethys. Shallow Tethys 5:101–135

Kozur H, Mostler H (1992) Erster paläontologischer Nachweis von Meliaticum und Süd-Rudabànyaicum in den nördlischen Kalkalpen (Oesterreich) und ihre Beziehungen zu den Abfolgen in den Westkarpaten. Geol Paläont Mitt Innsbruck 18:87–129

Kozur HW, Senel M, Tekin K (1998) First evidence of Hercynian Lower Carboniferous deep-water sediments in the Lycian nappes, SW Turkey. Geol. Croatica 51:15–22

Krijgsman W, Hilgen FJ, Raffi I, Sierro FJ, Wilson DS (1999) Chronology, causes and progression of the Messinian salinity crisis. Nature 400:652–655

Laubscher HP (1983) The late alpine (periadriatic) intrusions and the Insubric Line. Mem Soc Geol Ital 26:21–30

Laubscher HP, Bernoulli D (1977) Mediterranean and Tethys. In: Nairn A, Kanes W, Stehli F (eds) The ocean basins and margins. Plenum Press, New York, pp 1–28

Le Borgne E, Le Mouel JL, Le Pichon X (1971) Aeromagnetic survey of south-western Europe. Earth Plan Sci Lett 12:287–299

Le Pichon X (1982) Land-locked oceanic basins and continental collision: the Eastern Mediterranean as a case example. In: Hsu K (ed) Mountain building processes, Academic Press, London, pp 201–211

Le Pichon X, Pautot G, Auzende J-M, Olivet JL (1971) La Méditerranée occidentale depuis l'Oligocène, schéma d'évolution. Earth Plan Sci Lett 13:145–152

Letouzey J, Trémolières P (1980) Paleostress fields around the Mediterranean since the Mesozoic derived from microtectonics: comparisons with plate tectonics data. In: Aubouin J, Debelmas J, Latreille M (eds) Géologie des chaînes alpines issues de la Tethys. Coll C5, 26th Int Geol Congr, Paris, Mem BRGM 115: 261–273

Leveque J-J, Rivera L, Wittlinger G (1993) On the use of the checkerboard test to assess the resolution of tomographic inversions. Geophys J Int 115:313–318

Liati A, Gebauer D, Froitzheim N, Fanning M (2003) Origin and geodynamic significance of metabasic rocks from the Antrona ophiolites (western Alps): new insights from SHRIMP-dating. EGS-AGU-EUG Joint Assembly, abstract 12648

Linzer HG (1996) Kinematics of retreating subduction along the Carpathian arc. Geology 24:167–170

Lippitsch R, Kissling E, Ansorge J (2003) Upper mantle structure beneath the Alpine orogen from high-resolution teleseismic tomography. J Geophys Res 108 (B8), art no 2376

Loncke L, Gaullier V, Mascle J, Vendeville B (2003) Shallow structure of the Nile delta deep sea fan: interactions between structural heritage and salt tectonics; consequences on sedimentary dispersal. CIESM Workshop Series 17

Lonergan L, White N (1997) Origin of the Betic-Rif mountain belt. Tectonics 16(3):504–522

Lucente FP, Speranza F (2001) Belt bending driven by lateral bending of subducting lithospheric slab: geophysical evidences from the northern Apennines (Italy). Tectonophysics 337(1–2):53–64

Lucente FP, Chiarabba C, Cimini GB, Giardini D (1999) Tomographic constraints on the geodynamic evolution of the Italian region. J Geophys Res 104 (B9):20307–20327

Makris J, Morelli C, Zanolla C (1998) The Bouguer gravity map of the Mediterranean Sea (IBCM-G). Boll Geofis Teor Appl 39: 79–98

Malinverno A, Ryan WBF (1986) Extension in the Tyrrhenian Sea and shortening in the Apennines as result of arc migration driven by sinking of the lithosphere. Tectonics 5:227–245

Malpas J, Calon T, Squires G (1993) The development of a late Cretaceous microplate suture zone in SW Cyprus. In: Prichard HM, Alabaster T, Harris NBW, Neary CR (eds) Magmatic processes and plate tectonics. Geol Soc London Spec Publ 76:177–195

Mantovani E, Albarello D, Babbucci D, Tamburelli C (1994) Extrusion tectonics in the central mediterranean area. Boll Geofis Teor Appl 36:435–462

Mantovani E, Albarello D, Babbucci D, Tamburelli C, Viti M (2002) Trench-Arc-BackArc Systems in the Mediterranean area: Examples of Extrusion Tectonics. In: Rosenbaum G, Lister GS (eds) Reconstruction of the evolution of the Alpine-Himalayan orogen. J Virtual Explorer 8:31–47

Marchant RH, Stampfli GM (1997) Subduction of continental crust in the Western Alps. Tectonophysics 269:217–235

Mascle J, Sardou O, Boucher P, Felt V, Prismed 2-Fanil Team (2002) Evidences of fluid escape structures, mud volcanoes and gas chimneys on the Nile deep sea fan. CIESM Workshop Series 17, pp 57–59

Mascle J, Loncke L, Camera L, Gonthier E, Mulder T, Murat A, Nile Scientific Team (2003) An integrated study of the Nile continental margin (Eastern Mediterranean). Ocean Margin Res Conf, Paris, 15–17 September, Abs, 141

Mattei M, Cipollari P, Cosentino D, Argentieri A, Rossetti F, Speranza F, Di Bella L (2002) The Miocene tectono-sedimentary evolution of the southern Tyrrhenian Sea: stratigraphy, structural and paleomagnetic data from the on-shore Amantea basin (Calabrian arc, Italy). Basin Res 14:147–168

Mauffret A, Fail JP, Montadert L, Sancho J, Winnock E (1973) Northwestern Mediterranean sedimentary basin from seismic reflection profile. Am Ass Petrol Geol Bull 57:2245–2262

Maury R, Fourcade S, Coulon C, Azzouzi M, Bellon H, Coutelle A, Ouabadi A, Semroud B, Megarsi M, Cotten J, Belanteur O, Louni-Hacini A, Piqué A, Capdevila R, Hernandez J, Rehault JP (2000) Post-collisional Neogene magmatism of the Mediterranean Maghreb margin: a consequence of slab breakoff. C R Acad Sci Paris 331:159–173

Mazzoli S, Helman M (1994) Neogene patterns of relative plate motions for Africa-Europe: some implications for recent central Mediterranean tectonics. Geol Rund 83:464–468

McClusky S, Balassanian S, Barka A, et al. (2000) Global Positioning System constraints on plate kinematics and dynamics in the eastern Mediterranean and Caucasus. J Geophys Res 105(B3):5695–5719

Monaghan A (2001) Coeval extension, sedimentation and volcanism along the Cainozoic rift system of Sardinia. In: Ziegler PA, Cavazza W, Robertson AHF, Crasquin-Soleau S (eds) Peritethyan rift/wrench basins and passive margins. Mém Museum National Hist Nat 186, Paris, pp 707–734

Monod O (1977) Recherches géologiques dans le Taurus occidental au Sud de Beyşehir (Turquie). Thèse Univ. Paris-Sud (Orsay)

Montadert L, Letouzey J, Mauffret A (1977) Messinian event: seismic evidence. In: Ross DA, Neprochnov YP (eds) Init Rep, Deep Sea Dril Proj, 42, I, US Gov Print Off, Washington DC, pp 1037–1050

Moores EM, Fairbridge RW (eds) (1997) Encyclopedia of European and Asian Regional Geology. Chapman & Hall, London

Moretti I, Delhomme JP, Cornet F, Bernard P, Schmidt-Hattenberg C, Borm G (2002) The Corinth rift laboratory; monitoring of active faults. First Break 20:91–97

Morris A, Tarling DH (eds) (1996) Paleomagnetism and tectonics of the Mediterranean region. Geol Soc London Spec Publ 105

Muttoni G, Garzanti E, Alfonsi L, Cirilli S, Germani D, Lowrie W (2001) Motion of Africa and Adria since the Permian: paleomagnetic and paleoclimatic constraints from Northern Libya. Earth Plan Sci Lett 192:159–174

Neumann P, Zacher W (1996) Multistratigraphic investigations and Sedimentary Cycles in the Upper Cretaceous of the Olonos-Pindos Zone (Greece): Paleogeography, Facies and Basin Development. In: Reitner J, Neuweiler F, Gunkel F (eds) Global and Regional Controls on Biogenic Sedimentation. Cretaceous Sedimentation Research Reports. Göttinger Arb Geol Paläont Sb 3:123–126

Nikishin AM, Ziegler PA, Abbott D, Brunet MF, Cloetingh S (2002) Permo-Triassic intraplate magmatism and rifting in Eurasia: implications for mantle dynamics. Tectonophysics 351:3–39

Nikishin AM, Korotaev MV, Ershov AV, Brunet MF (2003) The Black-Sea basin: tectonic history and Neogene-Quaternary rapid subsidence modelling. In: Brunet MF, Cloetingh S (eds) Integrated Peri-Tethyan Basins Studies. Sediment Geol 156:149–168

Niko S, Pillevuit A, Nishida T (1996) Early Late Permian (Wordian) non-ammonoid cephalopods from the Hamrat Duru Group, central Oman Mountains. Trans Proc Palaeont Soc Japan, 183:522–527

Noomen R, Springer TA, Ambrosius BAC, Herzberger K, Kuijper DC, Mets GJ, Overgaauw B, Wakker KF (1996) Crustal deformations in the Mediterranean area computed from SLR and GPS observations. J Geodyn 21 (1):73–96

Norman S, Chase CG (1986) Uplift of the shores of the western Mediterranean due to Messinian desiccation and flexural isostasy. Nature 322:450–451

Oberhänsli R, Hunziker JC, Martinotti G, Sten WB (1985) Geochemistry, geochronology and petrology of Monte Mucrone: an example of eo-alpine eclogitisation of Permian granitoïds in the Sesia Lanzo zone, western Alps, Italy. Chem Geol 52:165–184

Okay AI, Sahintürk Ö (1997) Geology of the eastern Pontides. In: Robinson AG (eds) Regional and petroleum geology of the Black Sea and surrounding region. Am Ass Petrol Geol Mem 68, pp 291–311

Okay AI, Tansel I (1992) New data on the upper age of the Intra-Pontide ocean from north of Sarkoy (Thrace). Mineral Res Expl Bull 114:23–26

Okay AI, Tüyzüs O (1999) Tethyan Sutures of northern Turkey. In: Durand B, Jolivet L, Horvath F, Seranne M (eds) Mediterranean Basins: Tertiary extension within the Alpine Orogen. Geol Soc London Spec Publ 15, pp 475–515

Okay AI, Siyako M, Bürkan KA (1991) Geology and tectonic evolution of the Biga Peninsula, northwest Turkey. Bull Tech Univ Istanbul 44:191–255

Okay AI, Sengör, AMC, Görür N (1994) Kinematic history of the opening of the Black Sea and its effect on the surrounding regions. Geology 22:267–270

Okay AI, Tansel I, Tüysüz O (2001) Obduction, subduction and collision as reflected in the Upper Cretaceous-Lower Eocene sedimentary record of western Turkey. Geol Mag 138:117–142

Olbertz D, Wortel MJR, and Hansen U (1997) Trench migration and subduction zone geometry. Geophys Res Lett 24:221–224

Oldow JS, Ferranti L, Lewis DS, Campbell JK, D'Argenio B, Catalano R, Pappone G, Carmignani L, Conti P, Aiken CLV (2002) Active fragmentation of Adria, the north African promontory, central Mediterranean orogen. Geology 30:779–782

Pamic J (2002) The Sava-Vardar zone of the Dinarides and Hellenides. Ecl geol Helv 95:99–114

Pamic J, Bahlen D, Herak M (2002) Origin and geodynamic evolution of late Paleogene magmatic associations along the Periadriatic-Sava-Vardar magmatic belt. Geodinamica Acta 15:209–231

Pandeli E, Pasini M (1990) Fusulinidi permiani nella successione metamorfica del sottosuolo del M. Amiata, Toscana meridionale (Italia). Rivista Ital Paleont Strat 96:3–20

Panza GF, Mueller S, Calcagnile G (1980) The gross features of the lithosphere-asthenosphere system in Europe from seismic surface waves and body waves. Pure Appl Geophys 118:1209–1213

Panza GF, Pontevivo A, Chimera G, Raykova R, Aoudia A (2003) The lithosphere-asthenosphere: Italy and surroundings. Episodes 26(3):169–174

Pascal GP, Mauffret A, Patriat P (1993) The ocean-continent boundary in the Gulf of Lion from analysis of expanding spread profiles and gravity modelling. Geophys J Inter 113:701–726

Patacca E, Sartori R, Scandone P (1993) Tyrrhenian basin and Apennines: kinematic evolution and related dynamic constraints. In: Boschi E, Mantovani E, Morelli A (eds) Recent evolution and seismicity of the Mediterranean region. Kluwer Academic Publishers, pp 161–171

Pe-Piper G (1998) The nature of Triassic extension-related magmatism in Greece: evidence from Nd and Pb isotope geochemistry. Geol Mag 135:331–348

Perez-Belzuz F, Alonso B, Ercilla G (1997) History of mud diapirism and trigger mechanisms in the Western Alboran Sea. Tectonophysics 282:399–422

Pfiffner OA (ed)(1996) Deep structure of the Swiss Alps. Results of NFP/PNR 20. Basel, Birkhäuser AG

Pfiffner OA, Frei W, Finck P, Valasek P (1988) Deep seismic reflection profiling in the Swiss Alps: explosion seismology results of Line NFP-20 east. Geology 16:987–990

Pillevuit A (1993) Les blocs exotiques du Sultanat d'Oman, évolution paléogéographique d'une marge passive flexurale. Mémoires de Géologie (Lausanne), vol 17

Pillevuit A, Marcoux J, Stampfli GM, Baud A (1997) The Oman exotics: a key to the understanding of the Neotethyan geodynamic evolution. Geodinamica Acta 10:209–238

Piqué A (1989) Variscan terranes in Morocco In: Dallmeyer D (ed) Terranes in the circum-Atlantic Paleozoic orogens. Geol Soc Am Spec Paper 230, pp 115–129

Piromallo C, Morelli A, (1997) Imaging the Mediterranean upper mantle by P-wave travel time tomography. Ann Geofis XL: 963–979

Piromallo C, Morelli A (2003) P wave tomography of the mantle under the Alpine-Mediterranean area. J Geophys Res 108(B2): art no 2065

Plasienka D (1996) Mid-Cretaceous (120–80Ma) orogenic processes in the central Western Carpathians: brief review and interpretation of data. Slovak Geol Mag 3–4:319–324

Platt JP, Vissers RLM (1989) Extensional collapse of thickened continental lithosphere : a working hypothesis for Alboran Sea and Gibraltar Arc. Geology 17:540–543

Platt JP, Behrmann JH, Cunningham PC, Dewey JF, Helman M, Parish M, Shepley MG, Wallis S, Weston PJ (1989) Kinematics of the Alpine arc and the motion history of Adria. Nature 337:158–161

Platt IP, Soto JI, Comas MC, Leg 161 Shipboard Scientists (1996) Decompression and high temperature, low pressure metamorphism in the exhumed floor of an extensional basin, Alboran Sea, Western Mediterranean. Geology 24:447–450

Platt JP, Soto JI, Whitehouse MJ, Hurford AJ, Kelley SP (1998) Thermal evolution, rate of exhumation and tectonic significance of metamorphic rocks from the floor of the Alboran extensional basin, Western Mediterranean. Tectonics 17:671–689

Platt JP, Allerton S, Kirker A, Mandeville C, Mayfield A, Platzman ES, Rimi A (2003) The ultimate arc: differential displacement, oroclinal bending, and vertical axis rotation in the External Betic-Rif arc. Tectonics 22(3):art no 1017

Poisson A (1984) The extension of the Ionian trough into southwestern Turkey. In: Dixon JE, Robertson AHF (eds) The geological evolution of the eastern Mediterranean. Geol Soc Spec Publ 17, pp 241–249

Puga E, Diaz de Federico A, Demant A (1995) The eclogitized pillows of the Betic Ophiolitic Association: relics of the Tethys Ocean floor incorporated in the Alpine Chain after subduction. Terra Nova 7:31–43

Rabineau M, Berné S, Le Drezen E, Lericolais G, Marsset T, Rotunno M (1998) 3D architecture of lowstand and transgressive Quaternary sand bodies on the outer shelf of the Gulf of Lion, France. Marine Petrol Geol 15:439–452

Ramovs A (1968) Biostratigraphie der klastischen Entwicklung der Trogkofelstufe in den Karawanken und Nachbargebieten. Neues Jahrb Geol Paläont, Abhandlungen 131:72–77

Rampone E, Piccardo G (2000) The ophiolite-oceanic lithosphere analogue:new insights from the northern Apennines (Italy). In: Dilek Y, Moores EM, Elthon D, Nicolas A (eds) Ophiolites and oceanic crust: new insights from field studies and the Ocean Drilling Project. Geol Soc Am Spec Paper 349:21–34

Ratschbacher L, Frisch W, Linzer HG, Merle O (1991) Lateral extrusion in the Eastern Alps, part 2: Structural analysis. Tectonics 10:257–271

Rebai S, Philip H, Taboada A (1992) Modern tectonic stress field in the Mediterranean region: evidence for variation in stress directions at different scales. Geophys J Int 110:106–140

Rehault JP, Boillot G, Mauffret A (1984a) The Western Mediterranean Basin, geological evolution. Mar Geol 55:447–477

Rehault JP, Boillot G, Mauffret A (1984b) The Western Mediterranean basins. In: Stanley DJ, Wezel FC (eds) Geological evolution of the Mediterranean basins. Springer-Verlag, Berlin, pp 101–129

Reinecker J, Heidbach O, Mueller B (2003) The 2003 release of the World Stress Map (available online at www.world-stress-map.org)

Reischmann T (1998) Pre-Alpine origin of tectonic units from the metamorphic complex of Naxos, Greece, identified by single zircon Pb/Pb dating. 8th Intern Congr Geol Soc Greece, pp 101–111

Richter D, Müller C, Mihm A (1993b) Die Flysch-Zonen Griechenlands, V. Zur Stratigraphie des Flysches der Pindos-Zone im nördlichen Pindos-Gebirge zwischen der albanischen Grenze und der Querzone von Kastaniotikos (Griechenland). Neues Jahr Geol Paläont 5:257–291

Robertson AHF (2002) Overview of the genesis and emplacement of Mesozoic ophiolites in the Eastern Mediterranean Tethyan region. Lithos 651–67

Robertson AHF, Degnan PJ (1993) Sedimentology and tectonic implications of the Lamayuru complex: deep-water facies of the Indian passive margin, Indus suture zone, Ladakh Himalaya. In: Treloar PJ, Searle MP (eds) Himalaya tectonics. Geol Soc Spec Publ 74:299–321

Robertson AHF, Xenophontos C (1993) Development of concepts concerning the Troodos ophiolite and adjacent units in Cyprus In: Prichard HM, Alabaster T, Harris NB, Neary CR (eds) Magmatic processes and plate tectonics. Geol Soc Spec Publ 76, pp 85–119

Robertson AHF, Dixon JE, Brown S, Collins A, Morris A, Pickett E, Sharp I, Ustaömer T (1996) Alternative tectonic models for the Late Palaeozoic-Early Tertiary development of Tethys in the Eastern Mediterranean region. In: Morris A, Tarling DH (eds) Palaeomagnetism and Tectonics of the Eastern Mediterranean Region. Geol Soc London Spec Publ 105, pp 239–263

Robinson AG (1997) Regional and petroleum geology of the Black Sea and surrounding region. Am Ass Petrol Geol Mem 68

Roca E (2001) The Northwest Mediterranean Basin (Valencia trough, Gulf of Lions and Liguro-Provençal baisns): structure and geodynamic evolution. In: Ziegler PA, Cavazza W, Robertson AHF, Crasquin-Soleau S (eds) Peritethyan rift/wrench basins and passive margins. Mém Museum National Hist Nat vol 186, Paris, pp 671–706

Rosenbaum G, Lister GS, Duboz C (2002a) Reconstruction of the tectonic evolution of the western Mediterranean since the Oligocene. In: Rosenbaum G, Lister GS (eds) Reconstruction of the evolution of the Alpine-Himalayan orogen. J Virtual Explorer 8:107–126

Rosenbaum G, Lister GS, Duboz C (2002b) Relative motions of Africa, Iberia and Europe during Alpine orogeny. Tectonophysics 359(1–2):117–129

Rouchy JM (1980) La genèse des évaporites messiniennes de Méditerranée: un bilan. Bull Centre Recherche Pau 4:511–545

Roure F (ed) (1994) Peri-Tethan Platforms. Ed. Technip, Paris

Roure F, Choukroune P, Berastegui X, Munoz JA (1989) Ecors deep seismic data and balanced cross sections: geometric constraints on the evolution of the Pyrenees. Tectonics 8:41–50

Roure F, Heitzman P, Polino R (eds) (1990) Deep structure of the Alps. Mém Soc Géol France 156, Soc Géol Suisse 1, Soc Geol It 1

Roure F, Roca E, Sassi W (1993) The Neogene evolution of the outer Carpathian flysch units (Poland, Ukraine and Romania): kinematics of a foreland/fold-and-thrust belt system. Sed Geol 86:177–201

Roure F, Choukroune P, Polino R (1996) Deep seismic reflection data and new insights on the bulk geometry of mountain ranges. CR Acad Sc, Paris

Roure F, Nazaj S, Mushka K, Fili I, Cadet JP, Bonneau M (2004) Kinematic evolution and petroleum systems: an appraisal of the Outer Albanides. In: McClay K (ed) Thrust tectonics. AAPG Memoir (in press)

Royden LH (1993) Evolution of retreating subduction boundaries formed during continental collision. Tectonics 12:629–638

Royden LH, Horvath F (1988) The Pannonian Basin. Am Ass Petrol Geol Mem 45, Tulsa

Royden LH, Patacca E, Scandone P (1987) Segmentation and configuration of subducted lithosphere in Italy: an important control on thrust belt and foredeep basin evolution. Geology 15:714–717

Rubatto D (1998) Dating of pre-Alpine magmatism, Jurassic ophiolites and Alpine subductions in the Westen Alps, ETHZ: 173

Russia-ITLP (1997) The paleogeographic atlas of northern Eurasia (380–10 Ma). Institute of Tectonics of Lithospheric Plates, Russian Academy of Sciences, Moscow (26 maps)

Ryan WFB (1976) Quantitative evaluation of the depth of the Western Mediterranean before, during and after the Messinian salinity crisis. Sedimentology 23:791–813

Ryan WBF, Pitman W (1998) Noah's flood. Simon and Schuster, New York

Ryan WBF, Pitman W, Major CO, Shimkus K, Moskalenko V, Jones GA, Dimitrov P, Görür N, Saknç M, Yüce H (1997) An abrupt drowning of the Black Sea shelf. Mar Geol 138:119–126

Ryan WBF, Çagatay N, Major CO, Lericolais G (2003) Evidence for a Black Sea flooding event. Geol Soc of America Abstracts with Program, Seattle, Nov 2–5, paper no 189-1

Sage L, Letouzey J (1990) Convergence of the African and Eurasian Plates in the Eastern Mediterranean. In: Letouzey J (ed) Petroleum and Tectonics in Mobile Belts. Ed. Technip, Paris, pp 49–68

Salas R, Guimera J, Mas R, Martin-Closas C, Melendez A, Alonso A (2001) Evolution of the Mesozoic Central Iberian Rift System and its Cainozoic inversion (Iberian Chain). In: Ziegler PA, Cavazza W, Robertson AHF, Crasquin-Soleau S (eds) Peritethyan rift/wrench basins and passive margins. Mém Museum National Hist Nat vol 186, Paris, pp 145–185

Sans M, Sàbat F (1993) Pliocene salt rollers and syn-kinematic sediments in the northeast sector of the València trough (western Mediterranean). Bull Soc Géol France 164, 2:189–198

Sartori R (2003) Luigi Ferdinando Marsili founding father of oceanography. In: Vai GB, Cavazza W (eds) Four centuries of the word geology. Minerva Edizioni, Bologna, pp 169–177

Savelli C (2002) Time-space distribution of magmatic activity in the western Mediterranean and peripheral orogens during the past 30 Ma (a stimulus to geodynamic considerations) J Geodyn 34:99–126

Schettino A, Scotese C (2002) Global kinematic constraints to the tectonic history of the Mediterranean region and surrounding areas during the Jurassic and Cretaceous. In: Rosenbaum G, Lister GS (eds) Reconstruction of the evolution of the Alpine-Himalayan orogen. J Virtual Explorer 8:149–168

Schmid SM, Kissling E, (2000) The arc of the western Alps in the light of geophysical data on deep crustal structure. Tectonics 19:62–85

Schmid SM, Pfiffner OA, Schönborn G, Froitzheim N, Kissling E (1996) Integrated cross section and tectonic evolution of the lps along the eastern transect. In: Pfiffner OA (ed) Deep structure of the Swiss Alps, results of NFP-PNR 20. Birkhäuser AG, Basel

Schönlaub HP, Histon K (1999) The Paleozoic Evolution of the southern Alps. Mitt Öster Geol Gesell 92:15–34

Schwan W (1997) Europe. In: Moores EM, Fairbridge RW (eds) (1997) Encyclopedia of European and Asian Regional Geology. Chapman & Hall, London, pp 201–227

Seber D, Barazangi M, Tadili BA, Ramdani M, Ibenbrahim A, Sari DB (1996) Three-dimensional upper mantle structure beneath the intraplate Atlas and interplate Rif mountains of Morocco. J Geophys Res 101:3125–3138

Seidel E, Okrusch M, Kreuzer H, Raschka H, Harre H (1976) Eo-Alpine Metamorphism in the Uppermost Unit of the Cretan Nappe System – Petrology and Geochronology Part 1. The Lendas Area (Asteroussia Mountains). Contrib Mineral Petrol 57:259–275

Selvaggi G, Chiarabba C (1995) Seismicity and P-velocity image of the southern Tyrrhenian subduction zone. Geophys J Int 121: 818–826

Sengör AMC (1979) Mid-Mesozoic closure of Permo-Triassic Tethys and its implications. Nature 279:590–593

Sengör AMC (1984) The Cimmeride orogenic system and the tectonics of Eurasia. Geol Soc Am Spec Paper 195

Sengör AMC (1985) The story of Tethys: how many wives did Okeanos have? Episodes 8:3–12

Sengör AMC, Yýlmaz Y, Sungurlu O (1984) Tectonics of the Mediterranean Cimmerides: nature and evolution of the western termination of Paleo-Tethys. In: Dixon JE, Robertson AHF (eds) The geological evolution of the eastern Mediterranean. Geol Soc London Spec Publ 17, pp 77–112

Serri G, Innocenti F, Manetti P (2001) Magmatism from Mesozoic to Present: petrogenesis, time-space distribution and geodynamic implication. In: Vai GB, Martini IP (eds) Anatomy of an orogen: the Apennines and adjacent basins, Kluwer Academic Publ, Dordrecht, pp 77–104

Srivastava SP, Tapscott CR (1986) Plate kinematics of the North Atlantic. In: Vogt PR, Tucholke BE (eds) The Western North Atlantic region. Geol Soc Am, pp 379–404

Spadini G, Cloetingh S, Bertotti G (1995) Thermo-mechanical modeling of the Tyrrhenian Sea: Lithospheric necking and kinematics of rifting. Tectonics 14:629–644

Spadini G, Robinson A, Cloetingh S (1996) Western versus Eastern Black Sea tectonic evolution, pre-rift lithospheric controls on basin formation. Tectonophysics 266:139–154

Spakman W (1986a) Subduction beneath Eurasia in connection with the Mesozoic Tethys. Geologie en Mijnbouw 65:145–153

Spakman W (1986b) The upper mantle structure in the Central European-Mediterranean region. In: Freeman R, Mueller S, Giese P (eds) European Geotraverse (EGT) Project, the central segment, European Science Foundation, Strasbourg, pp 215–222

Spakman W (1990) Images of the upper mantle of central Europe and the Mediterranean. Terra Nova, 2:542–553

Spakman W (1991) Delay time tomography of the upper mantle below Europe, the Mediterranean, and Asia Minor. Geoph J Int 107:309–332

Spakman W (1993) Iterative strategies for nonlinear travel-time tomography using global earthquake data. In: Iyer HM, Hirahara K (eds) Seismic tomography: theory and applications. Prentice-Hall, London, pp 190–226

Spakman W, Bijwaard H (2001) Optimization of cell parameterization for tomographic inverse problems. Pure and Appl Geophys 158:1401–1423

Spakman W, Nolet G (1988) Imaging algorithms, accuracy and resolution in delay time tomography. In: Vlaar NJ, Nolet G, Wortel MJR, Cloetingh SAPL (eds), Mathematical Geophysics: a survey of recent developments in seismology and geodynamics, Reidel, Dordrecht, pp 155–188

Spakman W, Van der Lee S, van der Hilst RD (1993) Travel-time tomography of the European-Mediterranean mantle down to 1,400 km. Phys Earth Planet Inter 79:3–7

Speed RC (1991) Tectonic Section Display – North American Continent-Ocean Transects Program. Geological Society of America, Boulder

Speranza F, Villa IM, Sagnotti L, Florindo F, Cosentino D, Cipollari P, Mattei M (2002) Age of the Corsica-Sardinia rotation and Liguro-Provençal Basin spreading: new plaeomagnetic and Ar/Ar evidence. Tectonophysics 347:231–251

Sperner B, Ioane D, Lillie RJ (2004) Slab behaviour and its surface expression: new insights from gravity modelling in the SE Carpathians. Tectonophysics (in press)

Srivastava SP, Tapscott CR (1986) Plate kinematics of the North Atlantic. In: Tucholke BE, Vogt PR (eds) The Western Atlantic Region. The Geology of North America Vol M. Geol Soc America, Boulder, pp 379–404

Srivastava SP, Roest WR, Kovacs LC, Oakay G, Lévesque S, Verhoef J, Macnab R (1990) Motion of Iberia since the Late Jurassic: results from detailed aeromagnetic measurements in the Newfoundland Basin. Tectonophysics 184:229–260

Stampfli G. (1978) Etude géologique generale de l'Elbourz oriental au sud de Gonbad-e-Qabus (Iran NE). Ph.D. dissertation, Univ. Genève: 329 pp

Stampfli GM (1989) Late Paleozoic evolution of the eastern Mediterranean region. IGCP 276 Paleozoic geodynamic domains and their alpidic evolution in the Tethys

Stampfli GM (1993) Le Briançonnais, terrain exotique dans les Alpes? Eclogae geol Helv 86:1–45

Stampfli GM (1996) The Intra-Alpine terrain: a Paleotethyan remnant in the Alpine Variscides. Eclogae geol Helv 89:13–42

Stampfli GM (2000) Tethyan oceans In: Bozkurt E, Winchester JA, Piper JDA (eds) Tectonics and magmatism in Turkey and surrounding area. Geol Soc London Spec Publ 173:1–23

Stampfli GM (in press) Plate tectonic of the Apulia-Adria microcontinents. In: Finetti IR (ed) CROP deep seismic exploration of the Mediterranean region. Elsevier, Amsterdam

Stampfli GM, Borel GD (2000) West-Tethys oceans. Geology 2000 Geol Ver. 90th meeting, p 113

Stampfli GM, Borel GD (2002a) A plate tectonic model for the Paleozoic and Mesozoic constrained by dynamic plate boundaries and restored synthetic oceanic isochrons. Earth Planet Sci Lett 196:17–33

Stampfli GM, Borel GD (2002b) A plate tectonic model for the Paleozoic and Mesozoic of the Tethyan domain constrained by dynamic plate boundaries and restored synthetic oceanic isochrons. 1st Int. Symposium of the Faculty of Mines on Earth Science and Engineering, Istanbul Technical University, p 77

Stampfli GM, Marchant RH (1995) Plate configuration and kinematics in the Alpine region. Accademia Nazionale delle Scienze, Scritti e Documenti 14 (Atti del congresso "Rapporti tra Alpi e Appennino"): pp 147–166

Stampfli GM, Marchant RH (1997) Geodynamic evolution of the Tethyan margins of the Western Alps In: Pfiffner OA, Lehner P, Heitzman PZ, Mueller S, Steck A (eds) Deep structure of the Swiss Alps – Results from NRP 20. Birkhäuser AG, Basel, pp 223–239

Stampfli GM, Mosar J (1999) The making and becoming of Apulia. Mem. Sci. Geologiche (Univ. Padova) 51:141–154

Stampfli GM, Pillevuit A (1993) An alternative Permo-Triassic reconstruction of the kinematics of the Tethyan realm In: Dercourt J, Ricou L-E, Vrielinck B (eds) Atlas Tethys Palaeoenvironmental Maps. Explanatory Notes. Gauthier-Villars, Paris, pp 55–62

Stampfli GM, Marcoux J, Baud A (1991) Tethyan margins in space and time In: Channell JET, Winterer EL, Jansa LF (eds) Paleogeography and paleoceanography of Tethys. Palaeogeogr Palaeoclim Palaeoecol 87:373–410

Stampfli GM, Mosar J, Marchant R, Marquer D, Baudin T, Borel G (1998) Subduction and obduction processes in the western Alps In: Vauchez A, Meissner R (eds) Continents and their mantle roots. Tectonophysics 296:159–204

Stampfli GM, Borel GD, Cavazza W, Mosar J, Ziegler PA (2001a) The paleotectonic atlas of the Perithethyan domain (CD-ROM). European Geophysical Society

Stampfli GM, Borel GD, Cavazza W, Mosar J, Ziegler PA (2001b) Palaeotectonic and palaeogeographic evolution of the western Tethys and PeriTethyan domain. Episodes 24:222–228

Stampfli GM, Mosar J, Favre P, Pillevuit A, Vannay J-C (2001c) Permo-Mesozoic evolution of the western Tethyan realm: the Neotethys/East-Mediterranean connection. In: Ziegler PA, Cavazza W, Robertson AHF, Crasquin-Soleau S (eds) Peritethyan rift/wrench basins and passive margins. Mém Museum National Hist Nat vol 186, Paris, pp 51–108

Stampfli GM, von Raumer J, Borel GD (2002a) The Palaeozoic evolution of pre-Variscan terranes: From peri-Gondwana to the Variscan collision In: Martinez-Catalan JR, Hatcher RD, Arenas R, Diaz Garcia F (eds) Variscan Appalachian dynamics: the building of the Upper Paleozoic basement. Geol Soc Am Spec Paper 364, pp 263–280

Stampfli GM, Borel GD, Marchant R, Mosar J (2002b) Western Alps geological constraints on western Tethyan reconstructions. J Virtual Expl 8:77–106

Stampfli GM, Vavassis I, De Bono A, Rosselet F, Matti B, Bellini M (2003) Remnants of the Paleotethys oceanic suture-zone in the western Tethyan area In: Cassinis G, Decandia FA (eds) Stratigraphic and Structural Evolution of the Late Carboniferous to Triassic Continental and Marine Successions in Tuscany (Italy): Regional Reports and General Correlation. Boll Soc Geol Ital Vol Spec 2: pp 1–24

Stampfli GM, Kozur H, Borel GD (in press) Europe from Variscan to the Alpine cycles In: Gee DG, Stephenson R (eds) European lithosphere dynamics. Memoir of the Geological Society, London

Stanley DJ, Wezel F-C (eds) (1985) Geological Evolution of the Mediterranean Sea. Springer-Verlag, Heidelberg

Steckler MS, Lofi J, Mountain GS, Ryan WBF, Berné S, Gorini C (2003) Reconstruction of the Gulf of Lions margin during the Messinian salinity crisis. Geophysical Research Abs 5: 07319

Steiner CW, Hobson A, Favre P, Stampfli GM, Hernandez J (1998) The Mesozoic sequence of Fuerteventura (Canary islands): witness of an Early to Middle Jurassic sea-floor spreading in the Central Atlantic. Geol Soc Am Bull 110:1304–1317

Stöcklin J (1968) Structural history and tectonics of Iran: a review. Am Ass Petrol Geol Bull 52:1229–1258

Stöcklin J (1974) Possible ancient continental margin in Iran. In: Burk CA, Drake CL (eds) The geology of continental margins. Springer Verlag, Heidelberg, pp 873–887

Tapponnier P (1977) Evolution du système alpin en Méditerranée. Poinçonnement et écrasement rigide plastique. Bull Soc Géol France 19:437–460

Thöni M (1999) A review of geochronological data from the Eastern Alps. Schweizerische Mineral Petrogr Mitt 79:209–230

Thöni M, Jagoutz E (1992) Some new aspects of dating eclogites in orogenic belts: Sm-Nd, Rb-Sr and Pb-Pb isotopic results from the Austroalpine Saualpe and Koralpe type-locality (Carinthia/Styria, SE Austria). Geochim Cosmochim Acta 56:347–368

Torsvik TH, Van der Voo R, Meert JG, Mosar J, Walderhaug HJ (2001) Reconstructions of continents around the North Atlantic at about the 60th parallel. Earth Plan Sci Lett 187:55–69

Trampert J, Vacher P, Vlaar N (2001) Sensitivities of seismic velocities to temperature, pressure and composition in the lower mantle. Phys Earth Planet Int 124 (3–4):255–267

TRANSALP Working Group (2002) First deep seismic reflection images of the Eastern Alps reveal giant crustal wedges and transcrustal ramps. Geophys. Res. Letters 29/10:92-1–92-4

Trümpy R (1980) Geology of Switzerland, Part A. Wepf & Co

Trümpy R (1988) A possible Jurassic-Cretaceous transform system in the Alps and the Carpathians. Geological Society of America Special Paper 218, pp 93–109

Trümpy R (1992) Ostalpen und Westalpen – Verbindendes und Trennendes. Jb Geol B-A 135:875–882

UNESCO-IUGS (2000) International Stratigraphic Chart

Vai GB (1994) Crustal evolution and basement elements in the Italian area: Paleogeography and characterization. Boll Geofis Teor Appl 36:411–434

Vai GB (2003) Development of the palaeogeography of Pangaea from Late Carboniferous to Early Permian. Palaeogeogr Paleoecl 196: 125–155

Vai GB, Martini IP (eds) (2001) Anatomy of an Orogen: the Apennines and adjacent Mediterranean basins. Kluwer Academic Publishers, Dordrecht

van der Hilst R (1995) Complex morphology of subducted lithosphere in the mantle beneath the Tonga trench. Nature 374(6518): 154–157

van der Hilst RD, Engdahl ER, Spakman W, Nolet G (1991) Tomographic imaging of subducted lithosphere below northwest Pacific island arcs. Nature 353:37–43

van der Hilst RD, Widiyantoro S, Engdahl ER (1997) Evidence for deep mantle circulation from global tomography. Nature 386: 578–584

van der Meulen MJ, Meulenkamp JE, Wortel MJR (1998) Lateral shifts of Appennine foredeep depocentres reflecting detachment of subducted lithosphere. Earth Planet Sc Lett 154:203–219

van der Meulen MJ, Buiter SJH, Meulenkamp JE, Wortel MJR (2000) An Early Pliocene uplift of the central Apenninic foredeep, and its geodynamic significance. Tectonics 19:300–313

Van der Voo R, Spakman W, Bijwaard H (1999a) Mesozoic subducted slabs under Siberia, Nature, 397:246–249

Van der Voo R, Spakman W, Bijwaard H (1999b) Tethyan subducted slabs under India, Earth Planet Sci Lett 171:7–20

van de Zedde DMA, Wortel MJR (2001) Shallow slab detachment as a transient source of heat at mid-lithospheric depths. Tectonics 20:868–882

Vavassis I, De Bono A, Stampfli GM, Giorgis D, Valloton A, Amelin Y (2000) U-Pb and Ar-Ar geochronological data from the Pelagonian basement in Evia (Greece): geodynamic implications for the evolution of Paleotethys. Schweizerische Mineral Petrogr Mitt 80:21–43

Vergés J, Garcia-Senz J (2001) Mesozoic evolution and Cainozoic inversion of the Pyrenean Rift. In: Ziegler PA, Cavazza W, Robertson AHF, Crasquin-Soleau S (eds) PeriTethys memoir 6: Peritethyan rift/wrench basins and passive margins. Mém. Museum National Hist Nat vol 186, Paris, pp 187–212

Vially R, Trémolières P (1996) Geodynamics of the Gulf of Lions: implications for petroleum exploration. In: Ziegler PA, Horvath F (eds) Structure and prospects of Alpine basins and forelands. Mem Mus National Hist Nat, Paris, vol 170, pp 129–158

Vigliotti L, Langenheim VE (1995) When did Sardinia stop rotating? New palaeomagnetic results. Terra Nova 7:424–435

Vissers RLM, Platt JP, van der Wal D (1995) Late orogenic extension of the Betic Cordillera and Alboran domain: a lithospheric view. Tectonics 14:786–803

von Raumer J, Stampfli GM, Borel GD, Bussy F (2002) The organisation of pre-Variscan basement areas at the north-Gondwanan margin. Int J Earth Sci 91:35–52

Wagreich M (1996) Age and significance of Upper Cretaceous siliciclastic turbidites in the central Pindos Mountains, Greece. Geol Mag 133:325–331

Waldhauser F, Kissling E, Ansorge J, Mueller S (1998) Three-dimensional interface modelling with two-dimensional seismic data: the Alpine crust-mantle boundary. Geophys J Int 135(1): 264–278

Wilson M, Bianchini G (1999) Tertiary-Quaternary magmatism within the Mediterranean and surrounding regions. In: Durand B, Jolivet L, Horvàth F, Séranne M. (eds) The Mediterranean Basin: Tertiary Extension within the Alpine Orogen. Geol Soc London Spec Publ, vol 156, pp 141–168

Wong A Ton SYM, Wortel MJR (1997) Slab detachment in continental collision zones: an analysis of controlling parameters. Geophys Res Lett 24:2095–2098

Wortel MJR, Goes SDB, Spakman W (1990) Structure and seismicity of the Aegean subduction zone. Terra Nova, 2:554–562

Wortel MJR, Spakman W (1992) Structure and dynamics of subducted lithosphere in the Mediterranean region. Proceedings of the koninklijke nederlandse Akademie van Wetenschappen Series B: Palaeontology, Geology, Physics, Chemistry, Anthropology 95:325–347

Wortel MJR, Spakman W (1993) The dynamic evolution of the Apenninic-Calabrian, Helvetic, and Carpathian arcs: an unifying approach. Terra Abstracts 5:97

Wortel MJR, Spakman W (2000) Subduction and slab detachment in the Mediterranean-Carpathian region. Science 290:1910–1917

Wortel MJR, Spakman W (2001) Subduction and slab detachment in the Mediterranean-Carpathian region. Science 291:437–437 (erratum)

Wortmann UG, Weissert H, Fuink H, Hauck J (2001) Alpine plate kinematics revisited: the Adria problem. Tectonics 20:134–147

Wybraniec S, Zhou S, Thybo H, Forsberg R,, Perehuc E, Lee M, Demianov GD, Strakhov VN (2004) New Map Compiled of Europe's Gravity Field [*http://solid_earth.ou.edu/readings/europe_gravity.html*]

Yanev Y, Bardintzeff J-M (1997) Petrology, volcanology and metallogeny of Paleogene collision-related volcanism of the eastern Rhodope. Terra Nova 9:1–8

Yilmaz PO, Norton IO, Leary D, Chuchla RJ (1996) Tectonic evolution and paleogeography of Europe. In: Ziegler PA, Horvath F (eds) Structure and Prospects of Alpine Basins and Forelands. Mémoires du Muséum National d'Histoire Naturelle, Paris, vol 170, pp 47–60

Yoshioka S, Wortel MJR (1995) Three-dimensional numerical modeling of detachment of subducted lithosphere. J Geophys Res 100:20223–20244

Zanolla C, Morelli C, Marson I (1998) The magnetic anomalies of the Mediterranean Sea. Boll Geofis Teor Appl 39:1–46

Ziegler PA (1988) Evolution of the Arctic-North Atlantic and Western Tethys. American Association of Petroleum Geologists Memoir 43

Ziegler PA (1990) Geological Atlas of Western and Central Europe. Shell Internationale Petroleum Mij. B.V. and Geological Society, London

Ziegler P, Roure F (1996) Architecture and petroleum systems of the Alpine orogen and associated basins. In: Ziegler P, Horvath F (eds) Structure and prospects of Alpine basins and forelands, Mém Mus National Hist Nat, Paris, vol 170, pp 15–45

Ziegler PA, Stampfli GM (2001) Late Paleozoic Early Mesozoic plate boundary reorganisation: collapse of the Variscan orogen and opening of Neotethys. In: Cassinis R (eds) The continental Permian of the Southern Alps and Sardinia (Italy) regional reports and general correlations. Annali Museo Civico Science Naturali vol 25, Brescia, pp 17–34

Ziegler PA, Cloetingh S, van Wees J-D (1995) Dynamics of intraplate compressional deformation: the Alpine foreland and other examples. Tectonophysics 252:7–59

Ziegler PA, Cloetingh S, Guiraud R, Stampfli GM (2001a) PeriTethyan Platforms: constraints on dynamics of rifting and basin inversion. In: Ziegler PA, Cavazza W, Robertson AHF, Crasquin-Soleau S (eds) Peritethyan rift/wrench basins and passive margins. Mém Museum National Hist Nat vol 186, Paris, pp 9–49

Ziegler PA, Cavazza W, Robertson AHF, Crasquin-Soleau S (2001b) Peritethyan rift/wrench basins and passive margins. Mém Museum National Hist Nat vol 186, Paris

Ziegler PA, Bertotti GV, Cloetingh S (2002) Dynamic processes controlling foreland development. The role of mechanical (de)coupling of orogenic wedges and forelands. EGS Sp Publ, vol 1, pp 9–92

Zoback ML, Zoback MD (1980) State of stress in the conterminous United States J Geophys Res 85:6113–6156

Zoback MD, Zoback ML (1991) Tectonic stress field of North America and relative plate motions. In: Slemmons DB, Engdahl ER, Zoback MD, Blackwell DD (eds) Neotectonics of North America. Geol Soc Amer DNAG Map vol 1, Boulder

Zoback ML, Zoback MD, Adams J, Assumpcao M, Bell S, Bergman EA, Bluemling P, Denham D, Ding J, Fuchs K, Gregersen S, Gupta HK, Jacob K, Knoll P, Magee M, Mercier JL, Muller BC, Paquin C, Stephansson O, Udias A, Xu ZH (1989) Global patterns of intraplate stress: A status report on the World Stress Map Project of the International Lithosphere Program. Nature 341:291–298

Zonenshain LP, Le Pichon X (1986) Deep basins of the Black Sea and Caspian Sea as remnants of Mesozoic back-arc basins. In: Auboin J, Le Pichon X, Monin AS (eds) Evolution of the Tethys. Tectonophysics 123:181–211

References

CD-ROM

Transect I: Iberian Meseta – Guadalquivir Basin – Betic Cordillera – Alboran Sea – Rif – Moroccan Meseta – High Atlas – Sahara Domain

Ahmamou M, Chalouan A (1988) Distension synsédimentaire plio-quaternaire et rotation anti-horaire des contraintes au Quaternaire ancien sur la bordure nord du bassin du Saïss (Maroc). Bull Inst Sci Rabat 12:19–26

Aldaya F, García-Dueñas V, Navarro-Vilá F (1979) Los Mantos Alpujárrides del tercio central de las Cordilleras Béticas. Ensayo de correlación tectónica de los Alpujárrides. Acta Geológica Hispánica 14:154–166

Aldaya F, Alvarez F, Galíndo-Zaldívar J, González-Lodeiro F, Jabaloy A, Navarro-Vilá F (1991) The Malaguide-Alpujarride contact (Betic Cordilleras, Spain): a brittle extensional detachment. C R Acad Sci Paris, Série II, 313:1447–1453

Alonso-Chaves FM, García-Dueñas V, Orozco M (1993) Fallas de despegue extensional miocenas en el área de Sierra Tejeda (Béticas centrales). Geogaceta 14:116–118

Andrieux J (1971) La structure du Rif central. Etudes des relations entre la tectonique de compression et les nappes de glissement dans un tronçon de la chaîne alpine. Editions du Service Géologique du Maroc, Notes et Mémoires 235, 155 pp and a geological map (scale 1:200 000)

Andrieux J, Frizon de Lamotte D, Braud J (1989) A structural scheme for the Western Mediterranean area in Jurassic and Early Cretaceous times. Geodinamica Acta 3:5–15

Argles TW, Platt JP, Waters DJ (1999) Attenuation and excision of a crustal section during extensional exhumation: the Carratraca Massif, Betic Cordillera, Southern Spain. J Geol Soc London 156:149–162

Argus DF, Gordon RG, Demets C, Stein S (1989) Closure of the Africa-Eurasia-North America plate motion circuit and tectonics of the Gloria fault. J Geophys Res 94:5585–5602

Azañón JM, Crespo-Blanc A (2000) Exhumation during a continental collision inferred from the tectonometamorphic evolution of the Alpujarride Complex in the central Betics (Alboran Domain, SE Spain). Tectonics 19:549–565

Azañón JM, Goffé B (1997) High-pressure, low-temperature metamorphic evolution of the Central Alpujarrides, Betic Cordillera (SE-Spain). Eur J Min 9:1035–1051

Azañón JM, García-Dueñas V, Martínez-Martínez JM, Crespo-Blanc A (1994) Alpujarride tectonic sheets in the central Betics and similar eastern allochthonous units (SE Spain). C R Acad Sci ser II 318:667–674

Azañón JM, Crespo-Blanc A, García-Dueñas V, Sánchez-Gómez M (1995) The Alpujarride Complex structure and its contribution to the ESCI-Béticas 2 deep seismic reflection profile interpretation (Alboran Domain, Betic Chain). Rev Soc Geol España 8:491–502

Azañón JM, Crespo-Blanc A, García-Dueñas V, Sánchez-Gómez M (1996) Folding of metamorphic isograds in the Adra extensional unit (Alpujarride Complex, Central Betics). C R Acad Sci Ser II, 323:949–956

Azañón JM, Crespo-Blanc A, García-Dueñas V (1997) Continental collision, crustal thinning and nappe-forming during the pre-Miocene evolution of the Alpujarride Complex (Alboran Domain, Betics). J Struct Geol 19:1055–1071

Azañón JM, García-Dueñas V, Goffé B (1998) Exhumation of high-pressure metamorphic metapelites and coeval crustal extension in the Alpujarride Complex (Betic Cordillera). Tectonophysics 285:231–252

Azdimousa A, Bourgois J, Poupeau G, Montigny R (1998) Histoire thermique du massif de Ketama (Maroc): sa place en Afrique du Nord et dans les Cordillères bétiques. C R Acad Sci Paris 326:847–853

Bakker HE, De Jong K, Helmers H, Biermann C (1989) The geodynamic evolution of the Internal Zone of the Betic Cordilleras (South-East Spain): a model based on structural analysis and geothermobarometry. J Met Geol 7:359–381

Balanyá JC (1991) Estructura del Dominio de Alborán en la parte norte del Arco de Gibraltar. PhD thesis, University of Granada

Balanyá JC, García-Dueñas V (1988) El cabalgamiento cortical de Gibraltar y la tectónica de Béticas y Rif. Actas Segundo Congreso Geológico de España (Simposios), Sociedad Geológica de España, Granada, pp 35–44

Balanyá JC, Azañón JM, Sánchez-Gómez M, García-Dueñas V (1993) Pervasive ductile extension, isothermal decompression and thinning of the Jubrique unit during the Paleogene times (Alpujarride complex, western Betics). C R Acad Sci Ser. II 316:1595–1601

Balanyá JC, García-Dueñas V, Azañón JM, Sánchez-Gómez M (1997) Alternating contractional and extensional events in the Alpujarride nappes of the Alboran Domain (Betics, Gibraltar Arc). Tectonics 16:226–238

Balanyá JC, García-Dueñas V, Azañón JM, Sánchez-Gómez M (1998) Reply to the comment by Platt. Tectonics 17:977–981

Banda E, Gallarta J, García-Dueñas V, Dañobeitia J, Makris J, (1993) Lateral variation of the crust in the Iberian Peninsula: new evidence from the Betic Cordillera. Tectonophysics 221:53–66

Banks CJ, Warburton J (1991) Mid-crustal detachment in the Betic system of southeast Spain. Tectonophysics 191:275–289

Beauchamp W, Allmendinger RW, Baranzagi M, Demnati A, El Alji M, Dahmani M (1999) Inversion tectonics and the evolution of the High Atlas Mountains, Morocco, based on a geological-geophysical transect. Tectonics 18:163–185

Berástegui X, Banks CJ, Puig C, Taberner C, Waltham D, Fernàndez M (1998) Lateral diapiric emplacement of Triassic evaporites at the southern margin of the Guadalquivir Basin, Spain. In: Mascle A, Puigdefabregas C, Fernàndez M (eds) Cenozoic foreland basins of western Europe. Geol Soc London Spec Publ 134:49–68

Berkhil M, Vachard D, Paicheler JC, Tahiri A (2000) Modèle sédimentaire et évolution géodynamique du nord-est de la Méséta occidentale marocaine au cors du Carbonifère inférieur. C R Acad Sci Paris 331:251–256

Biju-Duval B, Letouzey J, Montadert L (1978) Structure and evolution of the Mediterranean basin. Proceedings of Ocean Drilling Program, Initial Report 42:951–984

Blankenship C (1992) Structure and palaeogeography of the External Betic Cordillera, southern Spain. Marine Petrol Geol 9:256–264

Boote DRD, Clark-Lowes DD, Traut MW (1998) Paleozoic petroleum systems of North Africa. In: Macgregor DS, Moody RTJ, Clark-Lowes DD (eds) Petroleum geology of North Africa. Geol Soc London Spec Publ 133:7–68

Bouillin JP (1986) Le bassin maghrébin: une ancienne limite entre l'Europe et l'Afrique à l'ouest des Alpes. Bull Soc Géol France 8:547–558

Bouillin JP, Durand-Delga M, Olivier P (1986) Betic-Rifian and Tyrrhenian Arcs : distinctive features, genesis, and development stages. In: Wezel FC (ed) The origin of arcs. Elsevier, Amsterdam, pp 281–304

Bourgois J (1978) La transversale de Ronda (Cordillères Bétiques, Espagne). Données géologiques pour un modèle d'évolution de l'Arc de Gibraltar: Annales Scientifiques de l'Université de Besançon, vol 30, 445 pp

Bouybaouene ML (1993) Etude pétrologique des métapélites des Sebtides supérieures, Rif interne, Maroc. PhD thesis, Univ. Mohamed V Rabat

Bouybaouene ML, Goffé B, Michard A (1998) High-pressure granulites on top of the Beni-Boussera peridotites, Rif belt, Morocco: a record of the ancient thickened crust in the Alboran domain. Bull Soc Géol France 169:153–162

Braga JC, Martin JM (1987) Sedimentación cíclica lagunar y bioconstrucciones asociadas en el Trías superior alpujarride. Cuadernos Geología Ibérica 11:459–473

Caritg-Monnot S. (2003) Géologie structurale de la chaîne de l'Anti-Atlas marocain durant l'orogenèse hercynienne. PhD thesis, University of Neuchâtel

Chalouan A, Michard A (1990) The Ghomaride nappes, Rif coastal range, Morocco: a Variscan chip in the alpine belt. Tectonics 9: 1565–1583

Chalouan A, Michard A (2004) The Alpine Rif belt (Morocco): a case of mountain building in a subduction-subduction-transform fault triple junction. Pure Appl Geophys 161, in press

Chalouan A, Ouazani-Touhami A, Mouhir, L, Saji R, Benmakhlouf M (1995) Les failles normales à faible pendage du Rif interne (Maroc) et leur effet sur l'amincissement crustal du domaine d'Alboran. Geogaceta 17:107–109

Chalouan A, Saji R, Michard A, Bally AW (1997) Neogene tectonic evolution of the southwestern Alboran Basin as inferred from seismic data off Morocco. Am Ass Petrol Geol Bull 81:1161–1184

Chalouan A, Michard A, Feinberg H, Montigny R, Saddiqi O (2001) The Rif mountain building (Morocco): a new tectonic scenario. Bull Soc Géol France 172:603–616

Comas MC, Soto JI (1999) Brittle deformation in the metamorphic basement at Site 976: implications for Middle Miocene extensional tectonics in the West Alboran Basin. In: Zahn R, Comas MC, Klaus A (eds) Proceedings of the Ocean Drilling Program, Scientific Results 161. College Station, TX, pp 331–344

Comas MC, García-Dueñas V, Jurado MJ (1992) Neogene tectonic evolution of the Alboran Sea from MCS data. Geo-Marine Lett 12:157–164

Comas MC, Dañobeitia JJ, Alvarez-Marrón J, Soto, JI (1995) Crustal reflections and structure in the Alboran basin: preliminary results of the ESCI-Alboran survey. Rev Soc Geol España 8:529–542

Comas MC, Zhan R, Klaus A et al. (eds) (1996) Leg 161, Western Mediterranean. Proc ODP, Init Repts 161. College Station TX (Ocean Drilling Program, USA)

Comas MC, Platt JP, Soto JI, Watts AB (1999) The origin and tectonic history of the Alboran Basin: insights from Leg 161 results. In: Zahn R, Comas MC, Klaus A (eds) Proceedings of the Ocean Drilling Program, Scientific Results 161, pp 555–580

Comas MC, Soto JI, Talukder AR and TTR12-Leg-3 (MARSIBAL 1) Scientific Party (2003) Discovering active mud volcanoes in the Alboran Sea (Western Mediterranean). In: Geological and biological processes at deep sea European margins and oceanic basins. IOC-UNESCO, Workshop Report Series 187:14–16

Coulon C, Megartsi M, Fourcade S, Maury R, Bellon H, Louni-Hacini A, Cotten J, Coutelle A, Hermitte D (2002) Post-collisional transition from calc-alkaline to alkaline volcanism during the Noegene in Oranie (Algeria): magmatic expression of a slab breakoff. Lithos 62:87–110

Crespo-Blanc A (1995) Interference of extensional fault systems: a case study of the Miocene rifting of the Alboran basement (North of Sierra Nevada, Betic Chain). J Struct Geol 17:1559–1569

Crespo-Blanc A, Campos J (2001) Structure and kinematics of the South Iberian palaeomargin and its relationship with the Flysch Trough units: Extensional tectonics within the Gibraltar Arc fold-and-thrust belt (western Betics). J Struct Geol 23:1615–1630

Crespo-Blanc A, Orozco M, García-Dueñas V (1994) Extension versus compression during the Miocene tectonic evolution of the Betic Chain. Late folding of normal fault systems. Tectonics 13: 78–88

Dallmeyer RD, Martinez-Garcia E (1990) Pre-Mesozoic geology of Iberia. Springer-Verlag, Berlin Heidelberg New York

De Jong K (1991) Tectonometamorphic studies and radiometric dating in the Betic Cordilleras (SE Spain) – with implications for the dynamics of extension and compression in the western Mediterranean area. PhD thesis, Free University Amsterdam

De Jong MR, Wortel MJR, Spakman W (1994) Regional scale tectonic evolution and the seismic velocity structure of the lithosphere and upper mantle. J Geophys Res 99:12091–12108

Dercourt J et al. (1986) Geological evolution of the Tethys belt from the Atlantic to the Pamirs since the Lias. Tectonophysics 123: 241–315

Dercourt J, Ricou LE, Vrielink B (eds) (1993) Atlas of Tethys environmental maps. Gauthier-Villars, Paris

Dickey JS, Lundeen MT, Obata M (1979) Geological map of the ultramafic complex, southern Spain. Geol Soc Am Map and Chart Series, MC29:1–4

Didon J, Durand-Delga M, Kornporbst J (1973) Homologies géologiques entre les deux rives du Détroit de Gibraltar. Bull Soc Géol France, 7/15:77–105

Doglioni C, Mongelli F, Pialli G (1998) Boudinage of the Alpine belt in the Apenninic back-arc. Mem Soc Geol It 52:457–468

Doglioni C, Fernandez M, Gueguen E, Sabat F (1999) On the interference between the early Apennines-Maghebides back-arc extension and the Alps-Betics orogen in the Neogene geodynamics of the Western Mediterranean. Bull Soc Geol It 118:75–89

Duggen S, Hoernle K, van den Boggard P, Rüpke L, Morgan JP (2003) Deep roots of the Messinian salinity crisis. Nature 422: 602–605

Durand-Delga M, Fontboté JM (1980) Le cadre structural de la Méditerranée. In: Les chaînes alpines issues de la Tethys. Mem BRGM 115:67–85

Durand-Delga M, Olivier P (1988) Evolution of the Alboran block margin from early Mesozoic to early Miocene time. In: Jacobshagen VH (ed) The Atlas system of Morocco. Lecture Notes in Earth Sciences 15, Springer, Berlin Heidelberg New York, pp 465–480

Durand-Delga M, Hottinger L, Marçais J, Mattauer M, Milliard Y, Suter G (1962) Données actuelles sur la structure du Rif. Mem Soc Géol France, hors série n°1, pp 241–247

Durand-Delga M, Rossi P, Olivier Ph, Puglisi D (2000) Situation structurale et nature ophiolitique de roches basiques jurassiques associées aux flyschs maghrébins du Rif (Maroc) et de Sicile (Italie). C R Acad Sci Paris 331:29–38

Echarfaoui H, Hafid M, Aït Salem A (2002) Structure sismique du socle paléozoïque du bassin des Doukkala, Môle côtier, Maroc occidental. Indication en faveur d'une phase éo-varisque. C R Géosciences 334:29–38

El Azzouzi M, Bernard-Griffiths J, Bellon H, Maury RC, Piqué A, Fourcade S, Cotten J, Hernandez J (1999) Evolution des sources du volcanisme marocain au cours du Neogene. C R Acad Sci Paris 329:95–102

El Hassani A, Huon S, Hoepffner C, Whitechurch H, Piqué A (1991) Une déformation d'âge Ordovicien moyen dans la zone des Sehoul (Meseta marocaine septentrionale). Regards sur les segments "calédoniens" au NW de l'Afrique. C R Acad Sci Paris, 312 série II:1027–1032

El Hassani A, Tahiri A, Walliser OH (2003) The Variscan crust between Gondwana and Baltica. Cour Forsch Inst Senckenberg 242:81–87

El Kadiri K, Linares A, Oloriz R (1992) La Dorsale calcaire rifaine (Maroc septentrional): évolution stratigraphique et géodynamique durant le Jurassique-Crétacé. Notes Mém Serv Géol Maroc 336:217–275

El Wartiti M, Broutin J, Freytet P (1986) Premières découvertes paléontologiques dans les séries rouges permiennes du bassin de Tiddas (Maroc central). C R Acad Sci 303:263–268

Elazzab D, Galdeano A, Feinberg H, Michard A (1997) Prolongement en profondeur d'une écaille ultrabasique allochtone: traitement des données aéromagnétiques et modélisation 3D des péridotites des Beni Malek (Rif, Maroc). Bull Soc Géol France 168:667–683

Ellouz N, Patriat M, Gaulier JM, Bouatmani R, Saboujni S (2002) From rifting to Alpine inversion and Cenozoic subsidence history of some Moroccan basins. Sed Geol 156:185–212

Errarhaoui K (1997) Structure du Haut Atlas: plis et chevauchements du socle et de couverture (interprétations des données géophysiques et géologiques). PhD thesis, Université Paris XI Orsay

Fabre J (1976) Introduction à la géologie du Sahara algérien et des régions voisines. SNED, Alger, 422 pp

Faugères JC (1978) Les Rides sud-rifaines: évolution paléogéographique et structurale d'un bassin atlantico-mésogéen de la marge africaine. Unpublish. PhD thesis, Université Bordeaux I

Favre P (1992) Géologie des massifs calcaires situés au front sud de l'unité de Kétama (Rif, Maroc), PhD thesis, University of Geneva

Favre P (1995) Analyse quantitative du rifting et de la relaxation thermique de la partie occidentale de la marge transformante nord-africaine: le Rif externe (Maroc). Comparaison avec la structure actuelle de la chaîne. Geodinamica Acta 8:59–81

Favre P, Stampfli G (1992) From rifting to passive margin: the example of the Red Sea, Central Atlantic and Alpine Tethys. Tectonophysics 215:67–97

Feinberg H (1986) Les séries tertiaires des zones externes du Rif (Maroc). Notes et Mémoires du Service Géologique, 315, 192 pp

Feinberg H, Maaté H, Bouhdadi S, Durand-Delga M, Maaté M, Magné J, Olivier P (1990) Significations des dépôts de l'Oligocène supérieur-Miocène inférieur du Rif interne (Maroc) dans l'évolution de l'Arc de Gibraltar. C R Acad Sci Paris 310:1487–1495

Feinberg H, Saddiqi O, Michard A (1996) New constraints of the bending of the Gibraltar Arc from paleo magnetism of the Ronda peridotite (Betic Cordilleras, Spain). In: Morris A, Tarling DH (eds) Paleomagnetism and tectonics of the Mediterranean region. Geol Soc Spec Publ 105:43–52

Fernàndez M, Berástegui X, Puig C, García-Castellanos D, Jurado MJ, Torné M, Banks C (1998) Geophysical and geological constraints on the evolution of the Guadalquivir foreland basin, Spain. In: Mascle A, Puigdefabregas C, Fernàndez M (eds) Cenozoic foreland basins of western Europe. Geol Soc Spec Publ 134: 29–48

Fernández-Soler JM (1996) El volcanismo calco-alcalino en el Parque Natural de Cabo de Gata-Níjar (Almeria): Estudio volcanologico y petrografico. Monografia Medio Natural (Sociedad Española de Historia Natural-Junta de Andalucia), Almeria

Fernandez-Soler JM, Comas MC (2001) Aportaciones al estudio petrológico de las rocas volcánicas de la cuenca del Mar de Alboran. Bol Soc Esp Min 24-A:126–128

Flinch JF (1996) Accretion and extensional collapse of the external Western Rif (Northern Morocco). In: Ziegler PA, Horvath F (eds) Structure and prospects of Alpine basins and forelands. Mem Mus National Hist Nat 170:61–85

Flinch J, Bally A, Wu S (1996) Emplacement of a passive-margin evaporitic allochthon in the Betic Cordillera of Spain. Geology 14:67–70

Fraissinet C, Zouine EM, Morel JL, Poisson A, Andrieux J, Faure-Muret A (1989) Structural evolution of the southern and northern Central High Atlas in Paleocene and Mio-Pliocene times. In: Jacobshagen VH (ed) The Atlas system of Morocco. Lecture Notes in Earth Sciences 15, Springer, Berlin Heidelberg New York, pp 273–292

Frizon de Lamotte D (1985) La structure du Rif oriental: rôle de la tectonique longitudinale et importance des fluides. Thèse, Université Pierre et Marie Curie, Mémoires des Sciences de la Terre n° 85-03, 436 pp and a structural map (scale 1:200 000)

Frizon de Lamotte D, Andrieux J, Guézou, JC (1991) Cinématique des chevauchements néogènes dans l'Arc bético-rifain: Discussion sur les modèles géodynamiques. Bull Soc Géol France, 162:611–626

Frizon de Lamotte D, Saint Bezar B, Bracène R, Mercier E (2000) The two main steps of the Atlas building and geodynamics of the western Mediterranean. Tectonics 19:740–761

Galindo-Zaldívar J, González-Lodeiro F, Jabaloy A (1989) Progressive extensional shear structures in a detachment contact in the Western Sierra Nevada (Betic Cordillera, Spain). Geodinamica Acta 3:73–85

Galindo-Zaldívar J, Jabaloy A, González-Lodeiro F (1996) Reactivation of the Mecina detachment in the western sector of Sierra Nevada (Betic Cordillera, SE Spain). C R Acad Sci Paris 323:615–622

García-Casco A, Torres-Roldán R (1996) Desequilibrium induced by fast decompression in St-Bt-Grt-Ky-Sil-And metapelites from the Betic Belt (Southern Spain). J Petrol 37:1207–1239

García-Casco A, Haissen F, Torres-Roldán R (1992) Termometría en metapelitas de grado medio de unidades Alpujárrides de la Zona Bética occidental, in Actas, Tercer Congreso Geológico de España (Simposios), Sociedad Geológica de España, Salamanca 1:338–342

García-Dueñas V, Balanyá JC (1991) Fallas normales de bajo ángulo a gran escala en las Béticas occidentales, Geogaceta, 9:29–33

García-Dueñas V, Martínez-Martínez JM (1988) Sobre el adelgazamiento mioceno del Dominio de Alborán: El despegue de los Filabres (Béticas orientales). Geogaceta 5:53–55

García-Dueñas V, Martínez-Martínez JM, Navarro-Vilá F (1986) La zona de falla de Torres Cartas, conjunto de fallas normales de bajo ángulo entre Nevado-Filábrides y Alpujárrides (Sierra Alhamilla, Béticas orientales). Geogaceta 1:17–19

García-Dueñas V, Martínez-Martínez JM, Soto JI (1988) Los Nevado-Filabrides, una pila de plieghemantos separados por zona de cizalla. II Cong. geol. España, 17–26

García-Dueñas V, Balanyá JC, Martínez-Martínez JM (1992) Miocene extensional detachments in the outcropping basement of the northern Alboran Basin and their tectonic implications. Geo-Marine Lett 12:88–95

García-Dueñas V, Balanyá JC, Martínez-Martínez JM, Muñoz M, Azañón JM, Crespo-Blanc A, Orozco M, Soto JI, Alonso FM, Sánchez-Gómez M (1993) Kinematics of the Miocene extension detachment faults and shear zones in the Betics and Rif chains. Documents du BRGM France 219:76–77

García-Dueñas V, Banda E, Torné M, Córdoba D. and ESCI Working Group (1994) A deep seismic reflection survey across the Betic Chain (southern Spain): first results. Tectonophysics 232:77–89

García-Hernández M, López-Garrido AC, Rivas P, Sanz de Galdeano C, Vera JA (1980) Mesozoic palaeogeographic evolution of the external zones of the Betic Cordillera. Geologie en Mijnbouw 59:155–168

Gasquet D, Stussi JM, Nachit H (1996) Les granitoïdes hercyniens du Maroc dans le cadre de l'évolution géodynamique régionale. Bull Soc Geol France 167:517–528

Giese P, Jacobshagen V (1992) Inversion tectonics of intracontinental ranges: High and Middles Atlas, Morocco. Geol Rund 81:249–259

Goffé B, Michard A, García-Dueñas V, González-Lodeiro F, Monié P, Campos J, Galindo-Zaldívar J, Jabaloy A, Martínez-Martínez JM, and Simancas F (1989) First evidence of high pressure, low temperature metamorphism in the Alpujarride nappes, Betic Cordilleras (SE Spain). Eur J Miner 1:139–142

Gomez F, Beauchamp W, Barazangi M (2000) Role of the Atlas mountains (northwest Africa) within the African-Eurasian plate boundary zone. Geology 28:775–778

Gomez F, Barazangi M, Beauchamp W (2002) Role of the Atlas mountains (northwest Africa) within the African-Eurasian plate boundary zone: reply. Geology 30:96

Gómez-Pugnaire MT, Fernández-Soler JM (1987) High-pressure metamorphism in the metabasites from the Betic Cordilleras (SE Spain) and its evolution during the alpine orogeny. Contrib Miner Petrol 95:231–244

Gómez-Pugnaire MT, Braga JC, Martín JM, Sassi FP, Del Moro A (2000) Regional implications of a Palaeozoic age for the Nevado-Filabride cover of the Betic Cordillera, Spain. Bull Suisse Minér Pétrogr 80/1:45–52

Gràcia E, Dañobeitia J, Vergés J, Bartolomé R (2003) Crustal architecture and tectonic evolution of the Gulf of Cadiz (SW Iberian margin) at the convergence of the Eurasian and African plates. Tectonics 22:1033, doi:10.1029/2001ITC901045

Guerrera F, Martin-Algarra A, Perrone V (1993) Late Oligocene-Miocene syn-/late-orogenic successions in western and central Mediterranean chains from the Betic cordilleras to the southern Apennines. Terra Nova 5:525–544

Guiraud R (1998) Mesozoic rifting and basin inversion along the northern African Tethyan margin: an overview. In: Macgregor DS, Moody RTJ, Clark-Lowes DD (eds) Petroleum geology of North Africa, Geol Soc London Spec Publ 133:217–229

Gutscher MA, Malod J, Rehault JP, Contrucci I, Klingelhoefer F, Mendes-Victor L, Spakman, W (2002) Evidence for active subduction beneath Gibraltar. Geology 30:1071–1074

Haddoum H, Guiraud R., Moussine-Pouchkine A (2001) Variscan compressional deformations of the Ahnet-Mouydir Basin, Algerian Sahara platform: far-field stress effects of the Late Palaeozoic orogeny. Terra Nova 13:220–226

Hadri M (1993) Un exemple de plate-forme carbonatée au Lias-Dogger dans le Haut Atlas Central au nord-ouest de Goulmima, Maroc. PhD thesis, Université Paris-Sud Orsay

Hafid M (2000a) Incidences de l'évolution du du Haut Atlas occidental et de son Avant-Pays septentrional sur la dynamique meso-Cenozoïque de la marge atlantique (entre Safi et Agadir), apport de la sismique réflexion et des données de forage. PhD thesis, Université Ibn Tofail (Kénitra, Morocco)

Hafid M (2000b) Triassic-early Liassic extensional systems and their Tertiary inversion, Essaouira Basin (Morocco). Marine Petrol Geol 17:409–429

Hafid M, Ait Salem A, Bally AW (2000) The western termination of the Jebilet-High Atlas system (Offshore Essaouira Basin, Morocco). Marine Petrol Geol 17:431–443

Horvath F, Berckhemer H (1982) Mediterranean backarc basins. In: Berkhemer H, Hsü K (eds) Alpine Mediterranean geodynamics. American Geophysical Union, Washington DC, pp 141–173

Hoyez B (1989) Le Numidien et les flyschs oligo-miocènes de la bordure sud de la Méditerranée occidentale. PhD thesis, University of Lille

I.G.M.E. (1987) Contribución de la exploración petrolífera al conocimiento de la Geología de España. IGME, Madrid, 481 pp

Jenny J, Le Marrec A, Monbaron M (1981) Les couches rouges du Jurassique moyen du Haut Atlas central (Maroc). Corrélations lithostratigraphiques, éléments de datation et cadre tectonosédimentaire. Bull Soc Géol France 23(6):627–639

Jolivet L, Faccenna C (2000) Mediterranean extension and the Africa-Eurasia collision. Tectonics 19:1095–1106

Jolivet L, Faccenna C, Goffé B, Burov E, Agard P (2003) Subduction tectonics and exhumation of high-pressure metamorphic rocks in the Mediterranean orogens. Am J Sci 303:353–409

Kerzazi K (1994) Etude biostratigraphique du Miocène sur la base des Foraminifères planctoniques et nannofossiles calcaires dans le Prérif et la marge atlantique du Maroc (site 547A du DSDP Leg 79) ; aperçu sur leur paléoenvironnement. Thèse, Univ. P. et M. Curie, Paris, 230 pp

Kornprobst J (1974) Contribution à l'étude pétrographique et structurale de la zone interne du Rif (Maroc Septentrional). Editions du Service Géologique du Maroc, Notes et Mémoires 251, 256 pp

Kornprobst J, Piboule M, Roden M, Tabit A (1990) Corundum-bearing garnet clinopyroxenites at Beni Bousera (Morocco): Original plagioclase-rich gabbros recrystallized at depth within the mantle? J Petrol 31:717–745

Lafuste J, Pavillon MJ (1976) Mise en évidence d'Eifélien daté au sein des terrains métamorphiques des zones internes des Cordillères bétiques. Intérêt de ce nouveau repère stratigraphique. C R Acad Sci Paris 283:1015–1018

Lagarde JL (1989) Granites tardi-Carbonifère et déformation crustale. L'exemple de la Meseta marocaine. Mem. et Doc. Centre armoricain d'étude structurale des socles, Rennes 26, 342 pp

Lagarde JL, Aït Omar S, Roddaz B (1990) Structural characteristics of granitic plutons emplaced during weak regional deformation: examples from late Carboniferous plutons, Morocco. J Struct Geol 12:805–821

Laville E (1985) Evolution sédimentaire, magmatique et tectonique d'une partie du versant sud du Haut Atlas (Maroc): modèle en relais multiples de décrochements. PhD thesis, Université Montpellier II

Laville E (1988) Multiple releasing and restraining step-over model for the Jurassic strike-slip basin of the central High Atlas. In: Mainspeizer W (ed) Triassic-Jurassic rifting: continental breakup and the origin of the Atlantic ocean and passive margins. Elsevier, New York, pp 499–523

Laville E (2002) Role of the Atlas mountains (northwest Africa) within the African-Eurasian plate boundary zone: comment. Geology 30:96

Laville E, Piqué A (1991) La distension crustale atlantique et atlasique au Maroc au début du Mésozoïque: le rejeu des structures hercyniennes. Bull Soc Géol France 162:1161–1171

Laville E, Piqué A (1992) Jurassic penetrative deformation and Cenozoic uplift in the Central High Atlas (Morocco): a tectonic model. Structural and orogenic inversions. Geol Rund 81:157–170

Lenoir X, Garrido CJ, Bodinier JL, Dautria JM, Gervilla F (2000) The recrystallization front of the Ronda peridotite: evidence for melting and thermal erosion of subcontinental lithospheric mantle beneath the Alboran basin. J Petrol 42:141–158

Le Pichon X, Bergerat F, Roulet MJ (1988) Plate kinematics and tectonic leading to Alpine belt formation: a new analysis. In: Processes in continental lithospheric deformation. Geol Soc Am Spec Paper 218:111–131

Levy RG, Tilloy R (1952) Maroc septentrional (chaîne du Rif), partie B. Livret-guide des excursions A31 et C31. Congrès Géologique International, XIX session, Alger, série Maroc 8, 65 pp

Litto W, Jaaidi EB, Medina F, Dakki M (2001) Etude sismo-structurale de la marge nord du bassin du Gharb (avant-pays rifain, Maroc): mise en évidence d'une distension d'âge miocène tardif. Eclogae geol Helv 94:63–73

Lonergan L, Platt JP, Gallagher L (1994) The internal-external zone boundary in the eastern Betic Cordillera, SE Spain. J Struct Geol 16:174–188

Lopez-Casado C, Sanz de Galdeano C, Molina-Palacios S, Henares-Romero J (2001) The structure of the Alboran sea: an interpretation from seismological and geological data. Tectonophysics 338:79–95

López-Sánchez-Vizcaíno V, Rubatto D, Gómez-Pugnaire MT, Trommsdorff V, Müntener O (2001) Middle Miocene high-pressure metamorphism and fast exhumation of the Nevado-Filábride Complex, SE Spain. Terra Nova 5:327–332

Luján M, Balanyá JC, Crespo-Blanc A (2000) Contractional and extensional tectonics in Flysch and Penibetic units (Gibraltar Arc, SW Spain): new constraints on emplacement mechanisms. C R Acad Sci Paris, 330:631–638

Makris J, Demnati A, Klussmann J (1985) Deep seismic soundings in Morocco and a crust and upper mantle model deduced from seismic and gravity data. Ann Géophys 3:369–380

Martínez del Olmo W, García-Mallo J, Serrano A, Suárez J (1984) Modelo tectonosedimentario del Bajo Guadalquivir. I Congreso Geol España, Segovia, pp 199–213

Martínez-Martínez JM, Azañón JM (1997) Mode of extensional tectonics in the southeastern Betics (SE Spain): Implications for the tectonic evolution of the peri-Alboran orogenic system. Tectonics 16:205–225

Martínez-Martínez JM, Soto JI, Balanyá JC (1995) Large scale structures in the Nevado-Filabride Complex and crustal seismic fabrics of the deep seismic reflection profile ESCI-Béticas 2. Rev Soc Geol España 8:477–490

Martínez-Martínez JM, Soto JI, Balanyá JC (2002) Orthogonal folding of extensional detachments: Structure and origin of the Sierra Nevada elongated dome (Betics, SE Spain). Tectonics 21:1–22

Mattauer M, Tapponnier P, Proust F (1977) Sur les mécanismes de formation des chaînes intracontinentales: l'exemple des chaînes atlasiques du Maroc. Bull Soc Géol France 7:521–526

Maury RC, Fourcade S, Coulon C, El Azzouzi M, Bellon H, Coutelle A, Ouabadi A, Semroud B, Megartsi M, Cotten J, Belanteur O, Louni-Hacini A, Piqué A, Capdevila R, Hernandez J, Rehault JP (2000) Post-collisional Neogene magmatism of the Mediterranean Maghreb margin: a consequence of slab breakoff. C R Acad Sci Paris 331:159–173

Meghraoui M, Morel JL, Andrieux J, Dahmani M (1996) Tectonique Plio-Quaternaire de la chaîne tello-rifaine et de la mer d'Alboran: une zone complexe de convergence continent-continent. Bull Soc Géol France 167:147–157

Michard A (1976) Eléments de Géologie Marocaine. Notes et Mémoires du Service Géologique du Maroc 252, 402 pp

Michard A, Yazidi A, Benziane F, Hollard H, Willefert S (1982) Foreland thrusts and olistostromes on the pre-Sahara margin of thed Variscan orogen, Morocco. Geology 10:253–256

Michard A, Cailleux Y, Hoepffner Ch (1989) L'orogène mésétien du Maroc: structure, déformation hercynienne et déplacements. Notes Mém Serv Géol Maroc 335:313–327

Michard A, Feinberg H, Elazzab D, Bouybaouene M, Saddiqi O (1992) A serpentinite ridge in a collisional paleomargin setting: the Beni Malek massif, External Rif, Morocco. Earth Plan Sci Lett 113:435–442

Michard A, Goffé B, Bouybaouene ML, Saddiqi O (1997) Late Variscan Mesozoic thinning in the Alboran domain. Metamorphic data from the northern Rif. Terra Nova 9:171–174

Michard A, Chalouan A, Feinberg H, Goffé B, Montigny R (2002) How does the Alpine belt end between Spain and Morocco? Bull Soc Géol France 173:3–15

Monié P, Frizon de Lamotte D, Leikine M (1984) Etude géologique préliminaire par la méthode $^{40}Ar/^{39}Ar$ du métamorphisme alpin dans le Rif externe (Maroc). Précision sur le calendrier tectonique tertiaire. Rev Géol dyn Géogr phys 25:307–317

Monié P, Gonzales-Lodeiro F, Goffé B, Jabaloy A (1991) $^{39}Ar/^{40}Ar$ geochronology of Alpine tectonism in the Betic Cordillera (Southern Spain). J Geol Soc London 148:289–297

Monié P, Torres-Roldán RL, García-Casco A (1994) Cooling and exhumation of the western Betic Cordilleras, $^{40}Ar/^{39}Ar$ thermochronological constraints on a collapsed terrane. Tectonophysics 238:353–379

Montel JM, Kornprobst J, Vielzeuf D (2000) Preservation of old U-Th-Pb ages in shielded monazites: examples from the Beni Bousera Variscan kinzigites (Morocco). J Met Geol 18:335–342

Morales J, Serrano I, Jabaloy A, Galindo-Zaldivar J, Zhao D, Torcal F, Vidal F, Gonzàlez-Lodeiro F (1999) Active continental subduction beneath the Betic Cordillera and the Alboran sea. Geology 27:753–738

Morel JL (1988) Evolution récente de l'orogène rifain et son avantpays depuis la fin de la mise en place des nappes. Mém Géodiffusion 4, 584 pp

Morel JL, Zouine M, Andrieux J, Faure-Muret A (2000) Déformations néogènes et quaternaires de la bordure nord haut atlasique (Maroc): rôle du socle et conséquences structurales. J Afr Earth Sci 30:119–131

Morley CK (1992) Tectonic and sedimentary evidence for synchronous and out-of-sequence thrusting Larache-Acilah area, Western Morocco, Rif. J Geol Soc London 149:39–49

Navarro-Vilá F, García-Dueñas V (1980) Geological map. Sheet La Peza 1010. Instituto Geológico y Minero de España, Madrid, scale 1:50 000

Nijhuis HJ (1964) Plurifacial Alpine metamorphism in the southeastern Sierra de los Filabres south of Lubrín, SE Spain. PhD thesis, Univ Amsterdam

Obata M (1980) The Ronda peridotite: Garnet-spinel and plagioclase-lherzolite facies and the P-T trajectories of a high-temperature mantle intrusion. J Petrol 21:533–572

Orozco M, Alonso-Chaves M, Nieto F (1998) Development of large north-facing folds and their relation to crustal extension in the Alborán domain (Alpujarras region, Betic Cordilleras, Spain). Tectonophysics 298:271–295

Ouanaimi H, Petit JP (1992) La limite sud de la chaîne hercynienne dans le Haut Atlas marocain:reconstitution d'un saillant non déformé. Bull Soc Géol France 163:63–72

Perconig E (1960–62) Sur la constitution géologique de l'Andalousie occidentale, en particulier du bassin du Guadalquivir (Espagne méridionale). In: Livre à la mémoire du Prof. Fallot. Mémoire hors-série, Soc Géol France 1:229–256

Piqué A (1994) Géologie du Maroc, Les domaines régionaux et leur évolution structurale. PUMAG, Rabat, 284 pp

Piqué A, Michard A (1989) Moroccan Hercynides: a synopsis of the Paleozoic sedimentary and tectonic evolution at the northern margin of west Africa. Am J Sci 289:286–330

Piqué A, Brahim L, El Azzouzi M, Maury RC, Bellon H, Semroud B, Laville E (1998) Le poinçon maghrebin: contraintes structurales et géochimiques. C R Acad Sci Paris, 326:575–581

Piqué A, Tricart P, Guiraud R, Laville E, Bouaziz S, Amrhar M, Aït Ouali, R (2002) The Mesozoic-Cenozoic Atlas belt (North Africa): an overview. Geodinamica Acta 15:185–208

Platt JP, Vissers RLM (1989) Extensional collapse of thickened continental lithosphere: a working hypothesis for the Alboran Sea and Gibraltar Arc. Geology 17:540–543

Platt JP, Whitehouse MJ (1999) Early Miocene high-temperature metamorphism and rapid exhumation in the Betic Cordillera (Spain): evidence from U-Pb zircon ages. Earth Plan Sci Lett 171:591–605

Platt JP, Soto JI, Whitehouse MJ, Hurford AJ, Kelley SP (1998) Thermal evolution, rate of exhumation, and tectonic significance of metamorphic rocks from the floor of the Alboran extensional basin, western Mediterranean. Tectonics 17/5:671–689

Platt JP, Allerton S, Kirker A, Mandeville C, Mayfield A, Platzman ES, Rimi A (2003a) The ultimate arc: differential displacement, oroclinal bending and vertical axis rotation in the External Betic-Rif arc. Tectonics, DOI 10.1029-2001TC001321

Platt JP, Argles TW, Carter A, Kelley SP, Whitehouse MJ, Lonergan L (2003b) Exhumation of the Ronda peridotite and its crustal envelope: constraints from thermal modelling of a P-T-time array. J Geol Soc London 160:655–676

Plaziat JC, Ahmamou M (1998) Les différents mécanismes à l'origine de la diversité des séismites, leur identification dans le Pliocène du Saïss de Fès et de Meknès (Maroc) et leur signification tectonique. Geodinamica Acta 11:183–203

Poisson A, Hadri M, Milhi A, Julien M, Andrieux J (1998) The central High Atlas (Morocco). Litho- and chrono-stratigraphic correlations during Jurassic times between Tinjdad and Tounfite: origin of subsidence. Peri-Tethys Memoir 4, Mém Mus National Hist Nat 179:237–256

Puga E, Díaz de Federico A, Fediukova E, Bondi M, Morten L (1989) Petrology, geochemistry and metamorphic evolution of the ophiolitic eclogites and related rocks from the Sierra Nevada (Betic Cordilleras, Southeastern Spain). Schweiz Miner Petrogr Mitt 69:435–455

Puga E, Díaz de Federico A, Demant A (1995) The eclogitized pillows of the Betic ophiolitic association; relics of the Tethys ocean floor incorporated in the Alpine chain after subduction. Terra Nova 7/1:31–43

Puga E, Nieto JM, Díaz de Federico A, Bodinier JL, Morten L (1999) Petrology and metamorphic evolution of ultramafic rocks and dolerite dykes of the Betic ophiolitic association (Mulhacen complex, SE Spain): evidence of eo-alpine subduction following an ocean-floor metasomatic process. Lithos 49:23–56

Puga E, Díaz de Federico A, Nieto JM (2002a) Tectonostratigraphic subdivision and petrological characterisation of the deepest complexes of the Betic zone: a review. Geodinamica Acta 15: 23–43

Puga E, Ruiz Cruz MD, Díaz de Federico A (2002b) Polymetamorphic amphibole veins in metabasalts from the Betic ophiolitic association at Cobdar, southern Spain: relics of the ocean-floor metamorphism preserved through the Alpine orogeny. Canad Mineral 40:67–83

Rehault JP, Boillot G, Mauffret A (1984) The western Mediterranean basin geological evolution. Marine Geol 55:447–477

Reuber I, Michard A, Chalouan A, Juteau T, Jermoumi B (1982) Structure and empacement of the Alpine-type peridotites from Beni Bousera, Rif, Morocco: a polyphase tectonic interpretation. Tectonophysics 82:231–251

Robert-Charrue CG (2001) Etude tectonique de l'Anti-Atlas oriental Tafilalet, Maroc. Mémoire pour l'obtention du dîplome de géologue, Institut de Géologie, Université de Neuchâtel, 62 pp

Rodríguez-Fernández J, Martín-Penela AJ (1993) Neogene evolution of the Campo de Dalías and the surrounding offshore areas (Northeastern Alboran Sea). Geodinamica Acta 6/4:255–270

Rodríguez-Fernández J, Comas MC, Soria J, Martín-Pérez JA, Soto JI (1999) The sedimentary record of the Alboran Basin: an attempt at sedimentary sequence correlation and subsidence analysis. In: Zahn R, Comas MC, Klaus A (eds) Proceedings of the Ocean Drilling Program, Scientific Results, vol 161, College Station, TX

Rolley JP (1978) Notice explicative de la carte géologique du Maroc au 1:100 000, feuille d'Afourer (Haut Atlas central). Notes et Mémoires Serv. carte géol. Maroc 247bis, 103 pp

Rosenbaum G, Lister GS, Duboz C (2002) Relative motions of Africa, Iberia and Europe during Alpine orogeny. Tectonophysics 359:117–129

Royden LH (1993) Evolution of the retreating subduction boundaries formed during continental collision. Tectonics 12:629–638

Saddiqi O, Reuber I, Michard A (1988) Sur la tectonique de dénudation du manteau infracontinental dans les Beni Bousera, Rif septentrional, Maroc. C R Acad Sci Paris 307:657–662

Saddiqi O, Feinberg H, Elazzab D, Michard A (1995) Paléomagnétisme des péridotites des Beni Bousera, Rif interne, Maroc: conséquences pour l'évolution miocène de l'arc de Gibraltar. C R Acad Sci Paris 321:361–368

Saïdi A, Tahiri A, Aït Brahim L, Saidi M (2002) Etats de contraintes et mécanismes d'ouverture et de fermeture des bassins permiens du Maroc hercynien. L'exemple du bassin des Jebilet et des Rehamna. C R Geoscience 334:221–226

Saint-Bézar B, Frizon de Lamotte D, Morel JL, Mercier E (1998) Kinematics of large scale tip line folds from the High Atlas thrust belt, Morocco. J Struct Geol 20:999–1011

Samaka F, Benyaich A, Dakki M, Hcaine M, Bally AW (1997) Origine et inversion des bassins miocènes supra-nappes du Rif Central (Maroc), Etude de surface et de subsurface. Geodinamica Acta 10:30–40

Sánchez-Rodriguez L, Gebauer D (2000) Mesozoic formation of pyroxenites and gabbros in the Ronda area (southern Spain), followed by Early Miocene subduction metamorphism and emplacement into the middle crust: U–Pb sensitive high-resolution ion microprobe dating of zircon. Tectonophysics 316:19–44

Sánchez-Rodriguez L, Gebauer D, Tubia JM, Gil Ibarguchi JL, Rubatto D (1996) First SHRIMP-ages on pyroxenite, eclogites and granites of the Ronda complex and its country rocks. Geogaceta 20: 487–489

Sandvol E, Seber D, Calvert A, Barazangi M (1998) Grid search modeling of receiver functions: Implications for crustal structure in the Middle East and North Africa. J Geophys Res 103: 26899–26918

Seber D, Barazangi M, Tadili BA, Ramdani M, Ibenbrahim A, Ben Sari D (1996) Three-dimensional upper mantle structure beneath the intraplate Atlas and interplate Rif mountains of Morocco. J Geophys Res 101 B2:3125–3138

Sierro FJ, González-Delgado JA, Dabrio CJ, Flores JA, Civis J (1996) Late Neogene depositional sequence in the foreland basin of Guadalquivir (SW Spain). In: Friend PF, Dabrio CJ (eds) Tertiary basins of Spain: the stratigraphic record of crustal kinematics. Cambridge University Press, pp 399–345

Soria JM (1994) Sedimentación y tectónica durante el Mioceno en la región de Sierra Arana-Mencal y su relación con la evolución geodinámica de la Cordillera Bética. Bol Soc Geol España 7:199–213

Soto JI (1991) Estructura y evolución metamórfica del complejo Nevado-Filábride en la terminación oriental de la Sierra de los Filabres (Cordilleras Béticas). PhD thesis, University of Granada

Soto JI, Platt JP (1999) Petrological and structural evolution of high-grade metamorphic rocks from the floor of the Alboran Sea basin, W. Mediterranean. J Petrol 40:21–60

Soto JI, Platt JP, Sánchez-Gómez M, Azañón JM (1999). Pressure-Temperature evolution of the metamorphic basement of the Alboran Sea: Thermobarometric and structural observations. In: Zahn R, Comas MC, Klaus A (eds), Proceedings of the Ocean Drilling Program, Scientific Results 161, pp 263–275

Stampfli GM, Borel GD (2000) A plate tectonic model for the Paleozoic and Mesozoic constrained by dynamic plate boundaries and restored synthetic oceanic isochrons. Earth Plan Sci Lett 196:17–33

Stampfli GM, Mosar J, Marquer D, Marchant R, Baudin T, Borel G (1998) Subduction and obduction processes in the Swiss Alps. Tectonophysics 296:159–204

Studer M, du Dresnay R (1980) Déformations sédimentaires en compression pendant le Lias supérieur et le Dogger au Tizi n'Irhill (Haut Atlas central de Midelt, Maroc). Bull Soc Géol France 22:391–397

Suter G (1980a) Carte Géologique de la Chaîne Rifaine, scale 1:500 000. Editions du Service Géologique du Maroc, Notes et Mémoires 245a

Suter G (1980b) Carte Structurale de la Chaîne Rifaine, scale 1:500 000. Editions du Service Géologique du Maroc, Notes et Mémoires 245b

Teixell A, Arboleya ML, Julivert M, Charroud M (2003) Tectonic shortening and topography in the central High Atlas (Morocco). Tectonics 22:1051, doi:10.1029/2002TC001460

Tejera de Leon J, Boutakiout M, Ammar A, Aït Brahim L, El Hatimi N (1995) Les bassins du Rif central (Maroc): marqueurs de chevauchements hors séquence d'âge miocène terminal au cœur de la chaîne. Bull Soc Geol France 166:751–761

Tendero JA, Martín-Algarra A, Puga E, Díaz de Federico A (1993) Lithostratigraphie des métasédiments de l'association ophiolitique Névado-Filabride (SE Espagne) et mise en évidence d'objets ankéritiques évoquants des foraminifères planctoniques du Crétacé: conséquences paléogéographiques. C R Acad Sci Paris, Sér. II, 316:1115–1122

Torné M, Banda E (1992) Crustal thinning from the Betic Cordillera to the Alboran Sea. Geo-Marine Letters 12:76–81

Torné M, Banda E, García-Dueñas V, Balanyá JC (1992) Mantle-lithosphere bodies in the Alboran crustal domain (Ronda peridotites, Beti-Rif orogenic belt). Earth Plan Sci Lett 110:163–171

Torné M, Fernàndez M, Comas MC, Soto JI (2000) Lithospheric structure beneath the Alboran Basin: results from 3D gravity modeling and tectonic relevance. J Geophys Res 105(B2): 3209–3228

Torres-Roldán RL (1981) Plurifacial metamorphic evolution of the Sierra Bermeja peridotite aureole (southern Spain). Estudios Geol 37:115–133

Torres-Roldán RL, Polli L, Peccerillo A (1986) An early Miocene arc-tholeitic magmatic dike event from the Alboran Sea: evidence for precollisional subduction and back-arc crustal extension in the westermost Mediterranean. Geol Rund 75:219–234

Tubía JM (1994) The Ronda peridotites (Los Reales nappe): an example of the relationship between lithospheric thickening by oblique tectonics and late extensional deformation within the Betic Cordillera (Spain). Tectonophysics 238:381–398

Tubía JM, Gil-Ibarguchi I (1991) Eclogites of the Ojén nappe: a record of a subduction in the Alpujarride Complex (Betic Cordilleras, southern Spain). J Geol Soc London 148:801–804

Tubía JM, Cuevas J, Gil-Ibarguchi I (1997) Sequential development of the metamorphic aureole beneath the Ronda peridotites and its bearing on the tectonic evolution of the Betic Cordillera. Tectonophysics 279:227–252

Van den Bosch J (1971) Carte gravimétrique du Maroc, scale 1:500 000. Editions du Service Géologique du Maroc, Notes et Mémoires 234

Van der Meijde M, van der Lee S, Giardini D (2003) Crustal structure beneath broad-band seismic stations in the Mediterranean region. Geophys J Int 152:729–739

Van der Wal D, Vissers RLM (1993) Uplift and emplacement of upper mantle rocks in the western Mediterranean. Geology 21: 1119–1122

Vegas R, Vázquez JT, Medialdea T, Suriñach E (1995) Seismic and tectonic interpretation of ESCI-Béticas and Alboran deep seismic reflection profiles: structure of the crust and geodynamic implications. Rev Soc Geol España 8:449–460

Vissers RLM, Platt JP, Van der Wal D (1995) Late orogenic extension of the Betic cordillera and Alboran domain: a lithospheric view. Tectonics 14:786–803

Watts AB, Platt JP, Bulh P (1993) Tectonic evolution of the Alboran sea basin. Basin Research 5:153–177

Weijermars R (1993) Estimation of palaeostress orientation within deformation zones between two mobile plates. Geol Soc Am Bull 105:1491–1510

Weijermars R, Roep TB, Van den Eeckhout B, Postma G, Kleverlaan K (1985) Uplift history of a Betic fold nappe inferred from Neogene-Quaternary sedimentation and tectonics (in the Sierra Alhamilla and Almería, Sorbas and Tabernas Basin of the Betic Cordilleras, SE Spain). Geol Mijnbouw 64:397–411

Wernli R (1988) Micropaléontologie du Néogène post-nappes du Maroc septentrional et description systématique des foraminifères planctoniques. Notes et Mémoires du Service Géologique du Maroc vol 331

Westerhoff AB (1977) On the contact relations of High-temperature peridotites in the Serranía de Ronda, Southern Spain. Tectonophysics 39:579–591

Wigger P, Asch, G, Giese P, Heinsohn WD, El Alami SO, Ramdani F (1992) Crustal structure along a traverse across the Middle and High Atlas mountains derived from seismic refraction studies. Geol Rund 81:237–248

Wildi W (1983) La chaîne tello-rifaine (Algérie, Maroc, Tunisie): structure, stratigraphie et évolution du Trias au Miocène. Rev Géol Dyn Géogr Phys 24:201–297

Wildi W, Nold M, Uttinger J (1977) La Dorsale calcaire entre Tetouan et Asifane (Rif Interne, Maroc). Eclogae geol Helv 70:371–416

Wildi W, Nold M, Uttinger J (1981) Géologie de la Dorsale calcaire entre Tetouan et Asifane (Rif Interne, Maroc). Editions du Service Géologique du Maroc, Notes et Mémoires vol 300 and a structural map (scale 1:100 000)

Zeck HP (1997) Mantle peridotites outlining the Gibraltar Arc – centrifugal extensional allochthons derived from the earlier Alpine, westward subducted nappe pile. Tectonophysics 281:195–207

Zeck HP, Monié P, Villa IM, Hansen BT (1992) Very high rates of cooling and uplift in the Alpine belt of the Betic Cordilleras, southern Spain. Geology 20:79–82

Zeyen H, Ayarza P, Fernàndez M, Rimi A (submitted) Lithospheric structure under the western African-European plate boundary: A transect across the Atlas mountains and the Gulf of Cadiz

Zeyen H, Ayarzda P, Fernàndez M, Rimi A (2003) Integrated thermal and density modelling of the lithosphere across the Iberian-African plate boundary: the Gulf of Cadiz-Moroccan Atlas traverse. Geophysical Research Abstracts, vol. 5, EGS-AGU-EUG Joint Assembly, EAE03-A-05509

Ziegler PA (1988) Evolution of the Arctic-North Atlantic and Western Tethys. Am Ass Petrol Geol Mem 43, 198 pp

Zizi M (2002) Triassic-Jurassic extensional systems and their neogene reactivation in northern Morocco. The Rides Prérifaines and Guercif basin. Notes et Mémoires du Service Géologique du Maroc, 416, 138 pp

Zouhri L, Lamouroux C, Vachard D, Piqué A (2002) Evidence of flexural extension of the Rif foreland: The Rharb-Mamora basin (northern Morocco). Bull Soc Géol France 173:509–514

Transect II: Aquitaine Basin – Pyrenees – Ebro Basin – Catalan Coastal Ranges – Valencia Trough – Balearic Promontory – Algerian Basin – Kabylies – Atlas – Saharan Domain

Ábalos B, Carreras J, Druguet E, Escuder J, Gómez-Pugnaire T, Lorenzo-Álvarez S, Quesada C, Rodríguez-Fernández LR, Gil-Ibarguchi JI (2002) Variscan and pre-Variscan tectonics. In: Gibbons W, Moreno T (eds) The Geology of Spain. Geol Soc London, pp 155–183

Acosta J, Muñoz A, Herranz P, Palomo C, Ballesteros M, Vaquero M, Uchupi E (2001) Geodynamics of the Emile Baudot Escarpment and the Balearic Promontory, western Mediterranean. Mar Petrol Geol 18:349–369

Addoum B (1995) L'Atlas Saharien Sud-Oriental. Cinématique des plis- chevauchements et reconstitution du bassin du Sud-Est constantinois (confins algéro-tunisiens). PhD thesis, Université Paris Sud, Orsay, 158 pp

Aifa T, Feinberg H, Derder MEM, Merabet NE (1992) Rotations paléomagnetiques récentes dans le bassin du Cheliff (Algérie). C R Acad Paris 314:915–922

Aït Ouali R (1991) Le rifting des Monts des Ksour au Lias: Organisation du bassin, diagénèse des assises carbonatées, place dans les ouvertures mésozoïques au Maghreb. PhD thesis, Université d'Alger (Algérie), 306 pp

Aïté MO (1994) Analyse de la microfracturation et paléocontraintes dans le Néogène post nappes de la Grande Kabylie. PhD thesis, Université Maine, Le Mans, 166 pp

Aïté MO, Gélard JP (1997) Distension néogène post-colisionelle sur le transect de Grande-Kabylie. Bull Soc Géol France 168(4):423–436

Anadón P, Colombo F, Esteban M, Marzo M, Robles S, Santanach P, Solé-Sugrañes Ll (1982) Evolución tectonoestratigráfica de los Catalánides. Acta Geol Hisp 14:242–270

Anadón P, Cabrera L, Guimerà J, Santanach P (1985) Paleogene strike-slip deformation and sedimentation along the southeastern margin of the Ebro Basin. In: Biddle KT, Christie-Blick N (eds) Strike-slip deformation, basin formation and sedimentation. Soc Econ Pal Min Spec Publ 37:303–318

Anadón P, Cabrera Ll, Colombo F, Marzo M, Riba O (1986) Syntectonic intraformational unconformities in alluvial fan deposits, eastern Ebro Basin margins (NE Spain). In: Allen PA, Homewood P (eds) Foreland basins. Int Ass Sedim Spec Publ 8:259–271

Arenas C, Millán H, Pardo G, Pocoví A (2001) Ebro Basin continental sedimentation associated with late compressional Pyrenean tectonics (north-eastern Iberia): controls on basin margin fans and fluvial system. Basin Res 13:65–89

Argus DF, Gordon RG, Demets C, Stein S (1989) Closure of the Africa-Eurasia-North America plate circuit and tectonics of the Gloria fault. J Geophys Res 94:5585–5602

Ashauer H, Teichmüller R (1935) Die variszische und alpidische Gebirgsbildung Kataloniens. Abh Ges Wiss Göttingen, Math Phys Kl, 3 F., 16, 78 pp

Auzende JM, Bonnin J, Olivet JL (1975) La marge nord-africaine con-sidérée comme marge active. Bull Soc Géol France 17: 486–495

Ayala C (2001) Modelització Conjunta d'Anomalies Gravimètriques i de Geoide. Aplicació a l'Estudi de l'Estructura Litosfèrica de la Mediterrània Occidental. PhD thesis, Universitat de Barcelona, 244 pp

Ayala C, Pous J, Torné M (1996) The lithosphere-asthenosphere boundary of the Valencia trough (western Mediterranean) deduced from 2D geoid and gravity modelling. Geophys Res Lett 22:3131–3134

Ayala C, Torne M et al. (2003) The lithosphere-astenosphere boundary in the western Mediterranean from 3D joint gravity and geoid modeling: tectonic implications. Earth Plan Sci Lett 209:275–290

Baby P, Crouzet G, Specht M, Déramond J, Bilotte M, Debroas EJ (1988) Rôle des paléostructures albo-cénomaniennes dans la géometrie des chevauchements frontaux nord-pyrénéens. C R Acad Sci Paris 306:307–313

Banda E (1988) Crustal parameters in the Iberian Peninsula. Phys Earth Plan Inter 51:222–225

Banda E, Ansorge J, Boloix M, Córdova D (1980) Structure of the crust and upper mantle beneath the Balearic islands (Western Mediterranean). Earth Plan Sci Lett 49:219–230

Banda E, Udías A, Mueller St, Mézcua J, Boloix M, Gallart J, Aparicio A (1983) Crustal structure beneath Spain from deep seismic sounding experiments. Phys Earth Plan Inter 31:277–280

Barnolas A, Chiron JC (eds) (1996) Synthèse géologique et géophysique des Pyrénées, vol 1: Introduction, géophysique, cycle hercynien. BRGM and IGME, Orléans and Madrid, 729 pp

Bartrina MT, Cabrera L, Jurado MJ, Guimerà J, Roca E (1992) Evolution of the central margin of the València trough (western Mediterranean). Tectonophysics 203:219–247

Bassoulet JP (1973) Contribution à l'étude stratigraphique du Mésozoïque de l'atlas saharien occidental (Algérie). PhD thesis, Univ. Paris, 2 tomes, 497 pp

Beaumont C, Muñoz JA, Hamilton J, Fullsack P (2000) Factors controlling the Alpine Evolution of the Central Pyrenees inferred from a comparison of observations and geodynamical models. J Geophys Res 105, B4:8121–8145

Bellon H (1981) Chronologie radiométrique (K-Ar) des manifestations magmatiques autour de la Méditerranée occidentale entre 33 et 1 MA. In: Wezel FC (ed) Sedimentary basins of Mediterranean margins. Tecnoprint, Bologna, pp 341–360

Berástegui X, García JM, Losantos M (1990) Structure and sedimentary evolution of the Organyà basin (Central South Pyrenean Unit, Spain). Bull. Soc. Géol. France 8:251–264

Berástegui X, Losantos M, Muñoz JA, Puigdefàbregas C (1993) Tall Geològic del Pirineu Central, 1:200 000. Servei Geològic de Catalunya, Institut Cartogràfic de Catalunya, Barcelona

Bertraneu J (1955) Contribution à l'étude géologique des Monts du Hodna: Le massif du Boutaleb. PhD thesis, Alger, Pub. Serv. de la Carte géol. Algérie, nouv . série 4, 190 pp

Bois C (1992) The evolution of the layered lower crust and Moho through Geological time in Western Europe: contribution of deep seismic reflection profiles. Terra Nova 4:99–108

Bois C, Pinet B, Roure F (1989) Dating lower crustal features in France and adjacent areas from deep seismic profiles. In: Mereu et al. (eds) Properties and processes of Earth's lower crust. AGU and IUGG Monogr 6:17–31

Bond RMG, McClay K (1995) Inversion of a Lower Cretaceous extensional basin, south central Pyrenees, Sapin. In: Buchanan JG, Buchanan PG (eds) Basin inversion. Geol Soc Spec Publ 88:415–431

Boote DRD, Clark-Lowes DD, Traut MW (1998) Paleozoic petroleum systems of North Africa. In: Macgregor DS, Moody RTJ, Clark-Lowes DD (eds) Petroleum geology of North Africa. Geol Soc Spec Publ 133:7–68

Boudiaf A, Philip H, Coutelle A, Ritz JF (1999) Découverte d'un chevauchement d'âge quaternaire au sud de la Grande Kabylie (Algérie). Geodin Acta 12:71–80

Boudjema A (1987) Evolution structurale du bassin pétrolier "triasique" du Sahara Nord-Oriental (Algérie). PhD thesis, Univ. Paris-Sud (Orsay), 290 pp

Bouillin JP (1977) Géologie Alpine de la Petite Kabylie dans les régions de Collo et d'El Milia. PhD thesis, Univ. Pierre et Marie Curie (Paris VI), 511 pp

Bouillin JP (1986) Le "bassin Maghrébin": une ancienne limite entre l'Europe et l'Afrique à l'ouest des Alpes. Bull Soc Géol France 8(2):547–558

Bouillin JP, Durand-Delga M, Gélard JP, Leikine M, Raoult JF, Raymond D, Tefiani M, Vila JM (1971) Définition d'un flysch massylien et d'un flysch maurétanien au sein des flyschs allochtones d'Algérie. C R Acad Sci Paris, 270:2249–2252

Boullin JP, Bossière G, Bourrouilh R, Coutelle A, Durand-Delga M, Gélard JP, Gery B, Raoult JF, Raymond D, Tefiani M (1984) Mise au point sur l'âge des socles métamorphiques kabyles (Algérie). C R Acad Sci Paris, 298, II:665–660

Bourrouilh R (1983) Estratigrafía, sedimentología y tectónica de la isla de Menorca y del noreste de Mallorca (Baleares). Mem Inst Geol Min España 99, Madrid, 672 pp

Bracène R (2002) Géodynamique du nord de l'Algérie: impact sur l'exploration pétrolière. PhD thesis, Univ. Cergy-Pontoise, tome 1, 110 pp

Bracène R, Frizon de Lamotte D (2002) The origin of intraplate deformation in the Atlas system of western and central Algeria: from Jurassic rifting to Cenozoic-Quaternary inversion. Tectonophysics 357:207–226

Bracène R, Bellahcene A, Bekkouche D, Mercier E, Frizon de Lamotte D (1998) The thin-skinned style of the South Atlas Front in central Algeria. In: Macgregor DS, Moody RTJ, Clark-Lowes DD (eds) Petroleum geology of North Africa. Geol Soc London Spec Publ 133:395–404

Bracène R, Patriat M, Ellouz N, Gaullier JM (2003) Subsidence history in basins of northern Algeria. Sed Geol 156:213–239

BRGM (1974) Géologie du basin d'Aquitaine. Éditions BRGM, Paris, 27 pp

Brunet MF (1986) The influence of the evolution of the Pyrenees on adjacent basins. Tectonophysics 129:345–354

Buis MG, Cugny P (1978) Les poudingues de Palassou entre l'Ariège et le Douctouyre (Pyrénées Ariégeoises). Bull Soc Hist Nat Toulouse 114:212–236

Buis MG, Rey J (1975) Une évolution sédimentaire de type deltaique: le passage du Tertiaire marin au Tertiaire continental entre l'Ariège et le Douctouyre (Pyrénées Ariégeoises). Bull Soc Hist Nat Toulouse 111:80–95

Buforn E, Sanz de Galdeano C, Udías A (1995) Seismotectonics of the Ibero-Maghrebian region. Tectonophysics 248:247–261

Burg JP, Matte P (1978) Cross section through the French Massif Central and the scope of its Variscan geodynamic evolution. Z dt Geol Ges Hannover 9:429–460

Cabrera L, Calvet F (1996) Onshore Neogene record in NE Spain: Vallès-Penedès and El Camp half-grabens (NW Mediterranean). In: Friend P, Dabrio CJ (eds) Tertiary basins of Spain: the stratigraphic record of crustal kinematics. Cambridge University Press, pp 97–105

Caby R, Saadallah A, Hammor D (1996) Alpine tectonometamorphic evolution of the Kabylian crystalline massifs, Algeria. In: Durand B, Jolivet L, Horváth F, Séranne M (eds) The Mediterranean basins: Tertiary extension within the Alpine orogen. Geol Soc London Spec Publ 156

Caire A (1957) Etude géologique de la région des Bibans. PhD thesis, Pub. Serv. Carte. géol. Algérie, nouv. Série 16, 2 vol, 818 pp

Cande SC, Kent DV (1992) A new geomagnetic polarity time-scale for the late Cretaceous and Cenozoic. J Geophys Res 97(B10): 13917–13951

Carminati E, Wortel MJR, Meijer PT, Sabadini R (1998) The two stage opening of the western-central Mediterranean basins: a forward modeling test to a new evolutionary model. Earth Plan Sci Lett 160:667–679

Cattanéo G, Gélard JP, Aïté MO, Mouterde R (1999) La marge septentrionale de la Téthys maghrébine au Jurassique (Djurdjura et Chellata, Grande Kabylie, Algérie). Bull Soc Géol France 170:173–188

Chamot-Rooke N, Jestin F, Gaulier JM (1997) Constraints on Moho depth and crustal thickness in the Liguro-Provençal Basin from a 3D gravity inversion: geodynamic implications. Rev Inst Français du Pétrole 52:557–583

Cheadle MJ, McGeary S, Warner MR, Matthews DH (1987) Extensional structures on the Western UK continental shelf: a review of evidence from deep seismic profiling. In: Coward MP, Dewey JF, Hancock PL (eds) Continental extensional tectonics. Geol Soc Spec Publ 28:223–246

Choukroune, P. (1976), Structure et évolution tectonique de la zone nord-pyrénéenne. Analyse de la deformation dans une portion de la chaîne à schistosité subverticale. Mém Soc géol France 127

Choukroune P, ECORS Team (1989) The ECORS Pyrenean deep seismic profile reflection data and the overall structure of an orogenic belt. Tectonics 8:23–39

Choukroune P, Mattauer M (1978) Tectonique des plaques et Pyrénées: sur le fonctionnement de la faille transformante nord-pyrénéenne. Comparaison avec des modèles actuels. Bull Soc Géol France 20:687–700

Choukroune P, Pinet B, Roure F, Cazes M (1990) Major Hercynian structures along the ECORS Pyrenees and Biscaye Lines. Bull Soc Géol France 8:313–320

Clavell E (1991) Geologia del Petroli de les Conques Terciàries de Catalunya. PhD thesis, Universitat de Barcelona, 437 pp

Clavell E, Berástegui X (1991) Petroleum geology of the Gulf of Valencia. In: Spencer AM (ed) Generation, accumulation and production of Europe's hydrocarbons. Oxford University Press, pp 355–368

Collier JS, Buhl P, Torné M, Watts AB (1994) Moho and lower crustal reflectivity beneath a young rift basin: results from a two-ship, wide-aperture seismic-reflection experiment in the Valencia Trough (western Mediterranean). Geophys J Int 118: 159–180

Colodrón I, Núñez A, Ruiz V, Cabañas I, Uralde MA, Nodal T (1978) Mapa Geológico de España, e. 1:50000. Hoja 445-Cornudella. Instituto Geológico y Minero de España, Servicio de Publicaciones del Ministerio de Industria y Energía, Madrid, 22 pp

Colombo F, Vergés J (1992) Geometría del margen SE de la Cuenca del Ebro: discordancias progresivas en el Grupo Scala Dei. Serra de la Llena (Tarragona). Acta Geol Hisp 27:33–53

Comas MC, García-Dueñas V, Jurado MJ (1992) Neogene tectonic evolution of the Alboran Sea from MCS data. Geo-Mar Lett 12: 157–164

Coney PJ, Muñoz JA, McClay K, Evenchick CA (1996) Syntectonic burial and post-tectonic exhumation of the southern Pyrenees foreland fold-thrust belt. J Geol Soc London 153:9–16

Coulon C, Megartsi M, Fourcade S, Maury R, Bellon H, Louni-Hacini A, Cotten J, Coutelle A, Hermitte D (2002) Post-collisional transition from calc-alkaline to alkaline volcanism during the Noegene in Oranie (Algeria): magmatic expression of a slab breakoff. Lithos 62:87–110

Courme-Rault MD (1985) Stratigraphie du Miocène et chronologie comparée des déformations suivant deux transversales des atlasides orientales (Algérie, Sicile). PhD thesis, Université Orléans

Coutelle A (1979) Géologie du sud-est de la Grande Kabylie et des Babors d'Akbou. PhD thesis, Université Paris, 567 pp

Daignières M, De Cabissole B, Gallart J, Hirn A, Suriñach E, Torné M (1989) Geophysical constraints on the deep structure along the ECORS Pyrenees line. Tectonics 8:1051–1058

Dallmeyer RD, Martinez-Garcia E (1990) Pre-Mesozoic geology of Iberia. Springer, Berlin, 550 pp

Danobeitia JJ, Arguedas M, Gallart F, Banda E, Makris J (1992) Deep crustal configuration of the Valencia trough and its Iberian and Balearic borders from extensive refraction and wide-angle reflection profiling. Tectonophysics 203:37–55

Darder B (1925) La tectonique de la région orientale de l'île de Majorque. Bull Soc Géol France 25:245–278

DeMets C, Gordon RG, Argus DF, Stein S (1994) Effect of recent revisions to the geomagnetic reversal time scale on estimate of current plate motions. Geophys Res Lett 21:2191–2194

Debroas EJ (1990) Le Flysch noir albo-cénomanien témoin de la structuration albienne à sénonienne de la zone nord-pyrénéenne en Bigorre (Hautes Pyrénées, France). Bull Soc Géol France 8(6):273–285

Delfaud J (1986) Organisation scalaire des événements sédimentaires majeurs autour de la Mésogée durant le Jurassique et le Crétacé. Conséquences sur les associations biologiques. Bull Centr Rech Expl Prod Elf-Aquitaine 10:509–535

Delfaud J, Douihasni M, Rolet L (1974) Mise en évidence de tectoniques superposées dans la région d'Aïn Ouarka (Monts des Ksour). C R Acad Sci Paris 278, série D:1817–1820

Déramond J, Souquet P, Fondecave-Wallez MJ, Specht M (1993) Relationships between thrust tectonics and sequence stratigraphy surfaces in foredeeps: model and examples from the Pyrenees (Cretaceous-Eocene, France, Spain). In: Williams GD, Dobb A (eds) Tectonics and seismic sequence stratigraphy. Geol Soc Spec Publ 71:193–219

Desegaulx P, Roure F, Villien A (1990) Structural evolution of the Pyrenees: tectonic inheritance and flexural behaviour in the continental crust. Tectonophysics 182:211–225

Dewey JF, Helman ML, Turco E, Hutton DHW, Knott SD (1989) Kinematics of the western Mediterranean. In: Coward MP, Dietrich D, Park (eds) Alpine tectonics. Geol Soc Spec Publ 45:265–283

Dinarés J, McClelland E, Santanach P (1992) Contrasting rotations within thrust sheets and kinematics of thrust-tectonics as derived from palaeomagnetic data: an example from the southern Pyrenees. In: McClay K (ed) Thrust tectonics. Chapman & Hall, London, pp 265–275

Doglioni C, Mongelli F, Pialli G (1998) Boudinage of the Alpine belt in the Apenninic back-arc. Mem Soc Geol It 52:457–468

Doglioni C, Fernandez M, Gueguen E, Sabat F (1999) On the interference between the early Apennines-Maghebides back-arc extension and the Alps-Betics orogen in the Neogene geodynamics of the Western Mediterranean. Bull Soc Geol It 118:75–89

Durand-Delga M (1966) Titres et Travaux scientifiques. Imp. Priester, Paris, 43 pp

Durand-Delga M, Fontboté JM (1980) Le cadre structural de la Méditerranée occidentale: 26th International Geological Congress, C5, Géologie des chaînes alpines issues de la Téthys, Paris, pp 67–85

Durand-Delga M, Rossi P, Olivier P, Puglisi D (2000) Situation structurale et nature ophiolitique de roches basiques jurassiques associées aux flyschs maghrébins du Rif (Maroc) et de Sicile (Italie). C R Acad Sci Paris 331:29–38

ECORS Pyrenees Team (1988) The ECORS deep reflection seismic survey across the Pyrenees. Nature 331:508–511

Elmi S, Almeras Y, Ameur M, Bassoulet JP, Boutaktiout M, Benhamou M, Marok A, Mekahli L, Mekaoui A, Mouterde R (1998) Stratigraphic and paleogeographic survey of the Lower and Middle Jurassic along a North- South transect in western Algeria. In: Crasquin-Soleau S, Barrier É (eds) Peri-Tethys memoir 5. Mém Mus National Hist Nat 179:145–211

El Robrini M (1986) Evolution morphostructurale de la marge algérienne occidentale (Mediterranée occidentale): influence de la néotectonique et de la sédimentation. PhD thesis, Université Paris 6, 164 pp

Erickson AJ (1978) The measurement and interpretation of heat flow in the Mediterranean and Black seas. Mass Inst Tech Report 70-5, NTIS AD709070, 272 pp

Fallot P (1922) Étude géologique de la Sierra de Majorque. PhD thesis. Libr Polytechnique, Ch. Béranger ed., Paris Liège, 420 pp

Favre P, Stampfli G (1992) From rifting to passive margin: the example of the Red Sea, Central Atlantic and Alpine Tethys. Tectonophysics 215:67–97

Fernàndez M, Torné M, Zeyen H (1990) Lithospheric structure of the NE Spain and the North Balearic basin. J Geodyn 12:253–267

Fernàndez M, Foucher JP, Jurado MJ (1995) Evidence fro the multistage formation of the south-western Valencia Trough. Mar Petrol Geol 12:101–109

Fischer MW (1984) Thrust tectonics in the North Pyrenees. J Struct Geol 6:721–726

Fitzgerald PG, Muñoz JA, Coney PJ, Baldwin SL (1999) Asymmetric exhumation across the central Pyrenees: implications for the tectonic evolution of a collisional orogen. Earth Plan Sci Lett 173:157–170

Flinch JF (1996) Accretion and extensional collapse of the external Western Rif (Northern Morocco). In: Ziegler PA, Horvath F (eds) Structure and prospects of Alpine basins and forelands. Mem Mus National Hist Nat 170:61–85

Fontboté JM, Guimerà J, Roca E, Sàbat F, Santanach P, Fernández-Ortigosa F (1990) The Cenozoic Geodynamic evolution of the València trough (western Mediterranean). Rev Soc Geol España 3:249– 259

Fornós JJ, Marzo M, Pomar L, Ramos-Guerrero E, Rodríguez-Perea A (1991) Evolución tectonosedimentaria y análisis estratigráfico del Terciario de la isla de Mallorca. In: Colombo F (ed) I Congreso del Grupo Español del Terciario, Vic 1991. Libro-Guía Excursión no 2. Dept. GDGP, Universitat de Barcelona, 145 pp

Frizon de Lamotte D, Andrieux J, Guézou JC (1991) Cinématique des chevauchements néogènes dans l'Arc bético-rifain, discussions sur les modèles géodynamiques. Bull Soc Géol France 162:611–626

Frizon de Lamotte D, Mercier E, Outtani F, Addoum B, Ghandriche H, Ouali J, Bouaziz S, Andrieux J (1998) Structural inheritance and kinematics of folding and thrusting along the front of the Eastern Atlas Mountains (Algeria and Tunisia). In: Barrier E, Crasquin S (eds) Peri-Tethyan memoir 3. Mém Mus National Hist Nat 117:237–252

Frizon de Lamotte D, Saint-Bezar B, Bracène R, Mercier E (2000) The two steps of the Atlas building and geodynamics of the western Mediterranean. Tectonics 19:740–761

Galdeano A, Rossignol JC (1977) Assemblage à altitude constante de cartes d'anomalies magnétiques couvrant l'ensemble du bassin occidental de la Méditerranée. Bull Soc Géol France 7:461–468

Gallart J, Rojas H, Díaz J, Dañobeitia J (1990) Features of the deep crustal structure and the onshore/offshore transition at the Iberian flank of the Valencia trough (Western Mediterranean). J Geodyn 12:233–252

Gallart J, Vidal N, Dañobeitia JJ, the ESCI-Valencia Trough Working Group (1994) Lateral variations in the deep crustal structure at the Iberian margin of the Valencia trough imaged from seismic reflection methods. Tectonophysics 232:59–75

Gallart J, Vidal N, Estévez A, Pous J, Sàbat F, Santisteban C, Suriñach E, ESCI-València Trough Group (1997) The ESCI-València Trough vertical reflection experiment: a seismic image of the crust from the NE Iberian Peninsula to the Western Mediterranean. Rev Soc Geol España 8 (1995):401–415

García-Senz J (2002) Cuencas extensivas del Cretácico inferior en los Pirineos centrales, formación y subsecuente inversión. PhD thesis, Universitat de Barcelona, 310 pp

Garrido-Megías A, Ríos LM (1972) Síntesis geológica del Secundario y Terciario entre los ríos Cinca y Segre (Pirineo central de la vertiente surpirenaica, provincias de Huesca y Lérida). Bol Geol Min 83:1–47

Gaspar-Escribano JM, Van Wees JD, Ter Voorde M, Cloetingh S, Roca E, Cabrera L, Muñoz JA, Ziegler PA, García-Castellanos D (2001) 3D flexural modeling of the Ebro Basin (NE Iberia). Geophys J Int 145:349–367

Gaspar-Escribano JM, ter Voorde M, Roca E, Cloetingh S (2003) Mechanical (DE-) coupling of the lithosphere in the Valencia trough (NW Mediterranean): What does it mean?. Earth Plan Sci Lett, in press

Gelabert B (1997) L'Estructura Geològica de la Meitat Occidental de l'Illa de Mallorca. PhD thesis, Universitat de Barcelona, 204 pp

Gelabert B, Sàbat F, Rodríguez-Perea A (1992) A structural outline of the Serra de Tramuntana of Mallorca (Balearic Islands). Tectonophysics 203:167–183

Gélard JP (1979) Géologie du nord-est de la grande Kabylie (un segment interne de l'orogène littoral nord Africain). PhD thesis, Université Dijon, 326 pp

Géry B, Feinberg H, Lorenz C, Magné J (1981) Définition d'une série type de "l'Oligo-Miocène kabyle" anté-nappes dans le Djebel Aïssa-Mimoun (Grand Kabylie, Algérie. C R Acad Sci Paris 292: 1529–1532

Ghandriche H (1991) Modalités de la superposition de structures de plissement-chevauchement d'âge alpin dans les Aurès (Algérie). PhD thesis, Université Paris Sud, Orsay, 190 pp

Giese P, Jacbshagen V (1992) Inversion Tectonics of intracontinental ranges: High and Middle Atlas, Morocco. Geol Rund 81:249–259

Glover P, Pous J, Queralt P, Muñoz JA, Liesa M, Hole M (2000) Integrated two dimensional lithosphere conductivity modelling in the Pyrenees using field-scale and laboratory measurements. Earth Plan Sci Lett 178:59–72

Goldberg JM, Maluski H (1988) Données nouvelles et mise au point sur l'âge du métamorphisme pyrénéen. C R Acad Sci Paris 306(2): 429–435

Gómez M, Guimerà J (1999) Estructura alpina de la Serra de Miramar y del NE de las Muntanyes de Prades (Cadena Costera Catalana). Rev Soc Geol España 12:405–418

Gueguen E, Doglioni C, Fernandez M (1998) On the post-25 Ma geodynamic evolution of the Western mediterranean. Tectonophysics 298:259–269

Guerrera F, Martin-Algarra A, Perrone V (1993) Late Oligocene-Miocene syn-/late-orogenic successions in western and central Mediterranean Chains from the Betic cordilleras to the southern Apennines. Terra Nova 5:525–544

Guimerà J, Álvaro M (1990) Structure et évolution de la compression alpine dans la Chaîne ibérique et la Chaîne côtière catalane (Espagne). Bull Soc Géol France 8, VI:339–349

Guiraud R (1973) Evolution post triasique de l'avant pays de la chaîne alpine d'Algérie, d'après l'étude du bassin du Hodna et des régions voisines. PhD thesis, Univ. Nice (nouv. édition 1990), 270 pp

Guiraud R (1975) L'évolution post-triasique de l'avant-pays de la chaîne alpine en Algérie, d'après l'étude du bassin du Hodna et des régions voisines. Rev Géogr Phys géol Dyn 17:427–446

Guiraud R, Bosworth W (1997) Senonian basin inversion and rejuvenation of rifting in Africa and Arabia: Synthesis and implications to plate scale tectonics. Tectonophysics 282:39–82

Hinz K (1972) Results of seismic refraction investigations (Project Anna) in Western Mediterranean, south and north of the island of Mallorca. Bull. Centre Rech. Pau-SNPA 6, 2:405–426

Hinz K (1973) Crustal structure of the Balearic Sea. Tectonophysics 20:295–302

Hsü KJ, Montadert L et al. (1978) Initial Report of the Deep Sea Drilling Project, Vol. 42, Part 1. Initial reports of the Deep Sea Drilling Project, 42, 1. U. S. Government Printing Office, Washington D. C

Janssen ME, Torné M, Cloetingh S, Banda E (1993) Pliocene uplift of the eastern Iberian margin: Inferences from quantitative modelling of the València Trough. Earth Planet Sci Lett 119:585–597

Jolivet L, Faccenna C (2000) Mediterranean extension and the Africa-Eurasia collision. Tectonics 19:1095–1106

Jolivet L, Faccenna C, Goffé B, Burov E, Agard P (2003) Subduction tectonics and exhumation of high-pressure metamorphic rocks in the Mediterranean orogens. Amer Jour Sci 303:353–409

Jurado MJ (1989) El Triásico del subsuelo de la Cuenca del Ebro. PhD thesis, Universitat de Barcelona, 259 pp

Kazi-Tani N (1986) Evolution géodynamique de la bordure nord-africaine: le domaine intraplaque nord-algérien. Approche mégaséquentielle. PhD thesis, Université Pau, 871 pp

Kieken M (1974) Etude géologique du Hodna, du Titteri et de la partie occidentale des Biban. Publi Serv Carte géol Algérie 46, I, II, 217 + 281 pp

Kireche O (1993) Evolution géodynamique de la marge tellienne des maghrébides d'après l'étude du domaine parautochtone schistosé (Massifs du Chélif et d'Oranie, de Blida- Bou Maad, des Babors et des Biban. PhD thesis., Université Alger, 297 pp

Klett TR (2000) Total petroleum systems of the Trias/Ghadames Province, Algeria, Tunisia, and Libya – the Tanezzuft-Oued Mya, Tanezzuft-Melrhir, and Tanezzuft-Ghadames. U.S. Geological Survey Bulletin 2202-C, Denver, 26 pp

Laffitte R (1939) Etude géologique de l'Aurès. PhD thesis, Université Paris, Bull., Paris, Pub. Serv. Carte géol. Algérie 46, 217 + 281 pp

Lanaja M (1987) Contribución de la exploración petrolífera al conocimiento de la geología en España. Instituto Geológico y Minero de España, Madrid, 465 pp

Le Pichon X, Bergerat F, Roulet MJ (1988) Plate kinematics and tectonics leading to the Alpine belt formation: a new analysis. In: Clark SP, Clark Burchfiel JB, Suppe J (eds) Processes in continental lithospheric deformation. Geol Soc Am Spec Pap 218: 111–131

Leikine M. (1971) Etude géologique des Babords occidentaux (Algérie). PhD thesis, Université Paris, 536 pp

Ledo J, Ayala C, Pous J, Queralt P, Marcuello A, Muñoz JA (2000) New geophysical constraints on the deep structure of the Pyrenees. Geophys Res Lett 27:1037–1040

Lentini F, Catalano S, Carbone S (1996) The External Thrust System in southern Italy: a target for petroleum exploration. Petrol Geosci 2:333–342

Lewis CJ, Vergés J, Marzo M (2000) High mountains in a zone of extended crust: Insights into the Neogene-Quaternary topographic development of northeastern Iberia. Tectonics 19:86–102

Linzer HG, Decker K, Peresson H, Dell'Mour R, Frisch W (2002) Balancing lateral orogenic float of the Eastern Alps. Tectonophysics 354:211–237

Llopis-Lladó N (1947) Contribución al conocimiento de la morfoestructura de los Catalánides. C.S.I.C., Inst "Lucas Mallada", Barcelona, 373 pp

Lonergan L, White N (1997) Origin of the Betic-Rif mountain belt. Tectonics 16:504–522

López-Blanco M, Marzo M, Burbank DW, Vergés J, Roca E, Anadón P, Piña J (2001) Tectonic and climatic controls on the development of foreland fan deltas: Montserrat and Sant Llorenç del Munt systems (Middle Eocene, Ebro Basin, NE Spain). Sed Geol 138:17–39

Maillard A, Mauffret A (1999) Crustal structure and riftogenesis of the Valencia Trough (north-western Mediterranean Sea). Basin Res 11:357–379

Maillard A, Mauffret A, Watts AB, Torné M, Pascal G, Buhl P, Pinet B (1992) Tertiary sedimentary history and structure of the Valencia trough (western Mediterranean). Tectonophysics 203:57–75

Marillier F, Mueller S (1985) The western Mediterranean region as an upper-mantle transition zone between two lithospheric plates. Tectonophysics 118:113–130

Martí J, Mitjavila J, Roca E, Aparicio A (1992) Cenozoic magmatism of the Valencia trough (western Mediterranean): relationships between structural evolution and volcanism. Tectonophysics 203: 145–165

Martínez A, Vergés J, Muñoz JA (1988) Secuencias de propagación del sistema de cabalgamientos de la terminación oriental del manto del Pedraforca y relación con los conglomerados sinorogénicos. Acta Geol Hisp 23:119–127

Martínez del Olmo W (1996) E3 Depositional sequences in the Gulf of Valencia Tertiary basin. In: Friend P, Dabrio CJ (eds) Tertiary basins of Spain: the stratigraphic record of crustal kinematics. Cambridge University Press, pp 55–67

Martínez-Peña B, Pocoví A (1988) El amortiguamiento frontal de la estructura de la cobertera surpirenaica y su relación con el anticlinal de Barbastro-Balaguer. Acta Geol Hisp 23:81–94

Marzo M, Muñoz JA, Vergés J, López-Blanco M, Roca E, Arbues P, Barberà X, Cabrera L, Colombo F, Serra-Kiel J (1998) Excursion B2 – Sedimentation and tectonics: case studies from Paleogene, continental to deep water sequences of the South Pyrenean foreland Basin (NE Spain). In: Field Trip Guidebook of the 15th International Sedimentological Congress, Alicante, April 1998. Instituto Tecnológico Geominero de España, Madrid, pp 197–252

Masana E (1994) Neotectonic features of the Catalan Coastal Ranges, Northeastern Spain. Acta Geol Hisp 29:107–121

Masana E, Villamarín JA, Santanach P (2001) Paleoseismic results from multiple thenching analysis along a silent fault: The El Camp fault (Tarragona, northeastern Iberian Peninsula). Acta Geol Hisp 36:329–354

Mauffret A (1976) Etude géodynamique de la marge des îles Baléares. PhD thesis, Université Paris, 137 pp

Mauffret A, Fail JP, Montadert L, Sancho J, Winnock E (1973) North-western Mediterranean sedimentary basin from seismic reflection profile. Am Ass Petrol Geol Bull 57:2245–2262

Mauffret A, Maldonado A, Campilo A (1992) Tectonic framework of the eastern Alboran and western Algerian basins (Western Mediterranean). In: Maldonado A (ed) The Alboran Sea, special issue. Geo-Mar Lett 12:104–110

Mauffret A, Pascal G, Maillard A, Gorini C (1995) Tectonics and deep structure of the north-western Mediterranean Basin. Mar Petrol Geol 12:645–666

Mauffret A, Maillard A, Gorini C (2004) Structure of the Algerian Basin and its Western Margin. Bull Soc Géol France (submitted)

Maury R, Fourcade S, Coulon C, Azzouzi M, Bellon H, Coutelle A, Ouabadi A, Semroud B, Megarsi M, Cotten J, Belanteur O, Louni-Hacini A, Piqué A, Capdevila R, Hernandez J, Rehault JP (2000) Post-collisional Neogene magmatism of the Mediterranean Maghreb margin: a consequence of slab breakoff. C R Acad Sci Paris 331:159–173

Mazzoli S, Helman M (1994) Neogene patterns of relative plate motion for Africa-Europe: some implications fro recent Mediterranean tectonics. Geol Rund 83:464–468

McClay K, Muñoz JA, García-Senz J (2004) Extensional raft faulting: A newly identified tectonic event in the Spanish Pyrenees. Geology: submitted

Megharoui M, Morel JL, Andrieux J, Dahmani M (1996) Tectonique plio-quaternaire de la chaîne tello-rifaine et de la mer d'Alboran. Une zone complexe de convergence continent-continent. Bull Soc Géol France 167:141–157

Meigs AJ, Vergés J, Burbank DW (1996) Ten-million-year history of a thrust sheet. Geol Soc Amer Bull 108:1608–1625

Melgarejo JC (1987) Estudi geològic i metallogenètic del Paleozoic del sud de les Serralades Costaneres Catalanes. PhD thesis, Universitat de Barcelona, 646 pp

Mey PHW, Nagtegaal PJC, Roberti KJ, Hartevelt JJA (1968) Lithostratigraphic subdivision of post-hercynian deposits in the south-central Pyrenees, Spain. Leidse Geol Mededelingen 41:221–228

Michard A, Chalouan A, Feinberg H, Goffé B, Montigny R (2002) How does the Alpine belt end between Spain and Morocco?. Bull Soc Géol France 173:3–15

Mickus K, Jallouli C (1999) Crustal structure beneath the Tell and Atlas mountains (Algeria and Tunisia) through the analysis of gravity data. Tectonophysics 314:373–385

Millán H, Den Bezemer T, Vergés J, Marzo M, Muñoz JA, Roca E, Cirés J, Zoetemeijer R, Cloetingh S, Puigdefàbregas C (1995) Palaeo-elevation and effective elastic thickness evolution at mountain ranges: inferences from flexural modelling in the Eastern Pyrenees and Ebro Basin. Mar Petrol Geol 12:917–928

Monié P, Saadallah A, Maluski H, Caby R (1988) New ^{39}Ar/^{40}Ar ages for Hercynian and Alpine thermotectonic events in Grande Kabylie (Algeria). Tectonophysics 152:53–69

Morley CK (1988) Out-of-sequence thrusts. Tectonics 7:539–561

Morley CK (1992) Tectonic and sedimentary evidence for synchronous and out-of-sequence thrusting Larache-Acilah area, Western Morocco, Rif. J Geol Soc London 149:39–49

Muñoz JA (1992) Evolution of a continental collision belt: ECORS-Pyrenees crustal balanced cross-section. In: McClay K (ed) Thrust tectonics. Chapman & Hall, London, pp 235–246

Muñoz JA (2002) The Pyrenees. In: Gibbons W, Moreno T (eds) The geology of Spain. Geol Soc Publ House, London, pp 370–385

Muñoz JA, Martínez A, Vergés J (1986) Thrust sequences in the eastern Spanish Pyrenees. J Struct Geol 8:399–405

Muñoz JA, Coney P, McClay K, Evenchick C (1997) Discussion on syntectonic burial and post-tectonic exhumation of the southern Pyrenees foreland fold-thrust belt. Geol Soc London 154:361–365

Mutti E, Seguret M, Sgavetti M (1988) Sedimentation and deformation in the Tertiary sequences in the southern Pyrenees. AAPG Mediterranean Basins Conference, Nice, Field Guide 7. Ist Geol Univ Parma, Italy

Negredo A, Fernàndez M, Torné M, Doglioni C (1999) Numerical modeling of simultaneous extension and compression: the Valencia trough (western Mediterranean). Tectonics 18:361–374

Nijman W (1998) Cyclicity and basin axis shift in a piggyback basin: towards modelling of the Eocene Tremp-Ager Basin, South Pyrenees, Spain. In: Mascle A, Puigdefàbregas C, Luterbacher HP, Fernàndez M (eds) Cenozoic foreland basins of Western Europe. Geol Soc Spec Publ 134:135–162

Núñez A, Colodrón I, Ruíz V, Cabañas I, Uralde MA, Abellán F (1980) Mapa Geológico de España, e. 1 : 50 000. Hoja 472-Reus. Instituto Geológico y Minero de España, Servicio de Publicaciones del Ministerio de Industria y Energía, Madrid, 33 pp

Obert D (1981) Etude géologique des Babors orientaux (domaine tellien, Algérie). PhD thesis, Université Paris VI, 645 pp

Olivera C, Susagna T, Roca A, Goula X (1992) Seismicity of the Valencia trough and surrounding areas. Tectonophysics 203:99–109

Olivet JL (1996) Kinematics of the Iberian plate. Bull Cent Rech Explor Prod Elf Aquitaine 20:131–195

Parés JM, Freeman R, Roca E (1992) Neogene structural development in the Valencia trough margins from palaeomagnetic data. Tectonophysics 203:111–124

Parson B, Sclater JG (1977) An analysis of the variation of ocean floor bathymetry and heat flow with age. J Geophys Res 82:803–827

Pascal G, Torné M, Buhl P, Watts AB, Mauffret A (1992) Crustal and velocity structure of the València trough (western Mediterranean), Part II: Detailed interpretation of five Expanded Spread Profiles. Tectonophysics 203:21–35

Pascal GP, Mauffret A, Patriat P (1993) The ocean-continent boundary in the Gulf of Lion from analysis of expanding spread profiles and gravity modelling. Geophys J Inter 113:701–726

Peucat JJ, Bossière G (1981) Age Rb-Sr sur micas du socle métamorphique kabyle (Algérie): mise en évidence d'événements thermiques alpins. Bull Soc Géol France 23:439–447

Piqué A (2001) Geology of Northwest Africa. Gebrüder Bontraeger, Berlin Stuttgart, 310 pp

Piqué A, Aït Brahim L, El Azzouzi M, Maury RC, Bellon H, Semroud B, Laville E (1998) Le poinçon maghrébin: contraintes structurales et géochimiques. C R Acad Sci Paris 326:575–581

Piqué A, Tricart P, Guiraud R, Laville E, Bouaziz S, Amrhar M, Aït Ouali R (2002) The Mesozoic-Cenozoic Atlas belt (North Africa): an overview. Geodin Acta 15:185–208

Piromallo C, Morelli A (2003) P wave tomography of the mantle under the Alpine-Mediterranean area. J Geophys Res 108, doi:10.1029/2002JB001757

Pocoví A (1978) Estudio geológico de las Sierras Marginales Catalanas (Prepirineo de Lérida). PhD thesis, Universitat de Barcelona, 218 pp

Polyak BG, Fernàndez M, Khutorskoy MD, Soto JI, Basov IA, Comas MC, Khain VYe, Alonso B, Agapova GV, Mazurova IS, Negredo A, Tochitsky VO, Bogdanov NA, Banda E (1996) Heat flow in the Alboran Sea (the Western Mediterranean). Tectonophysics 263:191–218

Pous J, Ledo J, Marcuello A, Dagnières A (1995a) Electrical resistivity model of the crust and upper mantle from magnetotelluric survey through the central Pyrenees. Geophys J Int 121:750–762

Pous J, Muñoz JA, Ledo J, Liesa M (1995b) Partial melting of the subducted continental lower crust in the Pyrenees. J Geol Soc London 152:217–220

Puigdefàbregas C, Souquet P (1986) Tecto-sedimentary cycles and depositional sequences of the Mesozoic and Tertiary from the Pyrenees. Tectonophysics 129:173–203

Puigdefàbregas C, Muñoz JA, Marzo M (1986) Thrust belt development in the eastern Pyrenees and related depositional sequences in the southern foreland basin. In: Allen PA, Homewood P (eds) Foreland basins. Int Ass Sediment, Spec Publ 8:229–246

Puigdefàbregas C, Muñoz JA, Vergés J (1992) Thrusting and foreland basin evolution in the southern Pyrenees. In: McClay K (ed) Thrust tectonics. Chapman & Hall, London, pp 247–254

Pujalte V, Baceta JI, Payros A, Orue-Etxebarria X (1994) The upper Maastrichtian-Middle Eocene of the western Pyrenees: a case-study for 2nd-order, transgressive/regressive facies cycles. Strata 6:162

Rangheard Y (1972) Étude géologique des îles d'Ibiza et de Formentera (Baléares). Mem Inst Geol Min Esp 82, 340 pp

Rehault JP, Boillot G, Mauffret A (1984) The western Mediterranean Basin, geological evolution. Mar Geol 55:447–477

Riba O (1976) Syntectonic unconformities of the Alto Cardener, Spanish Pyrenees: a genetic interpretation. Sed Geol 15:213–233

Riba O, Jurado MJ (1992) Reflexiones sobre la geología de la parte occidental de la Depresión del Ebro. Acta Geol Hisp 27:177–194

Riba O, Reguant S, Villena J (1983) Ensayo de síntesis estratigráfica y evolutiva de la cuenca terciaria del Ebro. In: Libro Jubilar J. M. Ríos, Geología de España. I.G.M.E., Madrid, pp 131–159

Roca E (1992) L'estructura de la conca Catalano-Balear: paper de la compressió i de la distensió en la seva gènesi. PhD thesis, Universitat de Barcelona, 330 pp

Roca E (1996) La evolución geodinámica de la Cuenca Catalano-Balear y áreas adyacentes desde el Mesozoico hasta la actualidad. Acta Geol Hisp 29:3–25

Roca E (2001) The Northwest-Mediterranean basin (Valencia Trough, Gulf of Lions and Liguro-Provencal basins): structure and geodynamic evolution. In: Ziegler PA, Cavazza W, Robertson AFH, Crasquin-Soleau S (eds) Peri-Tethyan rift/wrench basins and passive margins. Mém Mus National Hist Nat 186: 671–706

Roca E, Desegaulx P (1992) Analysis of the geological evolution and vertical movements in the València Trough area, western Mediterranean. Mar Petrol Geol 9:167–185

Roca E, Guimerà J (1992) The Neogene structure of the eastern Iberian margin: structural constraints on the crustal evolution of the València trough (Western Mediterranean). Tectonophysics 203:203–218

Roca E, Sans M, Cabrera L, Marzo M (1999) Oligocene to Middle Miocene evolution of the Central Catalan margin (North-western Mediterranean). Tectonophysics 315:209–229

Roest WR, Srivastava SP (1991) Kinematics of the plate boundaries between Eurasia, Iberia and Africa in the North Atlantic from the Late Cretaceous to the present. Geology 19:613–616

Rollet N, Déverchère J, Beslier MO, Guennoc P, Réhault JP, Sosson M, Truffert C (2002) Back-arc extension, tectonic inherence, and volcanism in the Ligurian Sea, Western Mediterranean. Tectonics 21(3), 10,1029/2001TC900027

Rosenbaum G, Lister GS, Duboz C (2002) Relative motions of Africa, Iberia and Europe during Alpine orogeny. Tectonophysics 359:117–129

Roure F, Choukroune P, Berástegui X, Muñoz JA, Villien A, Matheron P, Bareyt M, Seguret M, Cámara P, Deramond J (1989) ECORS Deep Seismic data and balanced cross-sections, geometric constraints to trace the evolution of the Pyrenees. Tectonics 8:41–50

Roure F, Howell DG, Guellec S, Casero P (1990) Shallow structures induced by deep-seated thrusting. In: Letouzey F (ed) Petroleum and tectonics in mobile belts. Technip, Paris, pp 15–30

Roure F, Choukroune P (1998) Contribution of the ECORS seismic data to the Pyrenean geology: crustal architecture and geodynamic evolution of the Pyrenees. Mém Soc Géol France 173:37–52

Royden L (1993) The tectonic expression of slab pull at continental convergent boundaries. Tectonics 12:303–325

Saadallah A, Caby R (1996) Alpine extensional detachment tectonics in the Grande Kabylie metamorphic core complex of the Maghrebides (Northern Algeria). Tectonophysics 267:257–274

Sàbat F, Muñoz JA, Santanach P (1988) Transversal and oblique structures at the Serres de Llevant thrust belt (Mallorca Island). Geol Rund 77:529–538

Sàbat F, Roca E, Muñoz JA, Vergés J, Santanach P, Masana E, Sans M, Estévez A, Santisteban C (1997) Role of extension and compression in the evolution of the eastern margin of Iberia: the ESCI-València trough seismic profile. Rev Soc Geol España 8: 431–448

Sandwell DT, Smith WHF (1992) Global Marine Gravity from ERS-1, Geosat and Seasat Reveals New Tectonic Fabric. Scripps Institute of Oceanography, La Jolla, California (map)

Sans M, Sàbat F (1993) Pliocene salt rollers and syn-kinematic sediments in the northeast sector of the València trough (western Mediterranean). Bull Soc Géol France 164, 2:189–198

Santanach P (ed) (1997) ESCI. Estudios Sísmicos de la Corteza Ibérica. Rev. Soc. Geol. España 8 (1995): 1–542

Savostin LA, Sibuet JC, Zonenshain LP, Le Pichon X, Roulet MJ (1986) Kinematic evolution of the Tethys belt from the Atlantic Ocean to the Pamirs since the Triassic. Tectonophysics 123:1–35

Seemann U, Pümpin VF, Casson N (1990) Amposta oil field. In: Beaumont EA, Foster NH (eds) Structural traps II: traps associated with tectonic faulting. AAPG, Tulsa, pp 1–20

Sénéchal G, Thouvenot F (1994) Seismic diffraction from the North Pyrenean Fault: a depth-migrated line-drawing of the ECORS profile. Tectonophysics 233:83–89

Simó A, Ramón X (1986) Análisis sedimentológico y descripción de las secuencias deposicionales del Neógeno postorogénico de Mallorca. Bol Geol Min XCVII:445–472

Solé Sabarís L, Solé Sugrañes L, Calvet J, Pocoví A (1975) Mapa Geológico de España, e. 1 : 50 000. Hoja 417-Espluga de Francolí. Instituto Geológico y Minero de España, Servicio de Publicaciones del Ministerio de Industria y Energía, Madrid, 32 pp

Soto JI, Comas MC, de la Linde J (1996) Espesor de sedimentos en la cuenca de Alborán mediante una conversión sísmica corregida. Geogaceta, 20:382–385

Soule GS, Spratt DA (1996) An échelon geometry and two-dimensional model of a triangle zone, Grease Creek syncline area, Alberta. Bull Can Petrol Geol 44:244–257

Souquet P (1967) Le Crétacé superieur sud-pyreneen en Catalogne, Aragon et Navarre. PhD thesis, Université Toulouse, 529 pp

Souriau A, Pauchet H (1998) A new synthesis of Pyrenean seismicity and its tectonics implications. Tectonophysics 290:221–244

Spakman W, van der Lee S, van der Hilst R (1993) Travel-time tomography of the European-Mediterranean mantle down to 1400 km. Phys Earth Planet Int 79:3–74

Srivastava SP, Tapscott CR (1986) Plate kinematics of the North Atlantic. In: Vogt PR, Tucholke BE (eds) The geology of North America: the Western North Atlantic Region. Geol Soc Am, pp 379–404

Srivastava SP, Roest WR, Kovacs LC, Oakey G, Lévesque S, Verhoef J, Macnab R (1990) Motion of Iberia since the Late Jurassic: Results from detailed aeromagnetic measurements in the Newfoundland Basin. Tectonophysics 184:229–260

Stampfli GM, Mosar J, Favre P, Pillevuit A, Vannay JC (2001) Permo-Mesozoic evolution of the western Tethys realm: the Neo-Tethys East Mediterranean Basin connection. In: Ziegler PA, Cavazza W, Robertson AFH, Crasquin-Soleau S (eds) Peri-Tethyan rift/wrench basins and passive margins. Mém Mus National Hist Nat 186:51–108

Stoeckinger WT (1976) Valencia gulf offer deadline nears. Oil Gas J March 29:197–204

Suriñach E, Marthelot JM, Gallart J, Daignières M, Hirn A (1993) Seismic images and evolution of the Iberian crust in the Pyrenees. Tectonophysics 221:67–80

Tefiani M (1970) Présence d'olistostromes à la base des nappes de flyschs reposant sur la Dorsale Kabyle au sud-est d'Alger. C R Soc Géol France 8:315

Teixell A (1996) The Ansó transect of the southern Pyrenees: basement and cover thrust geometries. Jour Geol Soc London 153: 301–310

Torné M, De Cabissole B, Bayer R, Casas A, Daignières M, Rivero A (1989) Gravity constraints on the deep structure of the Pyrenean belt along the ECORS profile. Tectonophysics 165:105–116

Torne M, Pascal G, Buhl P, Watts AB, Mauffret A (1992) Crustal structure of the Valencia Trough (Western Mediterranean). Part 1. A combined refraction/wide angle reflection and near-vertical reflection study. Tectonophysics 203:1–20

Torné M, Banda E, Fernàndez M (1996) The Valencia Trough: geological and geophysical constraints on basin formation models. In: Ziegler PA, Horvàth F (eds) Structure and prospects of Alpine basins and forelands. Mém Mus National Hist Nat 170: 103–128

Torné M, Fernandez M, Comas MC, Soto JI (2000) Lithospheric structure beneath the Alboran Basin: results from 3D gravity modeling and tectonic relevance. J Geophys Res 105:3209–3228

Torres J, Bois C, Burrus J (1993) Initiation and evolution of the Valencia Trough (western Mediterranean): constraints from deep seismic profiling and subsidence analysis. Tectonophysics 228:57–80

Vacher P, Souriau A. (2001) A three-dimensional model of the Pyrenean deep structure based on gravity modelling, seismic images and petrological constraints. Geophys J Int 145:460–470

Vergés J (1993) Estudi tectònic del vessant sud del Pirineu central i oriental – Evolució cinemàtica en 3D. PhD thesis, Universitat de Barcelona, 203 pp

Vergés J, Burbank DW (1996) Eocene-Oligocene Thrusting and Basin Configuration in the Eastern and Central Pyrenees (Spain). In: Friend P, Dabrio CJ (eds) Tertiary basins of Spain: the stratigraphic record of crustal kinematics. Cambridge University Press, pp 120–133

Vergés J, García-Senz J (2001) Mesozoic evolution and Cenozoic inversion of the Pyrenean Basin. In: Ziegler PA, Cavazza W, Robertson AFH, Crasquin-Soleau S (eds) Peri-Tethyan rift/wrench basins and passive margins. Mém Mus National Hist Nat 186:187–212

Vergés J, Muñoz JA (1990) Thrust sequences in the Southern Central Pyrenees. Bull Soc Géol France 8, VI 265–271

Vergés J, Sàbat F (1999) Constraints on the western Mediterranean kinematic evolution along a 1,000-km transect from Iberia to Africa. In: Durand B, Jolivet L, Horváth F, Séranne M (eds) The Mediterranean basins: Tertiary extension within Alpine orogen. Geol Soc Spec Publ 134:63–80

Vergés J, Millán H, Roca E, Muñoz JA, Marzo M, Cirés J, Den Bezemer T, Zoetemeijer R, Cloetingh S (1995) Eastern Pyrenees and related foreland basins: pre-, syn- and post-collisional crustal-scale cross-sections. Mar Petrol Geol 12:903–915

Vergés J, Fernàndez M, Martínez A (2002) The Pyrenean orogen: pre-, syn-, and post-collisional evolution. J Virtual Explorer 8: 55–84

Vially R, Letouzey J, Bernard F, Haddadi N, Desforges G, Askri H, Boudjema A (1994) A basin inversion along the North African Margin. The Saharan Atlas (Algeria). In: Roure F (ed) Peri-Tethyan platforms. Technip, Paris, pp 79–118

Vidal N (1995) Estructura litosférica en el margen oriental de la Península Ibérica a partir de datos de sísmica de reflexión vertical y de gran ángulo. PhD thesis, Universitat de Barcelona, 287 pp

Vidal N, Gallart J, Dañobeitia J, Díaz J (1995) Mapping the Moho in the Iberian Mediterranean margin by multicoverage processing and merging of wide-angle and near-vertical reflection data. In: Banda E, Torné M, Talwani M (eds) Rifted ocean continent boundaries. NATO ASI Series Series C, Mathematical and Physical Sciences 463:291–308

Vidal N, Gallart J, Dañobeitia JJ (1997) Contribution of the ESCI-València Trough wide-angle data to a crustal transect in the NE Iberian margin. Rev Soc Geol España 8:417–429

Vidal N, Gallart J, Danobeitia JJ (1998) A deep seismic transect from the NE Iberian Peninsula to the Western Mediterranean. J Geophys Res 103:12381–12396

Vielzeuf D, Kornprobst J (1984) Crustal spliting and the emplacement of Pyrenean lherzolites and granulites. Earth Planet Sci Lett 67:383–386

Vila JM (1980) La chaîne alpine d'Algérie orientale et des confins algéro-tunisiens. PhD thesis, Université Paris VI, 2 tomes, 665 pp

Watson HJ (1982) Casablanca field offshore Spain, a paleogeomorphic trap. Am Ass Petrol Geol Mem 32:237–250

Watts AB, Torné M (1992a) Subsidence history, crustal structure, and thermal evolution of the Valencia Trough. A young extensional basin in Western Mediterranean. J Geophys Res 97: 20021–20041

Watts AB, Torné M (1992b) Crustal structure and the mechanical propieties of extended continental lithosphere in the Valencia trough (western Mediterranean). J Geol Soc London 149:813–827

Watts AB, Torné M, Buhl P, Mauffret A, Pascal G, Pinet B (1990) Evidence for reflectors in the lower continental crust before rifting in the Valencia trough. Nature 348:631–635

Wildi W (1983a) La chaîne tello-rifaine (Algérie, Maroc, Tunisie): structure, stratigraphie et évolution du Trias au Miocène. Rev Géogr Phys géol Dyn 24:201–297

Wildi W (1983b) Carte structurale de la chaîne tello-rifaine, scale 1 : 10 000 000. In: Wildi W, La chaîne tello-rifaine (Algérie, Maroc, Tunisie): structure, stratigraphie et évolution du Trias au Miocène. Rev Géogr Phys géol Dyn 24:201–297

Yegorova TP, Starostenko VI, Kozlenko VG, Pavlenkova NI (1997) Three-dimensional gravity modelling of the European Mediterranean lithosphere. Geophys J Int 129:355–367

Yelles-Chaouche AK, Aït-Ouali R, Bracene R, Derder MEM, Djelit H (2001) Chronologie de l'ouverture du bassin des Ksour (Atlas Saharien, Algérie) au début du Mésozoïque. Bull Soc Géol France 172:285–293

Yielding G, Ouyed M, King GCP, Hatfeld D (1989) Active tectonics of the Algerian Atlas Mountains-evidence from aftershocks of the 1980 El Asnam earthquake. Geophys J Int 99:761–788

Zeyen H, Fernàndez M (1994) Integrated lithospheric modelling combining thermal, gravity, and local isostasy analysis: application to the NE Spanish geotransect. J Geophys Res 99:18089–18102

Ziegler PA (1988) Evolution of the Arctic-North Atlantic and Western Tethys. Am Ass Pet Geol Mem 43

Ziegler PA (1990) Geological Atlas of Western and Central Europe. London, Shell Internationale Petroleum Mij., distributed by Geol Soc London Publ House, Bath

Zoetemeijer R, Desegaulx P, Cloetingh S, Roure F, Moretti I (1990) Lithospheric dynamics and tectonic-stratigraphic evolution of the Ebro Basin. J Geophys Res 95:2701–2711

Transect III: Massif Central – Provence – Gulf of Lion – Provençal Basin – Sardinia – Tyrrhenian Basin – Southern Apennines – Apulia – Adriatic Sea – Albanian Dinarides – Balkans – Moesian Platform

Aiello G, De Alteriis G (1991) Il margine adriatico della Puglia: fisiografia ed evoluzione terziaria. Mem Soc Geol It 47:197–212

Alvarez W, Cocozza T, Wezel FC (1974) Fragmentation of the Alpine orogenic belt by microplate dispersal. Nature 248:309–314

Ansorge J, Blundell D, Mueller St (1992) Europe's lithosphere – seismic structure. In: Blundell D, Freeman R, Mueller S (eds) A continent revealed: the European Geotraverse. Cambridge University Press, pp 33–70

Argnani A, Savelli C (1999) Cenozoic volcanism and tectonics in the southern Tyrrhenian Sea: space-time distribution and geodynamic significance. Geodynamics 27:409–432

Argnani A, Bonazzi C, Evangelisti D, Favali P, Frugoni F, Gasperini M, Ligi M, Marani M, Mele G (1996) Tettonica dell'Adriatico meridionale. Mem Soc Geol It 51:227–237

Arsovski M (1997) Tectonics of Macedonia. Special edition of Faculty of Mining and Geology (in Macedonian)

Arthaud F, Matte P (1975) Les décrochements tardi-hercyniens du SW de l'Europe: géométrie et essai de reconstitution des conditions de la déformation. Tectonophysics 25:139–171

Arthaud F, Séguret M (1981) Les structures pyrénéennes du Languedoc et du Golfe du Lion (Sud de la France). Bull Soc Géol France 23:51–63

Arthaud F, Ogier M, Séguret M (1981) Géologie et géophysique du Golfe du Lion et de sa bordure nord. Bull du BRGM 1(3):175–193

Babushka V, Plomerova J, Spasov E (1986) Lithosphere thickness beneath the territory of Bulgaria – a model derived from teleseismic P-residuals. Geologica Balcanica 16:51–54

Babuska V, Plomerová J, Vecsey L, Granet M, Achauer U (2002) Seismic anisotropy of the French Massif Central and predisposition of Cenozoic rifting and volcanism by Variscan suture hidden in the mantlelithosphere. Tectonics 22:10.1029/2001TC901035

Balduzzi A, Casnedi R, Crescenti U, Tonna M (1982a) Il Plio-Pleistocene del sottosuolo del Bacino pugliese (Avanfossa appenninica). Geologica Romana 21:1–28

Balduzzi A, Casnedi R, Crescenti U, Mostardini F, Tonna M (1982b) Il Plio-Pleistocene del sottosuolo del Bacino lucano (Avanfossa appenninica). Geologica Romana 21:89–111

Balia R, Fais S, Klingelé E, Marson I, Porcu A (1991) Aeromagnetic constraints on the geostructural interpretation of the southern part of the Sardinian rift, Italy. Tectonophysics 195:347–358

Barchi M, Minelli G, Pialli G (1998) The CROP 03 profile: a synthesis of results on deep structures of the Northern Apennines. Mem Soc Geol It 52:383–400

Baudrimont AF, Dubois P (1977) Un bassin mésogéen du domaine péri-alpin: le sud-est de la France. Bull Centr Rech Expl Prod, Elf-Aquitaine 1:261–308

Beccaluva L, Brotzu P, Macciotta G, Morbidelli L, Serri G, Traversa G (1989) Cainozoic tectono-magmatic evolution and inferred mantle sources in the Sardo-Tyrrhenian area. In: Boriani A, Bonafede M, Piccardo GB, Vai GB (eds) The lithosphere in Italy. Advances in Earth Science Research, Atti dei Convegni Lincei 80:229–248

Beccaluva L, Bonatti E, Dupuy C et al. (1990) Geochemistry and mineralogy of volcanic rocks from the ODP Sites 650, 651, 655 and 654 in the Tyrrhenian Sea. Proceedings of the ODP, Scientific Results 107:49–74

Beccaluva L, Coltorti M, Premti I, Saccani E, Siena F, Zeda G (1994) Mid-ocean ridge and supra-subduction affinities in ophiolitic belts from Albania. In: Beccaluva L (ed) Albanian ophiolites: state of arts and perspectives. Ofioliti 23:77–96

Benedicto A (1996) Modèles tectono-sédimentaires de bassins en extension et style structural de la marge passive du Golfe du Lion (SE France). Doctorat thesis, Univ Montpellier 2, 242 pp

Benedicto AP, Labaume P, Séguret M, Séranne M (1996), Low-angle crustal ramp and basin geometry in the Gulf of Lion passive margin. The Oligocene-Aquitanian Vistrenque graben, SE France. Tectonics 15:1192–1212

Bernoulli D, Laubscher H (1972) The palinspastic problem in the Hellenides. Eclogae geol Helv 65:107–118

Bigi G, Cosentino D, Parotto M, Sartori R, Scandone P (1990) Structural Model of Italy, scale 1 : 500 000, sheets 3–6, CNR, P.F. Geodinamica, SELCA, Florence

Bilibajkic P, Mladenovic M, Mujagic S, Rimac N (1979) Explanatory note of the gravimetry map of SFRJ-Bouguer anomalies, 1 : 500 000. Savezni Geol. zav., Beograd (in Serbo-Croatian)

Blundell D, Freeman R, Mueller S (1992) A continent revealed: the European Geotraverse. Cambridge University Press, 275 pp

Bonardi G, Amore FO, Ciampo G, De Capoa P, Miconnet P, Perrone V (1988), Il Complesso Liguride auct.: stato delle conoscenze e problemi aperti sulla sua evoluzione pre-appenninica ed i suoi rapporti con l'Arco Calabro. Mem Soc Geol It 41:17–35

Bonatti E, Seyler M, Channell J, Girardeau J, Mascle G (1990) Peridotites drilled from the Tyrrhenian Sea, ODP Leg 107. In: Kastens KA, Mascle J et al. (eds) Proc ODP, Scientific Results 107:37–48

Bonev N, Stampfli GM (2003) New structural and petrologic data on Mesozoic schists in the Rhodope (Bulgaria): geodynamic implications. C R Geoscience 335:691–699

Bortolotti V, Kodra A, Marroni M, Mustafa F, Pandolfi L, Principi G, Saccani E (1996) Geology and Petrology of ophiolitic sequences in the Mirdita region (Northern Albania). Ofioliti 21:3–20

Boyanov I, Dabovski C, Gochev P, Harkovska A, Kostadinov V, Tzankov Tz, Zagorchev I (1989) A new view of the Alpine tectonic evolution of Bulgaria. Geologica Rhodopica 1:107–121

Boykova A (1999) Moho discontinuity in central Balkan Peninsula in the light of the geostatistical structural analysis. Phys Earth Plan Inter 114:49–58

Burg JP, Brun JP, Van den Driessche J (1990) Le Sillon Houiller du Massif Central français : Faille de transfert pendant l'amincissement crustal de la chaine varisque? Comptes rendus Acad Sci Paris 311(II):147–152

Burrus J (1989) Review of geodynamic models for extensional basins; the paradox of stretching in the Gulf of Lions (Northwest Mediterranean). Bull Soc géol Fr V: 377–393

Burrus J, Foucher JP (1986) Contribution to the thermal regime of the Provençal basin based on Flumed heat-flow surveys and previous investigation. Tectonophysics 128:303–334

Carbone S, Lentini F (1990) Migrazione neogenica del sistema catena-avampaese nell'Appennino meridionale: problematiche paleogeografiche e strutturali. Riv It Paleont Strat 96:271–296

Carbone S, Catalano S, Lentini F, Monaco C (1988) Le unità stratigrafico-strutturali dell'Alta Val d'Agri (Appennino Lucano) nel quadro dell'evoluzione del sistema catena-avanfossa. Mem Soc Geol It 41:331–341

Carmignani L, Carosi R, Di Pisa A, Gattiglio M, Musumeci G, Oggiano G, Pertusati PC (1994) The Hercynian chain in Sardinia (Italy). Geodinamica Acta 7:31–47

Carminati E, Doglioni C (2004) Mediterranean geodynamics, Encyclopedia of Geology, in press

Carminati E, Wortel MJR, Spakman W, Sabadini R (1998) The role of slab detachment processes in the opening of the western-central Mediterranean basins: some geological and geophysical evidence. Earth Plan Sci Lett 160:651–665

Carminati E, Giardina F, Doglioni C (2002) Rheological control of subcrustal seismicity in the Apennines subduction zone (Italy). Geophysical Research Letters 29, doi:10.1029/2001GL014084

Carrara G (2002) Evoluzione cinematica neogenica del margine occidentale del bacino Tirrenico. PhD thesis, Parma University, 160 pp

Casero P, Roure F, Endignoux L, Moretti I, Muller C, Sage L, Vially R (1988) Neogene geodynamic evolution of the Southern Apennines. Mem Soc Geol It 41:109–120

Casnedi R, Crescenti U, D'Amato C, Mostardini F, Rossi U (1981) Il Plio-Pleistocene del sottosuolo molisano. The Plio-Pleistocene of Molise. Geologica Romana 20:1–42

Casula G, Cherchi A, Montadert L, Murru M, Sarria E (2001) The Cenozoic graben system of Sardinia (Italy): geodynamic evolution from new seismic and field data. Mar Petr Geol 18:863–888

Catalano R, Di Stefano P, Kozur H (1991) Permian circumpacific deep-water faunas from the western Tethys (Sicily, Italy) – new evidences for the position of the Permian Tethys. In: Channell JET, Winterer EL, Jansa LF (eds) Paleogeography and paleoceanography of Tethys. Palaeogeogr Palaeoclim Palaeoecol 87:75–108

Catalano R, Doglioni C, Merlini S (2001) On the Mesozoic Ionian basin. Geophys J Int 144:49–64

Cella F, Fedi M, Florio G, Rapolla A (1998) Gravity modelling of the litho-asthenosphere system in the Central Mediterranean. Tectonophysics 287:117–138

Chamot-Rooke N, Maillard A, Gaulier JM, Pascal G (1996) Crustal structure of the Liguro-Provençal basin from gravity modelling: geodynamic implications. In: Jolivet L, Mascle A (eds) The Mediterranean basins: Tertiary extension within Alpine orogen. International workshop, session 3

Chamot-Rooke N, Gaulier JM, Jestin F (1999) Constraints on Moho depth and crustal thickness in the Liguro-Provençal basin from a 3D gravity inversion: geodynamic implications. In: Durand B, Jolivet L, Horváth F, Séranne M (eds) The Mediterranean basins: Tertiary extension within the Alpine orogen. Geol Soc London Spec Publ 156:37–68

Channell JET, D'Argenio B, Horvath F (1979) Adria, the African promontory, in Mesozoic Mediterranean paleogeography. Earth Sci Rev 15:213–292

Chantraine J, Autran A, Cavelier C, Alabouvette B, Barfety JC, Cecca F, Clozier L, Debrand-Passard S, Dubreuilh J, Feybesse JL, Guennoc P, Ledru P, Rossi P, Ternet Y (1996) Carte Géologique de la France à l'échelle du millionième, 6eme édition. BRGM, Orléans

Collaku A, Cadet JP, Bonneau M, Jolivet L (1993) L'édifice structurale de l'Albanie septentrionale: des éléments de réponse sur les modalités de la mise en place des ophiolites. Bull Soc Géol France 163:455–468

Committee of Geology (1989–1995) Geological map of Bulgaria on the scale 1 : 100 000. Sheets Kozloduj (eds. G. Cheshitev, L. Filipov), Bjala Slatina (L. Filipov), Vratsa (Tz. Tzankov), Berkovitsa (I. Haydutov, R. Dimitrova), Sofia (S. Yanev), Bosilegrad and Radomir (I. Zagorchev), Kriva Palanka and Kyustendil (I. Zagorchev)

Conti P, Carmignani L, Funedda A (2001) Change of nappe transport direction during the Variscan collisional evolution of central-southern Sardinia (Italy). Tectonophysics 332:255–273

Contrucci I, Nercessian A, Bethoux N, Mauffret A, Pascal G (2001) A Ligurian (Western Mediterranean Sea) geophysical transect revisited. Geophys J Int 146:74–97

Dabovski C, Boyanov I, Khrischev Kh, Nikolov T, Sapounov I, Yanev Y, Zagorchev I (2002) Structure and Alpine evolution of Bulgaria. Geologica Balcanica 32:2–4

Dachev Ch (1988) Structure of the Earth's crust in Bulgaria. Tehnika, Sofia, 334 pp (in Bulgarian)

Dal Piaz GV, Del Moro A, Di Sabatino B, Sartori R, Savelli C (1983) Geologia del Monte Flavio Gioia (Tirreno centrale). Mem Sci Geol Padova 35:429–452

De Alteriis G, Aiello G (1993) Stratigraphy and tectonics offshore of Puglia (Italy, southern Adriatic sea). Mar Geol 113:233–253

Debrand-Passard S, Courbouleix S (1984) Synthèse géologique du Sud-Est de la France. Atlas, Mém n°126 BRGM, Orléans

De Dominicis A, Mazzoldi G (1987) Interpretazione geologico-strutturale del margine orientale della Piattaforma Apula. Mem Soc Geol It 38:163–176

De Gori P, Cimini GP, Chiarabba C, De Natale G, Troise C, Deschamps A (2001) Teleseismic tomography of the Campanian volcanic area and surrounding Apenninic belt. J Volcan Geotherm Res 109:55–75

Della Vedova B, Bellani S, Pellis G, Squarci P (2001) Deep temperatures and surface heat flow distribution. In: Vai GB, Martini IP (eds). Anatomy of an orogen: the Apennines and adjacent Mediterranean basins. Kuwler Academic Publishers, Dordrecht, pp 65–76

Dercourt J, Gaetani M, Vrielynck B, Barrier E, Biju-Duval B, Brunet MF, Cadet JP, Crasquin S, Sandulescu M (2000) Atlas Peri-Tethys, Paleogeographical maps. Commission for the Geological Map of the World, CGMW, Paris

de Voogd B, Nicolich R, Olivet JL, Fanucci F, Burrus J, Mauffret A, Pascal G, Argnani A, Auzende JM, Bernabini M, Bois C, Carmignani L, Fabbri A, Finetti I, Galdeano A, Gorini CY, Labaume P, Lajat D, Patriat P, Pinet B, Ravat J, Ricci Lucchi F., Vernassa S. (1991) First deep seismic reflection transect from the gulf of Lions to Sardinia (Ecors-Crop profiles in Western Mediterranean). In: Union AG (eds) Continental lithosphere: deep seismic reflections, pp 265–274

Dimitrijevic M (1996) Geology of Yugoslavia. Belgrade, 205 pp (in Serbian)

Dimo-Lahitte A, Monié P, Vergély P (2001) Metamorphic soles from the Albanian ophiolites: Petrology ^{40}Ar/^{39}Ar geochronology, and geodynamic evolution. Tectonics 20:78–96

Doglioni C (1991) A proposal for the kinematic modelling of W-dipping subductions; possible applications to the Tyrrhenian-Apennines system. Terra Nova 3:423–434

Doglioni C (1992) Main differences between thrust belts. Terra Nova 4:152–164

Doglioni C, Mongelli F, Pieri P (1994) The Puglia Uplift (SE Italy); an anomaly in the foreland of the Apenninic subduction due to buckling of a thick continental lithosphere. Tectonics 13:1309–1321

Doglioni C, Busatta C, Bolis G, Marianini L, Zanella M (1996a) On the structural evolution of the eastern Balkans (Bulgaria). Mar Petrol Geol 13:225–251

Doglioni C, Harabaglia P, Martinelli G, Mongelli F, Zito G (1996b) A geodynamic model of the Southern Apennines accretionary prism. Terra Nova 8:540–547

Doglioni C, Mongelli F, Pialli GP (1998) Boudinage of the Alpine belt in the Apenninic back-arc. Mem Soc Geol It 52:457–468

Doglioni C, Harabaglia P, Merlini S, Mongelli F, Peccerillo A, Piromallo C (1999a) Orogens and slabs vs their direction of subduction. Earth Sci Rev 45:167–208

Doglioni C, Gueguen E, Harabaglia P, Mongelli F (1999b) On the origin of W-directed subduction zones and applications to the western Mediterranean. Geol Soc London Spec Publ 156:541–561

Dumurdzanov N (1985) Petrogenetic characteristics of the high metamorphic and magmatic rocks of the Central and Western part of Selecka Mountain (Pelagonian massif), SR Macedonia, Yugoslavia. Geologica Macedonica, T.II, Fasc.1

Dumurdzanov N, Petrov G (1990) Lithostratigraphical and chemical characteristics of the Vardar ocean crust in the territory of SR Macedonia. XII Yugoslavia Geological Congres. B.I.Ohrid, R. Macedonia

Dumurdzanov N, Stojanov R, Petrovski K (1979) Explanatory note of the General Geological Map of SFRJ, 1:100 000, sheet Krusevo (in Macedonian, with English and Russian summary)

Duschenes J, Sinha MC, Louden KE (1986) A seismic refraction experiment in the Tyrrhenian Sea. Geophys J R Astron Soc 85:139–160

Echtler H, Malavieille J (1990) Extensional tectonics, basement uplift and Stephano-Permian collapse basin in a late Variscan metamorphic core complex (Montagne Noire, Southern Massif Central). Tectonophysics 177:125–138

Egger A, Demartin M, Ansorge J, Banda E, Maiestrello M (1988) The gross structure of the crust under Corsica and Sardinia. Tectonophysics 150:363–389

Engel W, Feist R, Franke W (1980) Le Carbonifère anté-Stéphanien de la Montagne Noire: Rapports entre mise en place des nappes et sédimentation. Bulletin du BRGM, 2:341–389

EXXON Production Research Company (1994) Tectonic map of the world, panel 10, World Mapping Project, Am Ass Petr Geol Foundation

Fais S, Klingele EE, Lecca L (1996) Oligo-Miocene half graben structure in western Sardinian shelf (western Mediterranean): reflection seismic and aeromagnetic data comparison. Mar Geol 133:203–222

Frasheri A, Nishani P, Bushati S, Hyseni A (1996) Relationship between tectonic zones of the Albanides, based on results of geophysical studies. In: Ziegler PA and Horvath F (eds) Structure and prospects of Alpine basins and forelands. Mém Mus National Hist Nat 170:485–511

Gelati R, Diamanti F., Prence J, Cane H (1997) The stratigraphic record of Neogene events in the Tirana Depression. Riv It Paleont Strat 103:81

Geological map of SFR Yugoslavia (1971), 1:500 000, printed by Federal Geological Department, Belgrade

Georgiev G, Dabovski C, Stanisheva-Vassileva G (2001) East Srednogorie-Balkan Rift Zone. In: Ziegler PA, Cavazza W, Robertson AHF, Crasquin-Soleau S (eds) Peri-Tethyan rift/wrench basins and passive margins. Mem Mus National Hist Nat 186:259–293

Gorini C, Le Marrec A, Mauffret A (1993) Contribution to the structural and sedimentary history of the Gulf of Lions, (Western Mediterranean), from the ECORS profiles, industrial seismic profiles and well data. Bull Soc Géol France 164:353–363

Gorini C, Mauffret A, Guennoc P, Le Marrec A (1994) Structure of the Gulf of Lions (Northwestern Mediterranean sea): a review. In: Mascle A (ed) Hydrocarbon and petroleum geology of France. Europ Assoc Petrol Geol 4:223–243

Graf J (2001) Alpine tectonics in western Bulgaria: Cretaceous compression of the Kraiste region and Cenozoic exhumation of the crystalline Osogovo-Lisec Complex. Dissertation, ETH Zürich, No 14/238, 197 pp

Granet M, Stoll G, Dorel J, Achauer U, Poupinet G, Fuchs K (1995a) Massif Central (France): new constraints on the geodynamical evolution from teleseismic tomography. Geophys J Int 121:33–48

Granet M, Wilson M, Achauer U (1995b) Imaging a mantle plume beneath the French Massif Central. Earth Plan Sci Lett 136:281–296

Gueguen E (1995) La Méditerranée Occidentale: un véritable océan. PhD Dissertation, Université de Bretagne Occidentale, 315 pp

Gueguen E, Doglioni C, Fernandez M (1998) On the post 25 Ma geodynamic evolution of the western Mediterranean. Tectonophysics 298:259–269

Guennoc P, Debeglia N, Le Marrec A, Gorini C, Mauffret A (1994) Anatomie d'une marge passive jeune (Golfe du Lion, sud France): apports des données géophysiques. Bull Cent Rech Explo Prod Elf-Aquitaine 18:19–32

Guennoc P, Gorini C, Mauffret A (2000) Histoire géologique du golfe du Lion et cartographie du rift oligo-aquitanien et de la surface messinienne. Géologie de la France 3:67–97

Guerrera F, Martin-Algarra A, Perrone V (1993) Late Oligocene-Miocene syn-/-late-orogenic succession in Western and Central Mediterranean Chains from the Betic Cordillera to the Southern Apennines. Terra Nova 5:525–544

Harta Gjeologjikee RPS Teshqiperise (1983) Shkalla 1:200 000, 3 sheets. Geological Survey of Albania, Tirana

Haydutov I (2002) Peri-Gondwanan terranes in the pre-Palaeozoic basement of Bulgaria. Geologica Balcanica 32:2–4

Haydoutov I, Gochev P, Kozhoukharov D, Yanev S (1997) Terranes in the Balkan area. In: Papanikolaou D (ed) IGCP Project 276 – terrane map and terrane descriptions. Annales Geol Pays Helleniques, Athens, pp 479–494

Haydutov I, Yanev S (1997) The Protomoesian microcontinent of Balkan peninsula – a peri-Gondwanaland piece. Tectonophysics 272:303–313

Hoxha L (2001) The Jurassic-Cretaceous orogenic event sand its effect in the exploration of sulphide ores, Albanian ophiolites, Albania. Eclog geol Helv 94:339–350

Italiano F, Martelli M, Martinelli G, Nuccio PM (2000) Geochemical evidence of melt intrusions along lithospheric faults of the Southern Apennines, Italy; geodynamic and seismogenic implications. J Geophys Res B105:13,569–13,578

Ivanov T, Misar Z, Bowes DR, Dudek A, Dumurdzanov N, Jaros J, Jelinek E, Pacesova M (1987) The Demir Kapija-Gevgelija ophiolite massif, Macedonia, Yugoslavia. Ofioliti 12:457–478

Jancevski J, Popvasilev V (1984) Explanatory note of the General Geological Map of SFRJ, 1:100 000, sheet Skopje (in Macedonian with English and Russian summary)

Karajovanovic M, Hadzi-Mitrova S (1982) Explanatory notes of the General Geological Map of SFRJ, 1:100 000, sheet Titov Veles (in Macedonian with English and Russian summary)

Karamata E, Dimitrijevic MM, Dimitrijevic MN, Milovanovic D (2000) A correlation of ophiolitic belts and oceanic realms of the Vardar Zone and the Dinarides. Zvornik, pp 191–194

Kastens KA, Mascle J, Auroux C et al. (1987) Proceedings of the Ocean Drilling Program (Part A-Initial Reports, Sites 650–656), College Station, vol 107

Kastens KA, Mascle J et al. (1988) O.D.P., Leg 107 in the Tyrrhenian Sea: insights into passive margin and back-arc basin evolution. Geol Soc Am Bull 100:1140–1156

Kellici I, de Wever P (1994) Ouverture triasique du bassin de la Mirdita (Albanie) révélée par les radiolaires. C R Acad Sc Paris 318:1669–1676

Kodra A, Vergely P, Gjata K, Bakalli F, Godroli M (1993) La formation volcano-sédimentaire du Jurassique superieur: témoin de l'ouverture du domaine ophiolitique dans les Alnaides internes. Bull Soc Géol France 164:61–67

Knott SD (1987) The Liguride complex of southern Italy – a Cretaceous to Paleogene accretionary wedge. Tectonophysics 142:217–226

Le Douaran S, Burrus J, Avedik F (1984) Deep structure of the north-western Mediterranean basin: results of a two-ship seismic survey. Mar Geol 55:325–345

Ledru P, Lardeaux JM, Santallier D, Autran A, Quenardel JM, Floc'h JP, Lerouge G, Maillet N, Marchand J, Ploquin A (1989) Où sont les nappes dans le Massif Central Français. Bull Soc Géol France 8:605–618

Ledru P, Courrioux G, Dallain C, Lardeaux JM, Montel JM, Vanderhaeghe O, Vitel G (2001) The Velay dome (French Massif Central): melt generation and granite emplacement during orogenic evolution. Tectonophysics 342:207–237

Lilov P, Zagorchev I, Peeva I (1983) Rubidium-strontium isochrone data on the age of the metamorphism in the Ograzhdenian complex, Maleshevska Mountain. Geologica Balcanica 13:31–40 (in Russian, with English abstract)

Linnemann U, Romer RL (2002) The Cadomian Orogeny in Saxo-Thuringia, Germany; geochemical and Nd-Sr-Pb isotopic characterization of marginal basins with constraints to geotectonic setting and provenance. Tectonophysics 352:33–64

Maerten L, Séranne M (1995) Extensional tectonics in the Oligo-Miocene Hérault Basin (S. France), Gulf of Lion margin. Bull Soc Géol France 166:739–749

Malinverno A, Ryan WBF (1986) Extension in the Tyrrhenian sea and shortening in the Apennines as a result of arc migration driven by sinking of lithosphere. Tectonics 5:227–245

Marcucci M, Kodra A, Pirdeni A, Gjata Th (1994) Radiolarian assemblage in the Triassic and Jurassic cherts of Albania. Ofioliti 19:101–115

Marsella E, Bally AW, Cippitelli G, D'Argenio B, Pappone G (1995) Tectonic history of the Lagonegro domain and Southern Apennine thrust belt evolution. Tectonophysics 252:307–330

Mascle J, Rehault JP (1990) A revised seismic stratigraphy of the Tyrrhenian Sea: implications for the basin evolution. In: Kastens KA, Mascle J et al. (eds) Proc ODP, Scientific Results, vol 107, pp 617–636

Matte P (1991) Accretionary history and crustal evolution of the Variscan belt in Western Europe. In: Hatcher RD, Zonenshain L (eds) Accretionary tectonics and composite continents. Elsevier Science Publishers, Amsterdam, pp 309–337

Matte P (1998) Continental subduction and exhumation of HP rocks in Paleozoic orogenic belts: Uralides and Variscides, GGF 120:209–222

Matte P (2001) The Variscan collage and orogeny (480–290 Ma) and the tectonic definition of the Armorica microplate: a review. Terra Nova 13:122–128

Mauffret A, Durand de Grossouvre B, Dos Reis AT, Gorini C, Nercessian A (2001) Structural geometry in the eastern Pyrenees and western Gulf of Lion (Western Mediterranean). J Struct Geol 23:1701–1726

Mauffret A, Gorini C (1996) Structural style of the Camargue area and western Provençal basin (southeastern France), geodynamic consequences. Tectonics 15:356–375

Mauffret A, Pascal G, Maillard A, Gorini C (1995) Structure of the deep Northwestern Mediterranean basin. Mar Petrol Geol 12:645–666

Mauritsch HJ, Scholger R, Bushati SL, Ramiz H (1995) Palaeomagnetic results from southern Albania and their significance for the geodynamic evolution of the Dinarides, Albanides and Hellenides. Tectonophysics 242:5–18

Mazzoli S, Corrado S, De Donatis M, Scrocca D, Butler RWH, Di Bucci D, Naso G, Nicolai C, Zucconi V (2000) Time and space variability of "thin skinned" and "thick skinned" thrust tectonics in the Apennines. Rendiconti Lincei Scienze Fisiche e Naturali XI:5–39

Mazzotti AP, Stucchi E, Fradelizio GL, Zanzi L, Scandone P (2000) Seismic exploration in complex terrains; a processing experience in the Southern Apennines. Geophysics 65:1402–1417

Meco S, Aliaj S (2000) Geology of Albania. Gebr. Borntraeger, Berlin

Mele G, Rovelli A, Seber D, Barazangi M (1997) Shear wave attenuation in the lithosphere beneath Italy and surrounding regions; tectonic implications. J Geophys Res 102(B6):11863–11875

Menardi Noguera A, Rea G (2000) Deep structure of the Campanian-Lucanian Arc (Southern Apennine, Italy). Tectonophysics 324:239–265

Merle O, Michon L, Camus G, De Goer A (1998) L'extension Oligocène sur la transversale septentrionale du rift du Massif Central. Bull Soc Géol France 165:615–626

Miconnet P (1988) Evolution mesozoique du secteur de Lagonegro. Mem Soc Geol It 41:321–330

Monaco C, Tortorici L (1995) Tectonic role of ophiolite-bearing terranes in the devolopment of the Southern Apennines orogenic belt. Terra Nova 7:153–160

Monaghan A (2001) Coeval extension, sedimentation and volcanism along the Cainozoic rift system of Sardinia. In: Ziegler PA, Cavazza W, Robertson AHF, Crasquin-Soleen S (eds) Peri-Tethyan rift-wrench basins and passive margins. Mem Mus National Hist Nat 186:707–734

Mostardini F, Merlini S (1986) L'Appennino centro-meridionale. Sezioni geologiche e proposta di modello strutturale. Mem Soc Geol It 35:177–202

Murphy JB, Nance, RD, Keppie JD (2002) West African proximity of the Avalon Terrane in the latest Precambrian; discussion. Geol Soc Am Bull 114:1049–1050

Nance RD, Murphy JB, Keppie JD (2002) Massifs and correlations across the Cadomo-Avalonian orogens. Tectonophysics 352:11–31

Nehlig P (1999) Histoire géologique simplifiée du volcan du Cantal. In: Nehlig P (ed) Volcanismes, tectoniques et sédimentations Cénozoïques. BRGM, Orléans, pp 49–114

Nicolas A, Boudier F (1999) Slow spreading accretion and mantle denudation in the Mirdita ophiolite (Albania). J Geoph Res 104:15155–15167

Nicolich R (1981) Crustal structures in the Italian Peninsula and surrounding seas; a review of DSS data. In: Wezel, FC (ed) Sedimentary basins of Mediterranean margins. Pitagora, Bologna, pp 3–17

Nicolich R, Dal Piaz GV (1990) Moho Isobaths. In: Bigi G, Cosentino D, Parotto M, Sartori R, Scandone P (eds) Structural model of Italy, scale 1:500 000. C.N.R. Progetto Finalizzato Geodinamica

Ogniben L (1969), Schema introduttivo alla geologia del confine calabro-lucano. Mem Soc Geol It 8:453–763

Olivet JL (1996) La cinématique de la plaque ibérique. Bull Centr Rech Explor Prod Elf Aquitaine 20:131–195

Pamic J (2002) The Sava-Vardar Zone of the Dinarides and Hellenides versus the Vardar Ocean. Eclog Geol Helv 95:99–113

Pamic J (2003) The allochthonous fragments of the internal units in the western part of the South Pannonian Basin. Acta Geol Hungarica 46/1:41–62

Pamic J, Balen D (2001) Petrology and geochemistry of Egerian-Eggenburgian and Badenian tholeiite-calc-alkaline volcanics from the South Pannonian Basin (Croatia). N Jb Mineral Abh 176(3):237–267

Pamic J, Gusic J, Jelaska V (1998) Geodynamic evolution of the Central Dinarides. Tectonophysics 297:251–268

Panza GF, Scandone P, Calcagnile G, Mueller St, Suhadolc P (1990) The lithosphere-asthenosphere system in Italy and surrounding regions. In: Bigi G, Cosentino D, Parotto M, Sartori R, Scandone P (eds) Structural model of Italy, scale 1:500 000. C.N.R. Progetto Finalizzato Geodinamica

Papazachos BC (1988) Active tectonics in the Aegean and surrounding area. In: Bonnin J, Cara M, Cisternas A, Fantechi R (eds) Seismic hazard in Mediterranean regions. Proc Summer School, Strasbourg. Kluwer Acad Publ Dordrecht, pp 301–331

Papazachos BC, Comninakis PE (1977) Geotectonic significance of the deep seismic zones in the Aegean area. Thera and the Aegean World, Proc Second Int Scient. Congress

Pascal G, Mauffret A, Patriat P (1993) The ocean-continent boundary in the Gulf of Lion from analysis of expanding spread profiles and gravity modeling. Geophys J Int 113:701–726

Pascal G, Truffert C, Marquis G, Labaume P (1994) ECORS-Gulf of Lion deep seismic reflection profiles revisited: geodynamical implications. 6th Conference, Eur Ass Petrol Geosci Eng, Vienna, p 803

Patacca E, Scandone P (2001) Late thrust propagation and sedimentary response in the thrust-belt-foredeep system of the Southern Apennines (Pliocene-Pleistocene). In: Vai GB, Martini IP (eds) Anatomy of an orogen: the Apennines and the adjacent Mediterranean basins. Kluwer Academic Publishers, Dordrecht, pp 401–440

Patacca E, Sartori R, Scandone P (1990) Tyrrhenian basin and Appeninic Arcs: kinematic relations since late Tortonian times. Mem Soc Geol It 45:425–451

Patacca E, Scandone P, Bellatalla M, Perilli N, Santini U (1992a) The Numidian-sand event in the southern Apennines. Mem Sc Geol 18:297–337

Patacca E, Scandone P, Bellatalla M, Perilli N, Santini U (1992b) La zona di giunzione tra l'arco appenninico settentrionale e l'arco appenninico meridionale nell'Abruzzo e nel Molise. In: Tozzi M, Cavinato GP, Parotto M (eds) Studi preliminari all'acquisizione dati del profilo CROP 11 Civitavecchia-Vasto. Studi Geol Camerti vol spec 1992:417–441

Pescatore T, Renda P, Tramutoli M (1988) Rapporti tra le unità lagonegresi e le unità siciliidi nella media Valle del Basento, Lucania (Appennino meridionale). Mem Soc Geol It 41:353–361

Petkovski P, Ivanovski T (1980) Explanatory note of the General geological map of SFRJ, 1:100 000, sheet Kicevo (in Macedonian with English and Russian summary)

Petkovski P, Popovski S (1982) Explanatoru note of the General geological Map of SFRJ, 1:100 000, sheet Gostivar (in Macedonian with English and Russian summary)

Philip J, Masse PJL, Machour L (1987) L'évolution paléogéographique et structurale du front de chevauchement nord-toulonnais (Basse-Provence occidentale, France). Bull Soc Géol France 8:541–550

Piromallo C, Morelli A (2003) P wave tomography of the mantle under the Alpine-Mediterranean area. J Geophys Res 108(B2):2065, doi:10.1029/2002JB001757

Pollack HN, Hurter SJ, Johnson JR (1993) Heat Flow from the Earth's interior: analysis of the global data set. Rev Geophys 31:267–280

Pontevivo A, Panza GF (2002) Group velocity tomography and regionalization in Italy and bordering areas. Phys Earth Plan Int 134:1–15

Recq M, Rehault JP, Steinmetz L, Fabbri A (1984) Amincissement de la croute et accretion au centre du basin Tyrrhénien d'après la sismique refraction. Marine Geol 85:411–428

Rehault JP, Boillot G, Mauffret A (1984) The western Mediterranean Basin, geological evolution. Mar Geol 55:447–477

Rey P (1993) Seismic and tectono-metamorphic characters of the lower continental crust in Phanerozoic areas: a consequence of post-thickening extension. Tectonics 12:580–590

Ricchetti G, Ciaranfi N, Luperto Sinni E, Mongelli F, Pieri P (1988) Geodinamica ed evoluzione sedimentaria e tettonica dell'avampaese apulo. Mem Soc Geol It 41:57–82

Robertson A, Shallo M (2000) Mesozoic-Tertiary tectonic evolution of Albania in the regional Eastern Mediterranean context. Tectonophysics 316:197–254

Rollet N, Deverchère J, Beslier MO, Guennoc P, Réhault JP, Sosson M, Truffert C (2002) Back-arc extension, tectonic inheritance and volcanism in the Ligurian Sea, Western Mediterranean. Tectonics 21(3), 10.1029:2001TC900027

Roure F, Casero P, Vially R (1991) Growth processes and melange formation in the southern Apennines accretionary wedge. Earth Plan Sci Lett 102:395–412

Roure F, Nazaj S, Mushka K, Fili I, Cadet JP, Bonneau M (2003) Kinematic evolution and Petroleum systems: an appraisal of the outer Albanides. Mem AAPG, in press

Sandwell DT, Smith WHF (1997) Marine gravity anomaly from Geosat and ERS 1 satellite altimetry. J Geophys Res 102(B5):10039–10054

Sapin M, Hirn A (1974) Results of explosion seismology in the southern Rhône valley. Ann Geophys 30:181–202

Sartoni S, Crescenti U (1961) Ricerche biostratigrafiche nel Mesozoico dell'Appennino meridionale. Giorn Geol 29:161–302

Sartori R (1986) Notes on the geology of the acoustic basement in the Tyrrhenian Sea. Mem Soc Geol It 36:99–108

Sartori R, Carrara G, Torelli L, Zitellini N (2001) Neogene evolution of the southwestern Tyrrhenian Sea (Sardinia Basin and western Bathyal Plain). Marine Geol 175:47–66

Sartori R, Torelli L, Zitellini N, Carrara G, Magaldi M, Mussoni P (2003) Crustal characteristics of the Central Tyrrhenian Sea (Mediterranean). Tectonophysics, submitted

Savelli C. (1988) Late Oligocene to Recent episodes of magmatism in and around the Tyrrhenian Sea: implications for the processes of opening in the young inter-arc basin of intra-orogenic (Mediterranean) type. Tectonophysics 146:163–181

Savov I, Ryan J, Haydutov I, Schijf J (2001) Late Precambrian Balkan-Carpathian ophiolite – a slice of the Panafrican ocean crust?: geochemical and tectonic insights from the Cherni Vrah and Deli Jovan massifs, Bulgaria and Serbia. J Volcanol Geoth Res 110:299–318

Scandone P (1967) Studi di geologia lucana: la serie calcareo-silico-marnosa e i suoi rapporti con l'Appennino Calcareo. Boll Soc natur Napoli 76:301–469

Scandone P (1972) Studi di geologia lucana: la carta dei terreni della serie calcareo-silico-marnosa e note illustrative. Boll Soc natur Napoli 81:225–300

Scarascia S, Lozej A, Cassinis R (1994) Crustal structures of the Ligurian, Tyrrhenian and Ionian seas and adjacent onshore areas interpreted from wide-angle seismic profiles. Boll Geofis Teor Appl 36:5–19

Scrocca D, Doglioni C, Innocenti F, Manetti P, Mazzotti A, Bertelli L, Burbi L, D'Offizi S (eds) (2003a) CROP Atlas: seismic reflection profiles of the Italian crust. Memorie Descrittive della Carta Geologica d'Italia 62 (in press)

Scrocca D, Sciamanna S, Di Luzio E, Tozzi M, Nicolai C, Gambini R (2003b) Structural setting along the CROP-04 Deep Seismic Profile (Southern Apennines, Italy). Boll Soc Geol It Vol Spec (in press)

Selli R (1957) Sulla trasgressione del Miocene nell'Italia meridionale. Giorn Geol 26:1–54

Selli R (1962) Il Paleogene nel quadro della geologia dell'Italia meridionale. Mem Soc Geol It 3:733–789

Selvaggi G, Chiarabba C (1995) Seismicity and P-wave velocity image of the Southern Tyrrhenian subduction zone. Geophys J Int 121:818–826

Séranne M (1999) The Gulf of Lion continental margin (NW mediterranean) revisited by IBS: an overview. In: Durand B, Jolivet L, Horváth F, Séranne M (eds) The Mediterranean basins: Tertiary extension within the Alpine orogen. Geol Soc London Spec Publ 156:15–36

Séranne M, Merle O (1999) Cenozoic rifts basins of Western Europe, GeoFrance3D - résultats et perspectives. BRGM, Orleans, pp 112–115

Séranne M, Benedicto A, Labaume P, Truffert K, Pascal G (1995) Structural style and evolution of the Gulf of Lion Oligo-Miocene rifting: role of the Pyrenean Orogeny. Mar Petrol Geol 12:809–820

Séranne M, Camus H, Lucazeau F, Barbarand J, Quinif Y (2002) Surrection et érosion polyphasées de la bordure cévenole – un exemple de morphogenèse lente. Bull Soc Géol France, 173: 97–112

Serri G, Innocenti F, Manetti P (2001) Magmatism from Mesozoic to Present: petrogenesis, time-space distribution and geodynamic implications. In: Vai GB, Martini IP (eds) Anatomy of an orogen: the Apennines and adjacent Mediterranean basins. Kluwer Acad Publ, Dordrecht, pp 77–104

Shallo M (1990) Ophiolitic melange and flyschoidal sediments of the Tithonian-Lower Cretaceous in Albania. Terra Nova 2: 476–483

Shallo M (1992) Geological Evolution of the Albanian ophiolites and their platform periphery. Geol Rund 81:681–694

Smith AG (1993) Tectonic significance of the Hellanic-Dinaric ophiolites. In: Prichard HM et al. (eds) Magmatic processes and plate tectonics. Geol Soc London Spec Publ 76:213–243

Sobolev S, Zeyen H, Stoll H, Werling F, Altherr R, Fuchs K (1996) Upper mantle temperatures from teleseismic tomography of French Massif Central including effects of composition, mineral reactions, anharmonicity, anelasticity and partial melt. Earth Plan Sci Lett 139:147–163

Speranza F, Islami I, Kissel C, Hyseni A (1995) Paleomagnetic evidence for Cenozoic clockwise rotation of the external Albanides. Earth Planet Sci Lett 29:121–134

Speranza F, Villa IM, Sagnotti L, Florindo F, Cosentino D, Cipollari P, Mattei M (2002) Age of the Corsica-Sardinia rotation and Liguro-Provençal Basin spreading: new plaeomagnetic and Ar/Ar evidence. Tectonophysics 347:231–251

Stampfli GM (1996) The Intra-Alpine terrain: a Paleotethyan remnant in the Alpine Variscides. Eclogae geol Helv 89:13–42

Stampfli GM, Borel GD (2002) A plate tectonic model for the Paleozoic and Mesozoic constrained by dynamic plate boundaries and restored synthetic oceanic isochrons. Earth Plan Sci Lett 196:17–33

Stampfli GM, Borel G, Cavazza W, Mosar J, Ziegler PA (2001) The Paleotectonic Atlas of the PeriTethyan Domain. European Geophysical Society, CD-ROM

Stampfli GM, von Raumer J, Borel GD (2002) The Palaeozoic evolution of pre-Variscan terranes: From peri-Gondwana to the Variscan collision. In: Martinez-Catalan JR, Hatcher RD, Arenas R, Diaz Garcia F (eds) Variscan Appalachian dynamics: the building of the Upper Paleozoic basement. Geol Soc Am Special Paper 364:263–280

Stojanov R (1974) Petrological characteristics of magmatic and metamorphic rocks of Prilep area, Macedonia. Bulletin de l'Institut Geologique de la Republique Socialistique Macedonienne, Spec Edit, Skopje (in Macedonian with English summary)

Stojkovic et al. (1976) Explanatory note of the magnetic map of SFRJ, 1:500000

Tanguy de Rémur: Les Fonds de la Méditerranée, scale 1:4250000. Hachette, Guides Bleus, Paris

Tectonic Map of Albania (1999) Three sheets. Ministry of Mineral Resources and Energetics, Tirana

Thomas B, Gennesseaux M, Lecca L (1988) La structure de la marge occidentale de la Sardaigne et la fragmentation de l'île au Cenozoïque. Mar Geol 83:31–41

Vai GB (2001) Basement and Early (Pre-Alpine) History. In: Vai GB, Martini IP (eds) Anatomy of an orogen: the Apennines and adjacent Mediterranean basins. Kuwler Academic Publishers, Dordrecht, pp 65–76

Van Den Driessche J, Brun JP (1992) Tectonic evolution of the Montagne Noire (French Massif Central): a model of extensional gneiss dome. Geodynamica Acta 5(1–2)

Vanderhaeghe O, Burg JP, Teyssier C (1999) Exhumation of migmatites in two collapsed orogens: Canadian Cordillera and French Variscides. In: Ring U, Brandon MT, Lister GS, Willett SD (eds) Exhumation processes: normal faulting, ductile flow and erosion. Geological Society, London, pp 181–204

Vavassis I, De Bono A, Stampfli GM, Giorgis D, Valloton A, Amelin Y (2000) U-Pb and Ar-Ar geochronological data from the Pelagonian basement in Evia (Greece): geodynamic implications for the evolution of Paleotethys. Schweiz Min Petr Mitt 80:21–43

Vigliotti L, Kent DV (1990) Paleomagnetic results of Tertiary sediments from Corsica: evidence of post-Eocene rotation. Phys Earth Plan Int 62:97–108

Vigliotti L, Langenheim VE (1995) When did Sardinia stop rotating? New palaeomagnetic results. Terra Nova 7:424–435

Volvovsky IS, Dachev Ch, Popova OG, Babinets VA (1987) The Earth's crust of Bulgaria from the data of the Petrich-Nikopol wide-angle reflexion deep seismic sounding profile. Mezhduvedomstvenniy geofizicheskiy komitet AN SSSR, Moscow; 112 pp (in Russian)

Volvovsky IS, Starostenko VI (eds) (1996) Geophysical parameters of the lithosphere in the southern sector of the Alpine orogen. Kiev, Naukova dumka, 216 pp (in Russian)

von Raumer J, Stampfli GM, Borel GD, Bussy F (2002) The organisation of pre-Variscan basement areas at the north-Gondwanan margin. Int J Earth Sci 91:35–52

von Raumer J, Stampfli GM, Bussy F (2003) Gondwana-derived microcontinents – the constituents of the Variscan and Alpine collisional orogens. Tectonophysics 365:7–22

Wood AW (1981) Extensional tectonics and the birth of the Lagonegro Basin (southern Italian Apennines). Neues Jahrbuch Geol Palaeont 161:93–131

Wortel MJR, Spakman W (1992) Structure and dynamics of subducted lithosphere in the Mediterranean region. Proc Koninklijke Nederlandse Akademie van Wetenschappen 95:325–347

Wortel MJR, Spakman W (2000) Subduction and slab detachment in the Mediterranean-Carpathian region. Science 290:1910–1917

Yanev S (1993) Gondwana Palaeozoic Terranes in the Alpine Collage System of the Balkans. J Himalayan Geol 4:257–270

Yegorova T.P., Starostenko VI, Kozlenko VG, Pavlenkova NI (1997) Three-dimensional gravity modelling of the European Mediterranean lithosphere. Geophys J Int 129:355–367

Zagorchev I (1992) Neotectonics of the central parts of Balkan Peninsula: basic features and concepts. Geol Rund 81:635–654

Zagorchev I (1993) Multiphase crustal thickening in the central parts of the Balkan Peninsula. Bull Geol Soc Greece 28:87–97

Zagorchev I (1994) Alpine evolution of the pre-Alpine amphibolite-facies basement in South Bulgaria. Mitt Österr Geol Ges 86: 9–21

Zagorchev I (1998) Rhodope controversies. Episodes 21(3):159–166

Zagorchev I (1998) Pre-Triassic sections and units in West Bulgaria. IGCP Project No 276, Newsletter No 6. Spec Publ Geol Soc Greece 3:42–53

Zagorchev I (2001) Geology of SW Bulgaria: an overview. Geologica Balcanica 21:3–52

Zagorchev I, Moorbath S (1986a) Dating of the granitoid magmatism in Sushtinska Srednagora Mts by the Rb-Sr isochrone method. Rev Bulg Geol Soc 47(3):62–68 (in Bulgarian, with English abstract)

Zagorchev I, Moorbath S (1986b) Problems of metamorphism in the central Rhodope Mountains in the light of Sr/Rb isotopic data. Geologica Balcanica 16:61–78 (in Russian, with English abstract)

Transects IV, V and VI: The Alps and Their Forelands

Ansorge J (1968) Die Struktur der Erdkruste an der Westflanke der Zone von Ivrea. Schweiz Mineral Petrogr Mitt 48:247–254

Ansorge J, Blundell D, Müller S (1992) Europe's lithosphere – seismic structure. In: Blundell D, Freeman R, Müller S (eds). A continent revealed: the European Geotraverse. Cambridge University Press, pp 33–70

Antoine P (1971) La zone des Brèches de Tarantaise entre Bourg-Saint-Maurice (Vallée de l'Isère) et la frontière Italo-Suisse. Travaux Laboratoire Géologique Faculté des Sciences, PhD thesis, Université Grenoble

Argand E (1916) Sur l'arc des Alpes Occidentales. Eclogae geol Helv 14:145–204

Arlitt R, Kissling E, Ansorge J (1999) Three-dimensional crustal structure beneath the TOR array and effects on teleseismic wavefronts. Tectonophysics 314:309–319

Babuska V, Plomerova J, Granet M (1990) The deep lithosphere in the Alps: a model inferred from P residuals. Tectonophysics 176: 137–165

Bagnoud A, Wernli R, Sartori M (1998) Découverte de foraminifères planctoniques paléogènes dans la zone de Sion-Courmayeur à Sion. Eclogae geol Helv 91:421–429

Ballèvre M, Merle O. (1993) The Combin Fault: compressional reactivation of a Late Cretaceous-Early Tertiary detachment fault in the Western Alps. Schweiz Mineral Petr Mitt 73:205–227

Baudin T (1987) Etude géologique du massif du Ruitor (Alpes franco-italiennes): Evolution structurale d'un socle briançonnais. PhD thesis, Université de Grenoble, France

Bayer R, Carozzo MT, Lanza R, Miletto M, Rey D (1989) Gravity modelling along the ECORS-CROP vertical seismic reflection profile through the Western Alps. Tectonophysics 162:203–218

Bayer R, Lanza R, Truffert C (1996) Gravimetric and magnetic studies of the Western Alps in the context of the ECORS and ECORS-CROP projects; data and interpretations. Mém Soc géol France 170:61–72

Beaumont C, Fullsack PH, Hamilton J (1994) Styles of crustal deformation in compressional orogens caused by subduction of the underlying lithosphere. Tectonophysics 232:119–132

Berger A (1996) Geological-tectonic map of the Bergell pluton and surrounding country rocks (Southern Switzerland, Northern Italy). Schweiz Mineral Petrogr Mitt 76 (plate)

Berger A, Gieré R (1995) Structural observations at the eastern contact of the Bergell Pluton. Schweiz Mineral Petrogr Mitt 75: 241–258

Berger A, Rosenberg C, Schmid SM (1996) Ascent, emplacement and exhumation of the Bergell pluton within the Southern Steep Belt of the Central Alps. Schweiz Mineral Petrogr Mitt 76:357–382

Bergerat F, Mugnier JL, Guellec S, Truffert C, Cazes M, Damotte B, Roure, F (1990). Extensional tectonics and subsidence of the Bresse basin: an interpretation from ECORS data. In: Roure F, Heitzmann P, Polino R (eds) Deep structure of the Alps. Mém Soc geol Suisse 1:145–156

Bernoulli D, Laubscher H (1972) The palinspastic problem of the Hellenides. Eclogae geol Helv 65(1):107–118

Bernoulli D, Winkler W (1990) Heavy mineral assemblages from Upper Cretaceous South Alpine and Austro-alpine flysch sequences (N-Italy and S-Switzerland): source terranes and paleotectonic implications. Eclogae geol Helv 83(2):287–310

Bernoulli D, Heitzmann P, Zingg A (1990) Central and Southern Alps in southern Switzerland: tectonic evolution and first results of reflection seismics. In: Roure F, Heitzmann P, Polino R (eds) Deep structure of the Alps. Mém Soc geol Suisse 1:289–302

Bigi G, Castellarin A, Coli M, Dal Piaz GV, Sartori R, Scandone P (1990a) Structural model of Italy, sheet 1, 1:500000. Consiglio Nazionale delle Ricerche, Progetto Finalizzato Geodinamica, SELCA Firenze

Bigi G, Castellarin A, Coli M, Dal Piaz GV, Vai GB (1990b) Structural model of Italy, sheet 2, 1:500000. Consiglio Nazionale delle Ricerche, Progetto Finalizzato Geodinamica, SELCA Firenze

Bigi G, Coli M, Cosentino D, Dal Piaz GV, Parotto M, Sartori R, Scandone P (1992) Structural model of Italy, sheet 3, 1:500000. Consiglio Nazionale delle Ricerche, Progetto Finalizzato Geodinamica, SELCA Firenze

Bijwaard H, Spakman W (2000) Non-linear global P wave tomography by iterated linearized inversion. Geophys J Int 141:71–82

Bleibinhaus F (2003) Seismic profiling by the TRANSALP Working Group: refraction and wide-angle reflection seismic travel-time tomography. In: Nicolich R, Polizzi D, Furlani S (eds) Transalp Conference. Memorie di Scienze Geologiche 54:23–26

Blundell D, Freeman R, Müller S (eds) (1992) A continent revealed: the European Geotraverse. Cambridge University Press, 275 pp

Bousquet R, Oberhänsli R, Goffé B, Jolivet L, Vidal O (1998) High pressure-low temperature metamorphism and deformation in the Bündnerschiefer of the Engadine window: Implications for the regional evolution of the eastern Central Alps. J Metam Geol 16:657–674

Bousquet R, Goffé B, Vidal O, Patriat M, Oberhänsli R (2002) The tectono-metamorphic history of the Valaisan domain from the Western to the Central Alps: new constraints for the evolution of the Alps. Geol Soc Am Bulletin 114:207–225

Bouvier M et al. (1972) Blei und Zink in Österreich: Der Bergbau Bleiberg-Kreuth in Kärnten. Verlag des Naturhistorischen Museums Wien, 35 pp

Bucher S, Schmid SM, Bousquet R, Fügenschuh B (2003) Late-stage deformation in a collisional orogen (Western Alps): nappe refolding, back-thrusting or normal faulting? Terra Nova 15:109–117

Bucher S, Ceriani S, Fügenschuh B, Ulardic Ch, Bousquet R, Schmid SM (in press) Tectonic evolution of the Briançonnais units along a transect (ECORS-CROP) through the French-Italian Western Alps. Eclogae geol Helv

Buness H (1992) Collision structures in the crust at the boundaries of the NW Adriatic plate, PhD thesis, Freie Universität Berlin (in German)

Burkhard M, Sommaruga A (1998) Evolution of the western Swiss Molasse Basin: structural relations with the Alps and the Jura belt. In: Mascle C et al. (eds) Cenozoic foreland basins of Western Europe.Geol Soc Spec Publ 134:279–298

Caby R (1996) Low-angle extrusion of high-pressure rocks and the balance between outward and inward displacements of Middle Penninic units in the Western Alps. Eclogae geol Helv 89:229–268

Carminati E, Doglioni C, Argnani A, Carrara G, Dabovski Ch, Dumurdzanov N, Gaetani M, Georgiev G, Mauffret A, Nazai S, Sartori R, Scionti V, Scrocca D, Séranne M, Torelli L, Zagorchev I (this volume) TRANSMED Transect III: Massif Central – Provence – Gulf of Lion – Provençal Basin – Sardinia – Tyrrhenian Basin – Southern Apennines – Apulia – Adriatic Sea – Albanian Dinarides – Balkans – Moesian Platform

Caron Ch, Homewood P, Wildi W (1989) The original Swiss flysch: a reappraisal of the type deposits in the Swiss Prealps. Earth-Science Reviews 26:1–45

Carulli GB, Ponton M (1992) Interpretazione strutturale profonda del settore centrale Carnico-Friulano. Studi Geologici Camerti volume speciale (1992/2) CROP 1-1A: 275–284

Cassano E, Anelli L, Fichera R. Cappelli V (1986) Pianura Padana: Interpretazione integrata di dati geofisici e geologici. 73° Congresso Societa Geologica Italiana, Roma, AGIP, 27 pp

Castellarin A, Dal Piaz GV, Fantoni R, Vai GB, Nicolich R, TRANSALP Working Group (2003). Lower crustal style and models along the southern sector of the Transalp Profile. In: Nicolich R, Polizzi D, Furlani S (eds) Transalp Conference. Memorie di Scienze Geologiche 54:245–248

Cattaneo M, Eva C (1990) Propagation anomalies in Northwestern Italy by inversion of teleseismic residuals. Terra Nova 2:577–584

Ceriani S, Fügenschuh B Schmid SM (2001) Late-stage thrusting at the "Penninic Front" in the Western Alps between Mont Blanc and Pelvoux Massifs. Int J Earth Sciences 90:685–702

Csontos L, Nagymarosy A (1998) The Mid-Hungarian line: a zone of repeated tectonic inversions. Tectonophysics 297:51–71

Csontos L, Nagymarosi A, Horvath F, Kovac M (1992) Tertiary evolution of the Intra-Carpathian area: a model. Tectonophysics 208:221–241

Dal Piaz GV, Martin S (1998) Evoluzione litosferica e magmatismo nel dominio austro-sudalpino dall' orogenesi varisica al rifting permo-mesozoico. Riunione estiva S.G.I., Mem Soc Geol It 53: 43–62

Dal Piaz GV, Cortiana G, Del Moro A, Martin S, Pennacchioni G, Tartarotti P (2001) Tertary age and paleostructural inferences of the eclogitic imprint in the Austroalpine outliers and Zermatt-Saas ophiolite, Western Alps. Int J Earth Sci 90:668–684

Damotte BR, Nicholich M, Cazes M, Guellec S (1990) Mise en oeuvre, traitement et présentation du profil plaine du Po-Massif Central. In: Roure F et al. (eds) Deep structure of the Alps. Mém Soc géol France 156:65–76

Davidson C, Rosenberg C, Schmid SM (1996) Synmagmatic folding of the base of the Bergell pluton, Central Alps. Tectonophysics 265:213–238

Dewey JF Helman ML, Turco E, Hutton DHW Knott SD (1989) Kinematics of the western Mediterranean. In: Coward MP, Dietrich D, Park RG (eds) Alpine tectonics. Geol Soc London Spec Publ 45:265–283

Dèzes P, Schmid SM, Ziegler PA (submitted) Evolution of the European Cenozoic Rift System: interaction of the Alpine and Pyrenean orogens with their foreland lithosphere. Tectonophysics

Dobrin MB, Savit CH (1988) Introduction to Geophysical Prospecting, 4th ed. McGraw-Hill Book Company, 867 pp

Ebbing J, Braitenberg C Götze HJ (2001) Forward and inverse modelling of gravity revealing insight into crustal structures of the Eastern Alps. Tectonophysics 337:191–208

Elter G (1972) Contribution à la connaissance du Briançonnais interne et de la bordure Piémontaise dans les Alpes Graies nord-orientales et considérations sur les rapports entre les zones du Briançonnais et des schistes lustrés, Memorie Istituti di Geologia e Mineralogia Università Padova 28:1–19

Engi M, Berger A, Roselle,G (2001) The role of the tectonic accretion channel in collisional orogeny. Geology 29:1143–1146

Escher A, Masson H, Steck A (1988) Coupes géologiques des Alpes occidentales suisses. Mémoires de Géologie Lausanne 2

Escher A, Hunziker JC, Marthaler M, Masson H, Sartori M, Steck A (1997) Geologic framework and structural evolution of the western Swiss-Italian Alps. In: Pfiffner AO et al. (eds) Deep structure of the Swiss Alps: results from NRP 20. Birkhäuser, Basel, pp 205–221

Etter U (1987) Stratigraphische und strukturgeologische Untersuchungen im Gotthardmassivischen Mesozoikum zwischen dem Lukmanierpass und der Gegend von Ilanz. PhD thesis, Universität Bern

Exner Ch (1990) Erläuterugen zur Geologischen Karte des mittleren Lungaus. Mitt Ges Geol Bergbaustud Österr 36:1–38

Fabre J (1961) Contribution à l'étude de la zone Houllière en Maurienne et en Tarantaise (Alpes de Savoie). Mém Bur Rech Min 2

Faupl P, Wagreich M (1996) Basin analysis of the Gosau Group of the Northern Calcareous Alps (Turonian-Eocene, Eastern Alps). In: Wessely G, Liebl W (eds) Oil and gas in Alpidic thrust belts and basins of Central and Eastern Europe. EAGE Special Publication 5:127–135

Finetti IR, Boccaletti M, Bonini M, Del Ben A, Geletti R, Pipan M, Sani F (2001) Crustal section based on CROP seismic data across the North Tyrrhenian-Northern Apennines-Adriatic Sea. Tectonophysics 343:135–163

Florineth D, Froitzheim N (1994) Transition from continental to oceanic basement in the Tasna nappe (Engadine window, Graubünden, Switzerland): evidence for Early Cretaceous opening of the Valais Ocean. Schweiz Mineral Petrogr Mitt 74:437–448

Fodor L, Jelen B, Marton M, Skaberne D, Car J, Vrabec M (1998) Miocene-Pliocene tectonic evolution of the Slovenian Periadriatic fault: Implications for Alpine-Carpathian extrusion models. Tectonics 17:690–709

Frank W (1987) Evolution of the Austroalpine elements in the Cretaceous. In: Flügel HW, Faupl P (eds) Geodynamics of the Eastern Alps. Deuticke, Vienna, pp 379–406

Frank W, Esterlus M, Frey I, Jung G, Krohe A, Weber J (1983) Die Entwicklungsgeschichte von Stub- und Koralpenkristallin und die Beziehung zum Grazer Paläozoikum. Jber Hochschulschwerp S15/4, Graz: 263–293

Frey M, Desmons J, Neubauer F (1999) Metamorphic Maps of the Alps. CNRS (Paris) Swiss N.S.F. (Berne), BMfWFV and FWF (Vienna)

Frisch W (1975) Hochstegen-Fazies und Grestener Fazies – Ein Vergleich des Jura. N Jb Geol Paläont 2:82–90

Frisch W (1979) Tectonic progradation and plate tectonic evolution of the Alps. Tectonophysics 60:121–139

Frisch W, Gommeringer K, Kelm U, Popp F (1989) The Upper Bündner Schiefer of the Tauern Window – a key to understanding Eoalpine orogenic processes in the Eastern Alps. In: Flügel W, Faupl P (eds) Geodynamics of the Eastern Alps. Deuticke, Vienna, pp 55–69

Frisch W, Kuhlemann J, Dunkl I, Brügel A (1998) Palinspastic reconstruction and topographic evolution of the Eastern Alps during late Tertiary tectonic extrusion. Tectonophysics 297:1–15

Froitzheim N, Eberli GP (1990) Extensional detachment faulting in the evolution of a Tethys passive continental margin, Eastern Alps, Switzerland. Geol Soc Am Bull 102:1297–1308

Froitzheim N, Manatschal G (1996) Kinematics of Jurassic rifting, mantle exhumation, and passive-margin formation in the Austroalpine and Penninic nappes (eastern Switzerland). Geol Soc Am Bull 108:1120–1133

Froitzheim N, Schmid SM, Conti P (1994) Repeated change from crustal shortening to orogen-parallel extension in the Austroalpine units of Graubünden. Eclogae geol Helv 87:559–612

Froitzheim N, Schmid SM, Frey M (1996) Mesozoic paleogeography and the timing of eclogite-facies metamorphism in the Alps: a working hypothesis. Eclogae geol Helv 89:81–110

Froitzheim N, Conti P, Van Daalen M (1997) Late Cretaceous, synorogenic, low-angle normal faulting along the Schlinig fault (Switzerland, Italy, Austria) and its significance for the tectonics of the Eastern Alps. Tectonophysics 280:267–293

Fügenschuh B, Schmid SM (2003) Late stage of deformation and exhumation of an orogen constrained by fission-track data: a case study in the Western Alps. Bull Geol Soc Amer 115:1425–1440

Fügenschuh B, Seward D, Mancktelow N (1997) Exhumation in a convergent orogen: the western Tauern Window. Terra Nova 9: 213–217

Fügenschuh B, Loprieno A, Ceriani S, Schmid SM (1999) Structural analysis of the Subriançonnais and Valais units in the area of Moûtiers (Savoy, Western Alps). Paleogeographic and tectonic consequences. Int Jour Earth Sciences 88:201–218

Gawlick HJ, Frisch W, Vecsei A, Steiger T, Böhm F (1999) The change from rifting to thrusting in the Northern Calcareous Alps as recorded in Jurassic sediments. Int J Earth Sc 87:644–657

Gebauer D (1999) Alpine geochronology of the Central and Western Alps: new constraints for a complex geodynamic evolution. Schweiz Mineral Petr Mitt 79:191–208

Gee DG, Zeyen H (1996) EUROPROBE 1996 – Lithosphere dynamics: origin and evolution of continents. EUROPROBE Secretariate, Uppsala University, Uppsala, Sweden, 138 pp

Genser J, Neubauer F (1989) Low angle normal faults at the eastern margin of the Tauern window (Eastern Alps). Mitt Österr Geol Ges 81:233–243

Giamboni M, Ustaszewski K, Schmid SM, Schumacher ME, Wetzel A (2004) Plio-Pleistocene transpressional reactivation of Paleozoic and Paleogene structures in the Rhine-Bresse Transform Zone (Northern Switzerland and Eastern France). Int Jour Earth Sciences (in press)

Giese P, Buness H (1992) Moho depth. In: Freeman R, Müller S (eds) A continent revealed: the European Geotraverse. Atlas of compiled data, Cambridge University Press, pp 11–13 and Atlas Map 2

Gouffon Y (1993) Géologie de la nappe du Grand Saint-Bernard entre le Val de Bagnes et la frontière Suisse (Vallée d'Aoste, Italie). Mem Géol Lausanne 12:1–147

Guellec S, Mugnier M, Tardy JL, Roure F, (1990) Neogene evolution of the western Alpine foreland in the light of ECORS data and balanced cross section. In: Roure F et al. (eds) Deep structure of the Alps. Mém Soc géol France 156:165–184

Guyoton F (1991) Seismicity and lithospheric structure of the Western Alps. PhD thesis, University of Grenoble (in French)

Haas J, Kovacs S, Krystin L, Lein R (1995) Significance of Late Permian-Triassic facies zones in terrane reconstructions in the Alpine-North Pannonic domain. Tectonophysics 242:19–40

Haas J, Hamor G, Jambor A, Kovacs S, Nagymarosy A, Szenderkenyi T (2001) Geology of Hungary. Eötvös University Press, Budapest, 317 pp

Handy MR (1989) Deformation regimes and the rheological evolution of fault zones in the lithosphere: the effects of pressure, temperature, grain size and time. Tectonophysics 163:119–152

Handy MR, Zingg A (1991) The tectonic and rheological evolution of an attenuated cross section of the continental crust: Ivrea crustal section, Southern Alps, north-western Italy and southern Switzerland. Bull Geol Soc Am 103:236–253

Hirn A, Thouvenot F, Nicolich R, Pellis G, Scarascia S, Tabacco I, Castellano F, Merlanti F (1989) Mapping of the Moho of the Western Alps by wide-angle reflection seismics. Tectonophysics 162:193–202

Hitz L, Pfiffner OA (1994) A 3-D crustal model of the eastern external Aar massif interpreted from a network of deep seismic profiles. Schweiz Mineral Petrogr Mitt 74:405–420

Hofmann T, Mandl G, Peresson, H, Pestal, G, Pistotnik, J, Reitner J, Scharbert S, Schnabel W, Schönlaub HP (2002): Rocky Austria: Eine bunte Erdgeschichte von Östereich. Verlag Geol. Bundesanstalt, Wien, 63 pp

Hoinkes G, Koller F, Rantitsch G, Dachs E, Höck V, Neubauer F, Schuster R (1999) Alpine metamorphism of the Eastern Alps. Schweiz Mineral Petrogr Mitt 79:155–181

Holliger K (1991) Ray-based image reconstruction in controlled-source seismology with an application to seismic reflection and refraction data in the Central Swiss Alps, PhD thesis, ETH Zürich, no 9335, 156 pp

Holliger K, Kissling E (1991) Ray theoretical depth migration: methodology and application to deep seismic reflection data across the eastern and southern Swiss Alps. Eclogae geol Helv 84:369–402

Hurford AJ, Hunziker JC (1989) A revised thermal history for the Gran Paradiso massif. Schweiz Mineral Petrogr Mitt 69: 319–329

Hurford AJ, Stöckhert B, Hunziker JC (1991) Constraints on the late thermotectonic evolution of the Western Alps: evidence for episodic rapid uplift. Tectonics 10:758–769

Kerkhove C (1969) La zone du flysch dans les nappes de l'Embrunais-Ubaye. Géologie Alpine 45:5–204

Kissling E (1980) Krustenaufbau und Isostasie in der Schweiz. PhD thesis, ETH Zürich

Kissling E (1982) Aufbau der Kruste und des oberen Mantels in der Schweiz. Geodätisch-Geophysikalische Arbeiten in der Schweiz. Schweiz Geodätische Kommission 35:37–126

Kissling E (1984) Three-dimensional gravity model of the Northern Ivrea-Verbano zone. In: Wagner JJ, Müller St (eds) Geomagnetic and gravimetric studies of the Ivrea zone. Matér Géol Suisse, Géophys 21:55–61

Kissling E (1993) Deep structure of the Alps – what do we really know? Physics of the Earth and Planetary Interiors 79:87–112

Kissling E, Ansorge J, Baumann M. (1997) Methodological considerations of 3-D crustal structure modeling by 2-D seismic methods. In: Pfiffner OA, Lehner P, Heitzmann P, Müller S, Steck A (eds) Deep structure of the Swiss Alps. Birkhäuser, Basel Boston Berlin, pp 31–38

Kissling E, Schmid SM, Lippitsch R, Ansorge J, Fügenschuh B (submitted) Lithosphere structure and tectonic evolution of the Alpine arc: new evidence from high-resolution teleseismic tomography

Kollmann K (1977) Die Öl- und Erdgasexploration in der Molassezone Oberösterreichs und Salzburgs aus regional-geologischer Sicht. Erdöl-Erdgas-Z. 93. Sonderausgabe, Hammburg/Wien

Kummerow J, Kind R, Oncken O, Wylegalla K, Scherbaum F (2003) Seismic profiling by the TRANSALP working group: receiver function image and Upper Mantle anisotropy. In: Nicolich R, Polizzi D, Furlani S (eds) Transalp Conference. Memorie di Scienze Geologiche 54:27–28

Kurz W, Neubauer F, Dachs E (1998) Eclogite meso- and microfabrics: implications for the burial and exhumation history of eclogites in the Tauern Window (Eastern Alps) from P-T-d paths. Tectonophysics 285:183–209

Lammerer B (1986) Das Autochthon im westlichen Tauernfenster. Jb Geol B-A 129:51–67

Lammerer B, TRANSALP Working Group (2003) The crocodile model and balancing the seismic section. In: Nicolich R, Polizzi D, Furlani S (eds) Transalp Conference. Memorie di Scienze Geologiche 54:243–244

Lammerer B, Weger M (1998) Footwall uplift in an orogenic wedge: the Tauern window in the Eastern Alps of Europe. Tectonophysics 285:213–230

Laubscher HP (1970) Motion and heat in the Alpine orogeny. Schweiz Mineral Petrogr Mitt 50:565–596 (in German)

Laubscher HP (1971) The large-scale kinematics of the Western Alps and the northern Apennines and its palinspastic implications. Am J Science 271:193–226

Laubscher HP (1991) The arc of the Western Alps today. Eclogae geol Helv 84:631–659

Lemcke K (1988) Geologie von Bayern: I. Das bayrische Alpenvorland vor der Eiszeit - Erdgeschichte, Bau, Bodenschätze. E. Schweizerbart'sche Verlagsbuchhandlung, Stuttgart, 175 pp

Lemoine MP, Tricart P, Boillot G (1987) Ultramafic and gabbroic ocean floor of the Ligurian Tethys (Alps, Corsica, Apennines): in search of a genetic model. Geology 15:622–625

Liniger M (1992) Der ostalpin-penninische Grenzbereich im Gebiet der nördlichen Margna-Decke (Graubünden, Schweiz). PhD thesis, ETH Zürich

Lippitsch R (2002) Lithosphere and upper mantle structure beneath the Alps by high-resolution teleseismic tomography. PhD thesis, no 14726, ETH Zürich

Lippitsch R, Kissling E, Ansorge J (2003) Upper mantle structure beneath the Alpine orogen from high-resolution teleseismic tomography. J Geophys Res 108(B8):2376, doi:10.1029/2002JB002016

Litak RK, Marchant RH, Pfiffner OA, Brown LD, Sellami S, Levato L, Wagner JJ, Olivier R (1993) Crustal structure and reflectivity of the Swiss Alps from three-dimensional seismic modelling: 2. Penninic nappes. Tectonics 12:925–935

Loprieno A (2001) A combined structural and sedimentological approach to decipher the evolution of the Valais domain in Savoy, Western Alps. Dissertationen aus dem Geologisch-Paläontologischen Institut der Universität Basel 18, 285 pp

Lüschen E, Gebrande H, Millahn K, Nicolich R. (2003) Seismic profiling by the TRANSALP Working Group: deep crustal vibroseis and explosive seismic profiling. In: Nicolich R, Polizzi D Furlani S (eds) Transalp Conference. Memorie di Scienze Geologiche 54:1–14

Manatschal G, Bernoulli D (1999) Architecture and tectonic evolution of non-volcanic margins: present-day Galicia and ancient Adria. Tectonics 18:1099–1119

Manatschal G, Nievergelt P (1997) A continent-ocean transition recorded in the Err and Platta nappes (Eastern Switzerland). Eclogae Geol Helv 90:3–27

Mandl GW (2000) The Alpine sector of the Tethyan shelf – examples of Triassic to Jurassic sedimentation and deformation from the Northern Calcareous Alps. Mitt Österr Geol Ges 92: 61–77

Mandl GW (2001) Geologie der Dachstein-Region. In: Scheidler A, Boroviczeny F, Graf W, Hofmann T, Mandl G, Schubert G, Stichler W, Trimborn P, Kralik M (eds) Pilotprojekt "Karstwasser Dachstein": Karsthydrologie und Kontaminationsrisiko von Quellen. Archiv für Lagerstättenforschung 21:13–42

Mandl G, Ondrejickova A (1991) Über eine triadische Tiefwasserfazies (Radiolarite, Tonschiefer) in den nördlichen Kalkalpen - ein Vorbericht. Jb Geol B-A 134(2):309–318

Mandl G, Ondrejickova A (1993) Radiolarite und Conodonten aus dem Meliatikum im Ostabschnitt der NKA. Jb Geol B-A 136(4): 841–871

Marchant RH (1993) The Underground of the Western Alps. PhD thesis, Université Lausanne, 137 pp

Marchant RH, Stampfli GM (1997) Crustal and lithospheric structures of the Western Alps: geodynamic significance In: Pfiffner OA, Lehner P, Heitzman PZ, Mueller S, Steck A (eds) Deep structure of the Swiss Alps: results from NRP 20. Birkhäuser, pp 326–337

Miller C, Thöni M (1997) Eo-Alpine eclogitisation of Permian MORB-type gabbros in the Koralpe (Eastern Alps, Austria): new geochronological, geochemical and petrological data. Chemical Geology 137:283–310

Miller H, Gebrande H, Schmedes E (1977) An improved structural model for the Eastern Alps derived from seismic refraction data including the Alpine longitudinal profile. Geol Rdschau 66:289–308 (in German)

Milnes AG (1974) Structure of the Pennine zone (Central Alps): a new working hypothesis. Geol Soc Am Bull 85:1727–1732

Mosar J, Stampfli GM, Girod F (1996) Western Préalpes Médianes Romandes: Timing and structure. A review. Eclogae geol Helv 89:389–425

Müller S (1997) The lithosphere-asthenosphere system of the Alps. In: Pfiffner OA, Lehner P, Heitzmann P, Müller S, Steck A (eds) Deep structure of the Swiss Alps. Birkhäuser, Basel Boston Berlin, pp 338–347

Mugnier JL, Guellec S, Ménard G, Roure F, Tardy M, Vialon P (1990) A crustal scale balanced cross-section through the external Alps deduced from the ECORS profile In: Roure F et al. (eds) Deep Structure of the Alps. Mém Soc géol France 156:203–216

Mugnier JL, Bergerat F , Damotte B Guellec S, Nicolas A, Polino R, Roure F, Tardy M, Truffert C (1996) Crustal structure of the Western Alps and their forelands, Mém Soc géol France 170:73–97

Müntener O, Hermann J, Trommsdorff V (2000) Cooling history and exhumation of lower crustal granulites and upper mantle (Malenco, Eastern Central Alps). J Petrology 41:175–200

Nagel T, de Capitani C, Frey M, Froitzheim N, Stünitz H, Schmid, SM (2002) Structural and metamorphic evolution during rapid exhumation in the Lepontine dome (Southern Simano and Adula nappes, Central Alps, Switzerland). Eclogae geol Helv 95:301–321

Nicolas A, Hirn A, Nicolich R, Polino R, ECORS-CROP Working Group (1990) Lithospheric wedging in the Western Alps inferred from the ECORS-CROP traverse. Geology 18:587–590

Nicolas A, Polino R, Hirn A , Nicolich R, ECORS-CROP Working Group (1996) ECORS-CROP traverse and deep structure of the Western Alps: a synthesis. Mém Soc géol France 170:15–27

Nicolich R, Polizzi D, Furlani S (eds) (2003) Transalp Conference. Memorie di Scienze Geologiche 54, 268 pp

Nievergelt P, Liniger M, Froitzheim N, Ferreiro-Mählmann R (1996) Early Tertiary extension in the Central Alps: The Turba mylonite zone.Tectonics 15:329–340

Nussbaum Ch (2000) Neogene tectonics and thermal maturity of sediments of the easternmost Southern Alps (Friuli are, Italy). Unpublished PhD thesis, Université de Neuchâtel, Switzerland

Oberhänsli, R (1978) Chemische Untersuchungen an Glaukophanführenden basischen Gesteinen aus den Bündnerschiefern Graubündens. Schweiz Mineral Petrogr Mitt 58:139–156

Oberhauser R (1995) Zur Kenntnis der Tektonik und der Paläogeographie des Ostalpenraumes zur Kreide-, Paleozän- und Eocänzeit. Jb Geol B-A 138:369–432

Pamic J (2002) The Sava-Vardar Zone of the Dinarides and Hellenides versus the Vardar Ocean. Eclogae geol Helv 95:99–113

Pfiffner OA (1981) Fold-and-thrust tectonics in the Helvetic nappes (E Switzerland). In: Mc Clay KR, Price NJ (eds) Nappe tectonics. Geol Soc London Spec Publ 9:319–327

Pfiffner OA (1992) Alpine orogeny. In: Blundell D et al. (eds) A continent revealed: the European Geotraverse. Cambridge University Press, pp 180–190

Pfiffner OA, Hitz L (1997) Geologic interpretation of the seismic profiles of the Eastern Traverse (lines E1-E3, E7-E9): Eastern Swiss Alps. In: Pfiffner OA, Lehner P, Heitzmann P, Müller S, Steck A (eds) Deep structure of the Swiss Alps. Birkhäuser, Basel Boston Berlin, pp 73–100

Pfiffner OA, Klaper E, Mayerat AM, Heitzmann P (1990a) Structure of the basement-cover contact in the Swiss Alps. In: Roure F et al. (eds) Deep structure of the Alps. Mém Soc Geol France 156:247–262

Pfiffner OA, Frei W, Valasek P, Stäuble M, Levato L, Dubois L, Schmid SM, Smithson SB (1990b) Crustal shortening in the Alpine orogen: results from deep seismic reflection profiling in the Eastern Swiss Alps, line NFP 20-East. Tectonics 9: 327–1355

Pfiffner OA, Lehner P, Heitzmann P, Müller S, Steck A (eds) (1997) Deep Structure of the Swiss Alps: results from NRP 20. Birkhäuser, Basel Boston Berlin, 380 pp

Philippe Y, Colletta B, Deville E, Mascle A, (1996) The Jura fold-and-thrust belt: a kinematic model based on map-balancing. In: Ziegler PA, Horvàth F (eds) Structure and prospects of Alpine basins and forelands. Mém Mus National Hist Nat 170:235–261

Pieri M, Groppi G (1981) Subsurface geological structure of the Po Plain, Italy. Pubblicazione n° 414 del Progetto Finalizzato Geodinamica, Consiglio Nazionale delle Ricerche, AGIP, 13 pp

Piromallo C, Morelli A (2003) P wave tomography of the mantle under the Alpine Mediterranean area. J Geophys Res 108(B2): 2065

Plasienka D, Putis M, Kovac M, Sefara, Hrusecky J (1997) Zones of Alpidic subduction and crustal underthrusting in the Western Carpathians. In: Grecula P et al. (eds) Geological evolution of the Western Carpathians. Mineralia Slovaca-Monograph, Bratislava, pp 35–42

Prey S (1980) Die Geologie Österreichs in ihrem heutigen geodynamischen Entwicklungsstand sowie die geologischen Bauteile und ihre Zusammenhänge. In: Oberhauser R (eds) Der geologische Aufbau Österreichs. Springer, Wien: 8–117

Probst Ph (1980) Die Bündnerschiefer des nördlichen Penninikums zwischen Valser Tal und Passo di San Giacomo. Beitr geol Karte Schweiz NF 153

Ranalli G, Murphy DC (1987) Rheological stratification of the lithosphere. Tectonophysics 132:281–295

Ratschbacher L, Frisch W, Neubauer F, Schmid SM, Neugebauer J (1989) Extension in compressional orogenic belts: the Eastern Alps. Geology 17:404–407

Ratschbacher L, Frisch W, Linzer HG, Merle O (1991) Lateral extrusion in the Eastern Alps, part 2: Structural analysis. Tectonics 10:257–271

Ricou LE, Siddans AWB (1986) Collision tectonics in the Western Alps. In: Coward MP Ries AC (eds) Collision tectonics. Geological Soc Spec Publ 19:229–244

Roeder D, Bachmann G (1996) Evolution structure and petroleum geology of the German Molasse Basin. In: Ziegler PA, Horvàth F (eds) Structure and prospects of Alpine basins and forelands. Peri-Tethys Memoire 2. Mém Mus National Hist Nat 170: 263–284

Rosenberg CL, Berger A, Schmid SM (1995) Observations from the floor of a granitoid pluton: inferences on the driving force of final emplacement. Geology 23:443–446

Roure F, Polino R, Nicolich R (1990) Early Neogene deformation beneath the Po Plain, constraints on the post-collisional Alpine evolution. In: Roure F, Heitzmann P, Polino R (eds) Deep structure of the Alps. Mém Soc géol Suisse 1:309–322

Roure F, Choukroune P, Polino R (1996a) Deep seismic reflection data and new insights on the bulk geometry of mountain ranges. C. R. Acad. Sci. Paris 322 série IIa: 345–359

Roure F, Bergerat F, Damotte B, Mugnier JL, Polino R (eds) (1996b) The ECORS-CROP Alpine seismic traverse. Mém Soc Géol France 170, 113 pp

Roure F, Heitzmann P, Polino R (eds) Deep structure of the Alps. Mém Soc geol Suisse 1

Sandulescu M (1984) Geotectonica Romaniei. Editura Tehnica, Bucharest, 450 pp

Sandulescu M (1994) Overview on Romanian geology. ALCAPA II Field Guidebook, Supplement 2, Rom J Tectonics Reg Geol 75: 3–15

Sartori M (1990) L'unité du Barrhorn (Zone pennique, Valais, Suisse). Mém Géol (Lausanne) 6

Schmid SM (1975) The Glarus overthrust: field evidence and mechanical model. Eclogae geol Helv 68:251–284

Schmid SM (1993) Ivrea Zone and adjacent Southern Alpine basement. In: von Raumer JF, Neubauer F (eds) Pre-Mesozoic geology of the Alps. Springer, pp 567–583

Schmid SM (2000) Regional tectonics: from the Rhine graben to the Po Plain, a summary of the tectonic evolution of the Alps and their forelands. (http://comp1.geol.unibas.ch/groups/3_1/schmid/Text_Schmid.html)

Schmid SM, Froitzheim N (1994) Oblique slip and block rotation along the Engadine line. Eclogae geol Helv 86:569–593

Schmid SM, Haas R (1989) Transition from near-surface thrusting to intrabasement decollement, Schlinig Thrust, Eastern Alps. Tectonics 8:697–718

Schmid SM, Kissling E (2000) The arc of the Western Alps in the light of geophysical data on deep crustal structure. Tectonics 19/1:62–85

Schmid SM, Zingg A, Handy M (1987) The kinematics of movements along the Insubric line and the emplacement of the Ivrea zone. Tectonophysics 135:47–66

Schmid SM, Aebli HR, Heller F, Zingg A (1989) The role of the Periadriatic Line in the tectonic evolution of the Alps. In: Coward MP et al. (eds) Alpine tectonics. Geol Soc London Spec Publ 45:153–171

Schmid SM, Rück Ph, Schreurs G (1990) The significance of the Schams nappes for the reconstruction of the paleotectonic and orogenic evolution of the Pennine zone along the NFP 20 East traverse (Grisons, eastern Switzerland). Mém soc géol France 156:263–287

Schmid, SM, Pfiffner OA, Froitzheim N, Schönborn G, Kissling E (1996a) Geophysical-geological transect and tectonic evolution of the Swiss-Italian Alps. Tectonics 15:1036–1064

Schmid SM, Berger A, Davidson C, Gieré R, Hermann J, Nievergelt P, Puschnig A, Rosenberg C (1996b) The Bergell pluton (Southern Switzerland, Northern Italy): overview accompanying a geological-tectonic map of the intrusion and surrounding country rocks. Schweiz Mineral Petrogr Mitt 76:329–355

Schmid SM, Pfiffner OA, Schreurs G (1997) Rifting and collision in the Penninic zone of eastern Switzerland. In: Pfiffner AO et al. (eds) Deep structure of the Swiss Alps: results from NRP 20. Birkhäuser, pp 60–185

Schmid SM, Fügenschuh B, Lippitsch R (2003) The Western Alps-Eastern Alps transition: tectonics and deep structure. In: Nicolich R, Polizzi D Furlani S (eds) Transalp Conference. Memorie di Scienze Geologiche 54:257–260

Schmid SM, Fügenschuh B, Schuster R (in press) Tectonic map and overall architecture of the Alpine orogen. Eclogae geol Helv

Schönborn G (1992) Alpine tectonics and kinematic models of the central Southern Alps. Memorie di Scienze Geologiche 44: 229–393

Schönborn G (1999) Balancing cross sections with kinematic constraints: The Dolomites (northern Italy). Tectonics 18:527–545

Schreurs G (1993) Structural analysis of the Schamps nappes and adjacent tectonic units: implications for the orogenic evolution of the Penninic zone in eastern Switzerland. Bull Soc géol France 164:415–435

Schreurs G (1995) Structural analysis of the Schams nappes and adjacent tectonic units in the Pennine zone. In: The Schams nappes, part 2. Beitr Geol Karte Schweiz NF 167

Schumacher ME, Laubscher HP (1996) 3D crustal structure of the Alps-Apennines join – a new view on seismic data. Tectonophysics 260: 349–363

Schuster R (2003) Das Eo-Alpidische Ereignis in den Ostalpen: Plattentektonische Situation und interne Struktur des Ostalpinen Kristallins. Arbeitstagung Geol B-A 2003:141–159

Schuster R, Frank W (1999) Metamorphic evolution of the Austroalpine units east of the Tauern window: indications for Jurassic strike slip tectonics. Mitt Ges Geol Bergbaustud Österr 42: 37–58

Schuster R, Scharbert S, Abart R, Frank W (2001) Permo-Triassic extension and related HT/LP metamorphism in the Austroalpine-Southalpine realm. Mitt Ges Geol Bergbaustud Österr 45: 111–141

Sénéchal G, Thouvenot F (1991) Geometrical migration of line drawings: a simplified method apllied to ECORS data. In: Meissner R et al. (eds) Continental lithosphere: deep seismic reflection. Amer Geophys Union, Geodyn Ser 22:401–408

Sinclair HD, Allen PA (1992) Vertical versus horizontal motions in the Alpine orogenic wedge: stratigraphic response in the foreland basin. Basin Research 4:215–232

Slapansky P, Frank W (1987) Structural evolution and geochronology of the northern margin of the Austroalpine in the northwestern Schladming crystalline (NE Radstädter Tauern). In: Flügel HW, Faupl, P (eds) Geodynamics of the Eastern Alps. Deuticke, Wien, pp 244–262

Solarino S, Kissling E, Sellami S, Smriglio G, Thouvenot F, Granet M, Bonjer K, Slejiko D (1997) Compilation of a recent seismicity data base of the greater Alpine region from several seismological networks and preliminary 3D tomographic results. Annali di Geofisica 11:161–174

Spakman W (1991) Delay-time tomography of the upper mantle below Europe, the Mediterranean and Asia Minor. Geophys J Int 107:309–332

Spakman W, vander Lee S, vander Hilst R (1993) Traveltime tomography of the European-Mediterranean mantle down to 1400 km. Physics of the Earth and Planetary Interiors 79: 3–74

Spillmann P (1993) Die Geologie des penninisch-ostalpinen Grenzbereichs im südlichen Berninagebirge. PhD thesis, ETH Zürich, Nr. 10175

Stäuble M, Pfiffner OA (1991a) Processing, interpretation and modeling of seismic reflection data in the Molasse Basin of eastern Switzerland, Eclogae geol Helv 84:151–175

Stäuble M, Pfiffner OA (1991b) Evaluation of the seismic response of basement fold-and-thrust geometry in the Central Alps based on 2-D ray tracing. Ann Tectonicae 5:3–17

Stäuble M, Pfiffner OA, Smithson SB (1993) Crustal structure and reflectivity of the Swiss Alps from three-dimensional seismic modelling (1) Helvetic nappes. Tectonics 12:911–924

Stampfli G (1993) Le Briançonnais: Terrain exotique dans les Alpes? Eclogae geol Helv 86:1–45

Stampfli G (2000) Tethyan Oceans. In: Winchester JA et al. (eds) Tectonics and magmatism in Turkey and the surrounding area. Geol Soc London Spec Publ 173:1–23

Stampfli G, Marchant R (1997) Geodynamic evolution of the Tethyan margins of the Western Alps. In: Pfiffner OA, Lehner P, Heitzmann P, Müller S, Steck A (eds) Deep structure of the Swiss Alps. Birkhäuser, Basel Boston Berlin, pp 223–239

Stampfli GM, Mosar J, Marchant R, Marquer D, Baudin T, Borel G (1998) Subduction and obduction processes in the western Alps In: Vauchez A, Meissner R (eds) Continents and their mantle roots. Tectonophysics 296(1–2):159–204

Stampfli G, Mosar J, Favre Ph, Pillevuit A, Vannay JC (2001a) Permo-Mesozoic evolution of the western Tethys realm: the Neo-Tethys East Mediterranean Basin connection. In: Ziegler PA et al. (eds) Peri-Tethys memoir 6: Peri-Tethyan rift/wrench basins and passive margins. Mém Mus National Hist Nat 186:51–108

Stampfli GM, Borel G, Cavazza W, Mosar J, Ziegler PA (2001b) The paleotectonic atlas of the Peritethyan domain (CD-ROM). European Geophysical Society

Stampfli GM, Borel G, Marchant R, Mosar J (2002) Western Alps geological constraints on western Tethyan reconstructions. Journal Virtual Explorer 8:77–106

Steck A (1984) Structures de déformation tertiaires dans les Alpes centrales (transversale Aar-Simplon-Ossola). Eclogae geol Helv 77:55–100

Steck A (1990) Une carte des zones de cisaillement ductile des Alpes Centrales. Eclogae geol Helv 83:603–626

Steinmann, M (1994) Ein Beckenmodell für das Nordpenninikum der Ostschweiz. Jahrbuch der Geologischen Bundesanstalt 137: 675–721

Stipp M, Fügenschuh B, Gromet, LP, Stünitz H, Schmid SM (in press) Contemporaneous plutonism and strike-slip faulting along the easternmost segment of the Insubric line: the Tonale fault zone north of the Adamello pluton (Italian Alps). Tectonics

Sue C, Thouvenot F, Frechet J, Tricart P (1999) Widespread extension in the core of the Western Alps revealed by earthquake analysis. J Geophys Res 104(B11):25611–25622

Suhadolc P, Panza GF, Müller St. (1990) Physical properties of the lithosphere-asthenosphere system in Europe. Tectonophysics 176:123–135

Thöni M (1999) A review of geochronological data from the Eastern Alps. Schweiz Mineral Petr Mitt 79:209–230

Thouvenot F, Paul A, Sénéchal G, Hirn A, Nicolich, R (1990) ECORS-CROP wide-angle reflection seismics: constraints on deep intrefaces beneath the Alps. In: Roure F, Heitzmann P, Polino R (eds) Deep structure of the Alps. Mém Soc géol Suisse 1:97–106

Thouvenot F, Senechal G, Truffert C, Guellec S (1996) Comparison between two techniques of line-drawing migration (ray tracing and common tangent method). Mém Soc géol France 170:53–59

Tollmann A (1977) Geologie von Österreich, Band 1: Die Zentralalpen. Deuticke, Wien, 766 pp

Tomljenovic B (2002) Strukturne Znacajke Medvednice i Samoborskoj gorja. PhD thesis, University of Zagreb, 208 pp

TRANSALP Working Group (2001) European orogenic processes research transects the Eastern Alps. EOS Transactions of the American Geophysical Union 82(40):453, 460–461

TRANSALP Working Group (2002) First deep seismic reflection images of the Eastern Alps reveal giant crustal wedges and transcrustal ramps. Geophys Res Letters 29(10):92-1–92-4

Trommsdorff V, Piccardo GB, Montrasio A (1993) From magmatism through metamorphism to sea floor emplacement of subcontinental Adria lithosphere during pre-Alpine rifting (Malenco, Italy). Schweiz Mineral Petrogr Mitt 73:191–203

Trümpy R (1955) Remarques sur la corrélation des unités penniques externes antre la Savoie et le Valais et sur l'origine des nappes préalpines. Bull Soc géol France 6(5):217–231

Trümpy R (1960) Paleotectonic evolution of the Central and Western Alps. Bull Geol Soc America 71:843–908

Trümpy R (1992) Ostalpen und Westalpen – Verbindendes und Trennendes. Jb Geol B-A 135:875–882

Trümpy R (1999) Die tektonischen Grosseinheiten im Bereich AlpTransit. In: Löw S, Wyss R (eds) Vorerkundung und Prognose der Basistunnels am Gotthard und am Lötschberg. Balkema, Rotterdam, pp 21–29

Valasek P (1992) The tectonic structure of the Swiss Alpine crust interpreted from a 2D network of deep crustal seismic profiles and an evaluation of 3-D effects. PhD thesis, ETH Zürich No 9637, 195 pp and appendix

Valasek P, Müller S (1997) A 3D tectonic model of the Central Alps based on an integrated interpretation of seismic refraction and NRP20 reflection data. In: Pfiffner OA, Lehner P, Heitzmann P, Müller S, Steck A (eds) Deep structure of the Swiss Alps. Birkhäuser, Basel Boston Berlin, pp 305–325

Valasek P, Müller S, Frei W, Holliger K. (1991) Results of NFP20 seismic reflection profiling along the Alpine section of the European Geotraverse (EGT). Geophys J Int 105:85–102

Vearncombe, J.R. (1985) The structure of the Gran Paradiso basement massif and its enveloppe, Western Alps. Eclogae geol Helv 78:49–72

von Blanckenburg F, Davies, JH (1995) Slab breakoff: A model for syncollisional magmatism and tectonics in the Alps. Tectonics 14:120–131

Viel L, Berckhemer H, Müller S (1991) Some structural features of the Alpine lithospheric root. Tectonophysics 195:421–436

Wagner LR (1996) Stratigraphy and hydrocarbons in the Upper Austrian Molasse Foredeep (active margin). In: Wessely G, Liebel W (eds) Oil and gas in the Alpidic thrustbelt and basins of Central and Eastern Europe. EAGE Spec Publ 5:217–235

Wagreich M (2001) A 400-km-long piggyback basin (Upper Aptian-Lower Cenomanian) in the Eastern Alps. Terra Nova 13:401–406

Waldhauser F, Kissling E, Ansorge J, Müller S (1998) Three-dimensional interface modelling with two-dimensional seismic data: the Alpine crust-mantle boundary. Geophys J Int 135:264–278

Waldhauser F, Lippitsch R, Kissling E, Ansorge J (2002) High-resolution teleseismic tomography of upper-mantle structure using an a priori three-dimensional crustal model. Geophys J Int 150:403–414

Weh M, Froitzheim N (2000) Penninic cover nappes in the Prättigau half-window (Eastern Switzerland): Structure and tectonic evolution. Eclogae geol Helv 94:237–252

Wenk HR (1973): The structure of the Bergell Alps. Eclogae geol Helv 66:255–291

Wortel MJR, Spakman W (2002) Subduction and slab detachment in the Mediterranean-Carpathian region. Science 290:1910–1917

Ye S, Ansorge J, Kissling E, Müller S (1995) Crustal structure beneath the Swiss Alps derived from seismic refraction data. Tectonophysics 242:199–221

Ziegler PA (1990) Geological Atlas of Western and Central Europe, 2nd ed. Shell Internat Petrol Mij, dist Geol Soc Publ House, Bath, 239 pp, 56 enclosures

Ziegler PA, Dèzes P (2004) Crustal configuration of Western and Central Europe. In: Gee D, Stephenson R (eds) European lithosphere dynamics. Geol Soc London Memoir. See also Moho map available under http://www.unibas.ch/eucor-urgent

Ziegler PA, Stampfli GM (2001) Late Paleozoic Early Mesozoic plate boundary reorganisation: collapse of the Variscan orogen and opening of Neotethys. In: Cassinis R (ed) The continental Permian of the Southern Alps and Sardinia (Italy): regional reports and general correlations. Annali Museo Civico Science Naturali, Brescia 25:17–34

Ziegler PA, Bertotti G, Cloetingh S (2002) Dynamic processes controlling foreland development - the role of mechanical (de)coupling of orogenic wedges and forelands. In: Bertotti G, Schulmann K, Cloetingh S (eds) Continental collision and the tectono-sedimentary evolution of forelands. EGU Stephan Müller Special Publications Series 1:17–56

Zimmer W, Wessely G (1996) Hydrocarbon exploration in the Austrian Alps. In: Ziegler PA, Horvàth F (eds) Structure and prospects of Alpine basins and forelands. Mém Mus National Hist Nat 170:285–304

Transect VII: East European Craton – Scythian Platform – Dobrogea – Balkanides – Rhodope Massif – Hellenides – East Mediterranean – Cyrenaica

Altherr R, Kreuzer H, Wendt I, Lenz H, Wagner G, Keller J, Harre W, Hohndorf A (1982) A Late Oligocene/Early Miocene high-temperature belt in the Attic-Cycladic Crystalline Complex (SE Pelagonian, Greece). Geol Jb E 23:97–164

Andriessen PAM, Banga G, Helbeda EH (1987) Isotopic age study of pre-Alpine rocks in the basal units on Naxos, Sikinos and Ios, Greek Cyclades. Geol Mijnbouw 66:3–14

Antonescu E, Baltres A (1998) Palynostratigraphie de la Formation de Nalbant (Triasique-Jurassique) de la Dobrogea de nord et des formations jurassiques du sous-sol du Delta du Danube (Plateforme Scythienne). Geo-Eco-Marina 3: 159–188

Ardaens R, Colin J, Kozur H (1979) Sur la présence du Scythien supérieur fossilifère dans la chaîne du Vardousia (Grèce continentale). Conséquences paléogéographiques. CR Somm Soc Geol France, pp 132–135

Atanassov A, Bokov P, Georgiev G, Monakhov I (1984) Basic geological features of North Bulgaria relating to the oil and gas potential. Proc Res Inst Min Res 1:29–41 (in Bulgarian)

Aubouin J (1959) Contribution a l'étude géologique de la Grèce septentrionale: les confins de l'Epire et de la Thessalie. Ann Géol Pays Hellen 10:1–483

Aubouin J (1965) Geosynclines. Develop Geotectonics, 1, Elsevier, Amsterdam

Aubouin J (1977) Alpine tectonics and plate tectonics: thoughts about the Eastern Mediterranean. In: Ager DV, Brook M (eds) Europe from crust to core. Wiley, New York, pp 143–158

Aubouin J, Le Pichon X, Winterer E, Bonneau M (1977) Les Hellenides dans l'optiques de le tectonique des plaques. 6th Colloq Geol Aegean Region, Athens, 3:1333–1354

Avram E, Szasz L, Antonescu E, Baltres A, Iva M, Melinte M, Neagu T, Rădan S, Tomescu C (1993) Cretaceous terrestrial and shallow marine deposits in northern South Dobrogea (SE Rumania). Cret Res 14:265–305

Avram E, Costea I, Dragastan O, Mutiu R, Neagu T, Sindilar V, Vinogradov C (1996) Distribution of the Middle-Upper Jurassic and Cretaceous facies in the Romanian Eastern Part of the Moesian Platform. Rev Roum Geologie 39–40:3–33

Babushka V, Plomerova J, Spasov E (1986) Lithosphere thickness beneath the territory of Bulgaria – a model derived from teleseismic P-residuals. Geol Balc 16:51–54

Baltres A (1993) Somova Formation (North Dobrogea) sedimentological study. Unpublished PhD thesis, Univ. Bucharest

Barbu C, Ali Mehmed N, Paraschiv C (1969) The Paleozoic in the East Carpathian foreland between Buzaului Valley and the northern border of Romania. Petrol si Gaze XX(12):863–868 (in Romanian)

Baud A, Jenny C, Papanikolaou D, Sideris C, Stampfli G (1990) New observations on Permian stratigraphy in Greece and geodynamic interpretation. Bull Geol Soc Greece 25/1:187–206

Baumgartner PO (1985) Jurassic sedimentary evolution and nappe emplacement in the Argolis Peninsula (Peloponesus, Greece). Birkhäuser, Basel

Bebien J (1982) L'association ignée de Guevgueli (Macédoine, Grèce). Expression d'un magmatism ophiolitique dans une déchirure continentale. PhD thesis, Université Nancy

Bebien J, Mercier J (1977) Le cadre structural de l'association ophiolites-migmatites-granites de Guevqueli. (Macédoine, Grèce) – une croûte de bassin interarc. Bull Soc Geol France 19:927–934

Beccaletto L (2003) Geology, correlations, and geodynamic evolution of the Biga peninsula (NW Turkey). PhD thesis, Univ. Lausanne

Besenecker H, Dürr St, Herget G, Jacobshagen V, Kauffmann G, Lüdke G, Roth W, Tietze KW (1968) Geologie von Chios (Ägäis). Geologica et Paleontologica 2:121–150

Besutiu L (1997) Contributii la elaborarea unui model al zonei de tranzitie de la uscatul nord-dobrogean la platoul continental al marii negre, pe baza interpretarii datelor geofizice cu ajutorul modelelor de simulare. Unpubl PhD thesis, Bucharest Univ (in Romanian)

Biju-Duval B, Dercourt J, Le-Pichon X (1976) From the Tethys ocean to the Mediterranean seas: A plate tectonic model of the evolution of the Western Alpine system. In: Biju-Duval B, Montadert L (eds) Structural history of the Mediterranean basins. Ed Technip, Paris, pp 143–164

Boccaletti M (1979) Mesogea and Mesoparatethys: their development at the Tethyan continental margins and their importance on the later evolution of the Mediterranean and Paratethys. Ann Geol Pays Hell hors serie I, 139–148

Boccaletti M, Manetti P, Peccerillo A (1974) Hypothesis on the plate tectonic evolution of the Carpatho-Balkan areas. Earth Planet Sci Lett 23:193–198

Boccaletti M, Manetti P, Peccerillo A, Stanisheva-Vassileva G (1978) Late Cretaceous high-potassium volcanism in Eastern Srednogorie, Bulgaria. Geol Soc Am Bull 89:39–447

Bogdanova SV (1993) Segments of the East European Craton. In: Gee DG, Beckholmen M (eds) Europrobe Symposium in Jablonna 1991, Inst Geophys, Polish Acad Sci - Eur Sci Found, Warszawa, pp 33–38

Bombiță G (1964) Contributions to the stratigraphic study of the Eocene deposits from Dobrogea with special regard of nummulites and asilinas fauna. An Com Geol 32:381–438 (in Romanian)

Bombiță G (1987) Etages nummulitiques dans la couverture de la Plateforme moldave. D S Inst Geol Geofiz 72–73/4

Bonhoff M, Makris J, Papanikolaou D, Stavrakakis G (2001) Crustal investigation of the Hellenic subduction zone using wide aperture seismic data. Tectonophysics 343:236–262

Bonneau M (1973) Sur les affinités ioniennes des "calcaires en plaquettes" epimetamorphiques de la Crète, le charriage de la série de Gavrovo-Tripolitsa et la structure de l'arc Egéen. CR Acad Sci Paris 277:2453–2456

Bonneau M (1976) Esquisse structurale de la Crète alpine. Bull Soc Geol France 18:351–353

Bonneau M (1984) Correlation of the Hellenide nappes in the south-east Aegean and their tectonic reconstruction. Geol Soc London Spec Publ 17:517–527

Borsi S, Ferrara G, Mercier J (1964) Détermination de l'age des séries métamorphiques du Massif Serbo-Macedonien au Nordest de Thessalonique (Grèce) par les méthodes Rb/Sr et K/Ar. Ann Soc Geol Nord LXXXIV:223–225

Borsi S, Ferrara G, Mercier J (1966) Mesures d'age par la méthode Rb/Sr des granites filoniens de Karathodoro (Zones internes des Hellenides, Macédoine centrale Grèce). Savez Geol Drus Szr Jugoslaviye, pp 5–10

Boyanov I, Rouseva M, Toprakchieva V, Dimitrova E (1990) Lithostratigraphy of the Mesozoic rocks in the Eastern Rhodopes. Geol Balc 20:3–28

Boykova A (1999) Moho discontinuity in central Balkan Peninsula in the light of the geostatistical structural analysis. Phys Earth Planet Interiors 114:49–58

Bröcker M, Franz L (1998) Rb-Sr isotope studies on Tinos Island, (Cyclades, Greece). Additional time constraints for metamorphism, extent of infiltration-controlled overprinting and deformation activity. Geol Mag 135:369–382

Brunn J (1956) Contribution a étude Géologique du Pinde Septentrional et d'une partie de la Macédoine Occidental. Ann Geol Pays Hellen 7:1–358

Burchfiel BC (1980) Eastern European alpine system and the Carpathian orocline as an example of collision tectonics. Tectonophysics 63:31–61

Byrne P, Georgiev G, Todorov I, Marinov E (1995) Structure, depositional setting and petroleum geology of Eastern Balkanides, Bulgaria. 57th Conference and Technical Exhibition of EAGE, Glasgow, extended abstracts, vol 2, E-021

Campbell AS (1968) The Barce (Al Marj) earthquake of 1963. In: Barr FT (ed) The geology and archaeology of northern Cyrenaica, Libya. Petrol Explor Soc Libya, 10th annual field conference, pp 183–195

Capedri S, Lekkas E, Papanikolaou D, Skarpelis N, Venturelli G, Gallo F (1985) The ophiolite of the Koziakas range, Western Thessaly (Greece). N Jb Miner Abh 152:45–64

Célet P (1962) Contribution a étude géologique du Parnasse-Kiona et d'une partie des régions méridionales de la Grèce continentale. Ann Geol Pays Hellen 7:1–358

Célet P, Ferrière J (1978) Les Hellenides internes: Le Pélagonien. Ecl Geol Helv 71(3):467–495

Célet P, Clement B, Ferrière J (1976) La zone béotienne en Grece: implications paléogéographiques et structurales. Eclogae geol Helv 69:577–599

Chatalov G (1990) Geology of Strandza zone in Bulgaria. Publishing House of the Bulg. Acad. of Sciences, Sofia, Geologica Balcanica, Series Operum Singulorium 4 (in Bulgarian)

Clift PJ, Robertson AHF (1989) Evidence of a Late Mesozoic ocean basin and subduction-accretion in the southern Greek Neo-Tethys. Geology 17:559–563

Cogulu E, Krummenacher D (1967) Problèmes géochronometriques dans la partie NW de l'Anatolie Centrale (Turquie). Schweiz Miner Petrograph 47:825–833

Committee of Geology (1989–1995) Geological map of Bulgaria on the scale 1:100 000: sheets Shokaricho and Kroushari, Novi Pazar, Razgrad, Shumen, Sliven, Nova Zagora, Topolovgrad, Svilengrad, Ivaylovgrad, Kroumovgrad

Constantinescu L, Constantinescu P, Cornea I, Lazarescu V (1976) Recent seismic information on the lithosphere in Romania. Rev Roum Geol Geophys Geogr Geoph 20:33–40

Cristea P, Stanchievici B, Spanoche S, Pompilian A, Radulescu F (1994) Seismic information related to the crust in Dobrogea. XVIIth Symposium on Earth Physics & Applied Geophysics, Abstracts volume, A 13, Bucharest (in Romanian)

Crowley QG, Marheine D, Winchester JA, Seghedi A (2000) Recent geochemical and geochronological studies in Dobrogea, Romania. TESZ & PACE Conference, Zakopane, Poland, Abstracts volume 88

Dabovski C, Chatalov G, Savov S (1991) The Strandzha Cimmerides in Bulgaria. In: Proc Int Earth Sci Congr Aegean Regions, 1–6 Oct 1990, 2:92–101

Dabovski C, Boyanov I, Khrischev K, Nikolov T, Sapounov I, Yanev Y, Zagorchev I (2002) Structure and Alpine evolution of Bulgaria. Geol Balc 32:2–4

Dachev C (1988) Structure of the Earth's crust in Bulgaria. Tehnika, Sofia (in Bulgarian)

De Bono A, Martini Z, Zaninetti L, Hirsch I, Stampfli G, Vavassis I (2001) Permo-Triassic stratigraphy of the Pelagonian zone in central Evia Island (Greece). Eclog Geol Helv 94(3):289–311

Demirtasli F (1989) Stratigraphic correlation forms of Turkey. Rend Soc Geol It 12:183–211

Dercourt J (1970) L'expansion océanique actuelle et fossile, ses implications geotectoniques. Bull Geol Soc France 12:261–317

Dercourt J (1972) The Canadian cordillera, the Hellenides and the sea-floor spreading theory. Canad J Earth Sci 9709–743

Dermitzakis M, Papanikolaou D (1979) Paleogeography and geodynamics of the Aegean region during the Neogene. Ann Geol Pays Hell, Proceedings IV–VII Int Congress Med Neogene, 1981, pp 245–289

Dewey JF, Bird JM (1970) Mountain belts and the new global tectonics. J Geoph Res 75(14):2625–2647

Dewey JF, Sengör AMC (1979) Aegean and surrounding region: Complex multiplate and continuum tectonics in convergent zone. Geol Soc Am Bull 90:84–92

Dewey JF, Pitman WC, Ryan WBF, Bonnin J (1973) Plate tectonics and the evolution of the Alpine system. Bull Geol Soc Am 84: 3137–3180

Dimitrievic MD (1974) The Dinarides: a model based on the new global tectonics. In: Karamata S (ed) Metallogeny and concepts of the geotectonic development of Yugoslavia, Belgrad, pp 141–178

Dimitriu RG (2001) Geological structure of the Romanian continent-sea transition zone as inferred by geophysical data modelling. PhD thesis, Bucharest Univ (in Romanian)

Dinter D, Royden V (1993) Late Cenozoic extension in north-eastern Greece: Strymon valley detachment system and Rhodope metamorphic core complexes. Geology 21:45–48

Dixon JE, Dimitriadis S (1984) Metamorphosed ophiolitic rocks from the Serbo-Macedonian Massif, near Lake Volvi, North-East Greece. Geol Soc London Sp Publ 17:603–618

Dürr ST, Altherr R, Keller J, Okrusch M, Seidel E (1978) The Median Aegean Crystalline Belt: stratigraphy, structure, metamorphism, magmatism. In: Cloos H, Roeder D, Schmidt K (eds) Alps, Apennines, Hellenides. Schweizerbart, Stuttgart, pp 455–477

El Arnauti A, Shelmani M (1985) Stratigraphic and structural setting. In: Thusu B, Owens B (eds) Palynostratigraphy of Northeast Libya. J Brit Micropal Soc London 4/1:1–10

El Arnauti A, Shelmani M (1988) A contribution to the northeast Libya subsurface stratigraphy with emphasis on Pre-Mesozoic. In: El Arnauti A, Owens B, Thusu B (eds) Subsurface palynostratigraphy of Northeast Libya. Res Centre Garyounis Univ, Benghazi 1–17

El-Hawat AS (1997) Sedimentary basins of Egypt: an overview of dynamic stratigraphy. In: Selley RC (ed) Sedimentary basins of the world, vol 3: African basins. Elsevier, Amsterdam, pp 39–85

El-Hawat AS, Shelmani MA (1993) Short notes and guidebook on the geology of Al Jabal la Akhdar, Cyrenaica NE Libya. Interprint, Malta

El-Hawat AS, Missallati AA, Bezan AM, Taleb MT (1996) The Nubian Sandstone in Sirt Basin and its correlatives. In: Salem MJ, El-Hawat AS, Sbeta AM (eds) The geology of Sirt Basin. Elsevier, Amsterdam, pp 3–30

Ferrière J (1976) Sur la signification de séries du massif d'Othris (Grèce continentale centrale): la zone isopique maliaque. Ann Soc Geol Nord 96/2:121–134

Ferrière J (1982) Paléogéographie et tectoniques superposées dans les Hellenides internes: les massifs de l'Othrys et du Pélion. Soc Geol Nord Publ 8

Ferrière J (1985) Nature et développement des ophiolites helléniques du secteur Othrys et du Pelion (Grèce continentale). Ofioliti 10:225–278

Finetti I (1982) Structure, stratigraphy and evolution of Central Mediterranean. Boll Geol Teor Appl XXXIV, 96:296–298

Finetti I, Papanikolaou D, Del Ben A, Karvelis P (1990) Preliminary geotectonic interpretation of the East Mediterranean chain and the Hellenic Arc. Bull Geol Soc Greece 25/1:509–526

Fleury J (1980) Les zones de Gavrovo-Tripolitza et du Pinde-Olonos (Grece continentale et Peloponnese du Nord). Evolution d'une platforme et d'un basin dans le cadre alpin. Publ Soc Geol Nord 4

Gautier P, Brunn JP, Jolivet L (1993) Structure and kinematics of Upper Cenozoic extensional detachment on Naxos and Paros (Cyclades islands, Greece). Tectonics 12:1180–1194

Georgiev G, Dabovski C, Stanisheva-Vassileva G (2001) Eastern Srednogorie-Balkan rift zone. In: Ziegler PA, Cavazza W, Robertson AHF, Crasquin-Soleau S (eds) Peri-Tethyan rift/ wrench basins and passive margins. Mem Mus National Hist Nat 186:259–293

Giusca D (1977) Contributions to the petrography of the crystalline schists from Palazu Mare (Dobrogea). Acad R S România, Rev Géol Géophys Géogr Géologie 15:139–147 (in Romanian)

Giuscă D, Ianovici V, Mânzatu S, Soroiu E, Lemne M, Tănăsescu A, Ioncică M (1967) The absolute ages of the metamorphic rocks from the Carpathian foreland. St cerc geol geof geogr ser geol 12:287–297 Bucuresti (in Romanian)

Giuscă D, Vasiliu C, Medesan A, Udrescu C (1976) Formatiunea feruginoasă careliană din Dobrogea (La formation ferrugineuse carelienne de Dobrogea) Acad RSR, Stud Cerc Geol Geofiz Geogr ser geol 21:3–20

Godfriaux I (1968) Etude géologique de la région de l'Olympe (Grèce). Ann Geol Pays Hellen 19:1–281

Goodchild RG (1968a) Graeco-Roman Cyrenaica. In: Barr FT (ed): The geology and archaeology of northern Cyrenaica, Libya. Petrol Explor Soc Libya, 10th annual field conference, pp 23–40

Goodchild RG (1968b) Earthquakes in ancient Cyrenaica. In: Barr FT (ed) The geology and archaeology of northern Cyrenaica, Libya. Petrol Explor Soc Libya, 10th annual field conference, pp 41–44

Gorbatschev R, Bogdanova S (1993) Frontiers in the Baltic Shield. Prec Res 64:3–22

Grigorescu D, Avram E, Pop G, Lupu M, Anastasiu N, Radan S (1990) Nonmarine Cretaceous Correlation (Project 245), Tethyan Cretaceous Correlation – Terrestrial and shallow marine clastics (Project 262). Guide to Excursions A+B. International Symposium, Bucharest

Hallett D (2002) Petroleum geology of Libya. Elsevier, Amsterdam

Harre W, Kockel F, Kreuzer H, Lenz H, Mulder P, Walther HW (1968) Über Rejuvenationen im Serbo-Mazedonischen Massiv (Deutung radiometrischer Alters-Bestinmungen). Geologica et Palaentologica 2:193–194

Hatzipanagiotou K (1988) Einbindung der Oberste Einheit von Rhodos und Karpathos (Griechenland) in den alpidischen Ophiolith-Gürteln. N Jb Geol Paläont Abh 176:395–422

Hatzipanagiotou K, Pe-Piper G (1995) Ophiolitic and sub-ophiolitic metamorphic rocks of the Vatera area, southern Lesvos (Greece): geochemistry and geochronology. Ofioliti 20:17–29

Hausmann J FL (1845) Beiträge zur Oryktographic von Syra und ein neues Mineral, der Glaucophan. Göttingen Geol Anz, pp 193–198

Haydoutov I (1989) Precambrian ophiolites, Cambrian island arc and Variscan suture in the south Carpathian – Balkan region. Geology 17:905–908

Haydoutov I (2002) Peri-Gondwanan terranes in the pre-Palaeozoic basement of Bulgaria. Geol Balc 32:2–4

Haydoutov I, Yanev S (1997) The Protomoesian microcontinent of the Balkan Peninsula – a peri-Gondwanaland piece. Tectonophysics 272:303–313

Haydoutov I, Gochev P, Kozhoukharov D, Yanev S (1997) Terranes in the Balkan area. In: Papanikolaou D (ed) IGCP Project 276: terrane map and terrane descriptions. Annales Geol Pays Helleniques, pp 479–494

Hecht J (1972) Lesbos Island: Geological map at 1:50 000 scale. IGME, Athens, Greece

Henzes-Kunst F, Kreuzer H (1982) Isotopic dating of the pre-Alpidic rocks from the island of Ios (Cyclades Greece), Contrib Mineral Petrol 80:245–253

Hippolyte J-C (2002) Geodynamics of Dobrogea (Romania): new constraints on the evolution of the Tornquist-Teisseyre Line, the Black Sea and the Carpathians. Tectonophysics 357:33–53

Hippolyte J-C, Badescu D, Constantin P (1999) Evolution of the transport direction of the Carpathian belt during its collision with the east European Platform. Tectonics 18:1120–1138

Huguen C, Mascle J (2001) La Margie continentale libyenne, entre 23° 30' et 25° 30' de longitude est. CR Acad Sci Paris, Earth Pl Sci 332:553–560

Huguen C, Mascle J, Chaumillon E, Woodside JM, Benkhelil J, Kopf A, Volkonskaia A (2001) Deformational styles of the eastern Mediterranean Ridge and surroundings from combined swath mapping and seismic reflection profiling. Tectonophysics 343: 21–47

Hynes AJ, Nisbet EG, Smith AG, Welland MDJ, Rex DC (1972) Spreading and emplacement ages of some ophiolites in the Othris region (eastern central Greece). Z dt Geol Ges 123:455–468

Ianovici V, Giugci D (1961) New data regarding the crystalline basement of the Moldavian Highland and Dobrogea. Acad R SR, St cerc geol VI(1):153–159 (in Romanian)

Iliescu V (1974) Preliminary results of the palyno-protistological study of the pre-Silurian deposits from the basement of the Moldavian Highland. D S Inst Geol LX/3:225–234 (in Romanian)

Ioane D, Cristea P, Stanchievici B, Seghedi A, Neaga V, Moroz V, Romanov L (1995) Deep structure of the contact zone between the Predobrogea Deppression (Scythian Paltform) and North Dobrogea Orogen, based on geological and geophysical studies. Archives Geol Inst Romania, Bucharest (in Romanian)

Ion J, Iva M, Melinte M, Florescu C (1996) Subsurface Santonian-Maastrichtian deposits in north-eastern South Dobrogea (Romania). Rom J Stratigr 77

Ionesi L (1994) Geology of the platform units and the North Dobrogea Orogen. Ed Tehn, Bucuresti (in Romanian)

Iordan M (1974) Study of the Lower Devonian fauna from the Bujoarele Hills (Măcin Unit, North Dobrogea). D S Inst Geol LX/3:33–70, Bucuresti (in Romanian)

Iordan M, Mirăută E, Antonescu E, Iva M, Gheorghian D (1987) The Devonian and Jurassic Formations in the Viroaga borehole – South Dobrogea (South-Eastern part of the Moesian Platform). DS Inst Geol Geofiz 72–73/4:5–18

Ivanov Z, Moskovski S, Kolceva K (1979) Basic features of the structure of the central parts of the Rhodope massif. Geol Balc 9:3–50

Ivanov Z (1985) Position tectonique, structure géologique et évolution alpidique du massif des Rhodopes. Reun Extr Soc Geol France en Bulgarie, Guide, pp 1–31

Ivanovski T (1971) Tectonics of the region situated between Vardar River, Strumiva valley and Yugoslav-Greek frontier. Contribution to the knowledge of the Vardar zone. Posed ni uizdanjia 3: 3–98, Skopje

Jacobshagen V (1979) Structure and geotectonic evolution of the Hellenides. Proc 6th Coll Geol Aegean Region, Athens 1977, 3: 1355–1367

Jacobshagen V (1986) Geologie von Griechenland. Gebr. Bornträger

Jolivet L, Goffé B, Monié P, Truffert-Luxey C, Patriat M, Bonneau M (1996) Miocene detachment in Crete and exhumation P-T-t paths of high pressure metamorphic rocks. Tectonics 15:1129–1153

Jongsma D (1991) The Medina and Sirt wrenches: a quantitative estimate of strike slip deformation within the Sirt Rise over the past 5 Ma. In: Salem MJ, Sbeta AM, Bakbak MR (eds) 3rd symposium on the geology of Libya, vol 6, Elsevier, Amsterdam, pp 2331–2352

Juranov SG, Pimpirev H (1989) Lithostratigraphy of the Upper Cretaceous and the Palaeogene in the coastal part of East Stara Planina. Rev Bulg Geol Soc 50:1–17 (in Bulgarian)

Kahle HG, Cocard M, Peter Y, Geiger A, Reilinger R, Barka A, Veis G (2000) GPS-derived strain rate fixed within the boundary zones of the Eurasian, African and Arabian Plates. J Geophys Res 105:23353–23370

Katsikatsos G, Mercier J, Vérgely P (1976) L'Eubée méridionale: une double fenêtre polyphasée dans les Hellenides internes. CR Acad Sci Paris 283:459–462

Katsikatsos G, Migiros G, Triantaphyllis M, Mettos A (1986) Geological structure of internal Hellenides (E. Thessaly – SW Macedonia – Euboea – Attica – northern Cyclades Islands and Lesvos). Geol & Geoph Res Spec Vol 191–212

Kilias AA, Tranos MD, Orozco M, Alonso-Chaves FM, Soto JI (2002) Extensional collapse of the Hellenides. A review. Rev Soc Geol España 15:129–139

Kober L (1928) Der Bau der Erde. Berlin

Kober L (1929) Beiträge zur Geologie von Attika. Sitz Ber Akad Wiss, Wien, 138:299–327

Kober L (1931) Das Alpine Europa. Gebr. Bornträger, Berlin

Kockel F, Mollat H, Walther HW (1971) Geologie des Serbo-Mazedonischen Massivs und seines mesozoischen Rahmens (Northgriechenland). Geol Jb 89:529–551

Kockel F, Mollat H, Walther HW (1977) Erläuterungen zur Geologischen Karte der Chalkidiki und angrenzender Gebiete 1:100 000 (Nordgriechenland). Bundesanstalt für Geowissenschaften und Rohstoffe, Hannover

Koepke J, Kreuzer H, Seidel T (1985) Ophiolites in the southern Aegean arc (Crete, Karpathos, Rhodes) – linking the ophiolite belts of the Hellenides and Taurides. Ophiolite 10:343–354

Kokkinakis A (1978) Das intrusivgebiet des Symvolon-Gebirges und von Kavalla in Ostmakedonien, Griechenland. Unpubl Dr thesis, München

König H, Kuss S (1980) Neue Daten zur Biostratigraphie des permotriadischen Autochtonous der Insel Kreta (Grieschland). N Jb Geol Paläont Abh. 9:525–540

Kozhoukharov D (1984) Lithostratigraphy of the Precambrian metamorphics of the Rhodopian Supergroup in the Central Rhodopes. Geol Balc 14:43–88

Kozhukharov D (1988) Precambrian in the Rhodope Massif. In: Zoubek V et al. (eds) Precambrian in younger fold belts. Wiley, New York, pp 723–745, 762–765

Kozhoukharov D, Timofeev B (1989) Microphytofossil data on the Precambrian age of the Rhodope Supergroup (Sitovo and Asenovgrad Groups) in the Central and Western Rhodopes. Geol Balc 19:13–31

Kozhoukharova E (1984) Origin and structural position of the serpentinized ultrabasic rocks of the Precambrian ophiolite association in the Rhodope massif. I. Geologic position and composition of the ophiolite association. Geol Balc 14:9–36

Krahl J, Kauffmann G, Kozur H, Richter D, Forster O, Heinritzi F (1983) Neue Daten zur Biostratigraphie und zur tektonischen lagerung der Phyllit-Gruppe und der Trypali-Gruppe auf der Insel Kreta (Griechenland). Geol Rund 72:1147–1166

Krautner HG, Muresan M, Seghedi A (1988) Precambrian of Dobrogea. In: Zoubek V et al. (eds) Precambrian in younger fold belts. Wiley, New York, pp 361–379

Kröner S, Sengör C (1990) Archaean and Proterozoic ancestry in Late Precambrian to Early Paleozoic crustal elements of southern Turkey, as revealed by single zircon dating. Geology 18:1186–1190

Ktenas C (1924) Formations primaires semi-metamorphiques au Peloponnese centrale. CR Soc Geol France 61–63

Le Pichon X, Angelier J (1979) The Hellenic arc and trench system: a key to the neotectonic evolution of the Eastern Mediterranean area. Tectonophysics 60:1–42

Le Pichon X, Angelier J (1981) The Aegean Sea. Phil Trans R Soc London A300:357–372

Le Pichon X, Chamot-Rooke N, Lallement S, Noomen B, Veis G (1995) Geodetic determination of the kinematics of central Greece with respect to Europe: Implications for Eastern Mediterranean tectonics. J Geophys Res 100:12675–12690

Lekkas E (1988) Geological structure and geodynamic evolution of the Koziakas mountain range (Western Thessaly). Geol Monographs, Univ Athens

Liati A (1986) Regional metamorphism and overprinting contact metamorphism of the Rhodope zone, near Xanthi (N. Greece): petrology, geochemistry, geochronology. Unpubl. PhD thesis, Universität Braunschweig

Liati A, Gebauer D (1999) Constraining the prograded and retrograde P-T-t path of Eocene HP rocks by SHRIMP dating of different zircon domains: inferred rates of heating, burial, cooling and exhumation for central Rhodope, northern Greece. Contr Miner Petrol 135:340–354

Lilov P (1990) Rb-Sr and K-Ar dating of the Sakar granitoid pluton. Geol Balc 20:53–60

Lister GS, Banga G, Feenstra A (1984) Metamorphic core complexes of cordilleran type in the Cyclades, Aegean Sea, Greece. Geology 12:221–225

Macarovici N (1971) La faune silurienne du fondament du Plateau Moldave (les forages de Jassy et de Todireni-Botosani). –Anal St Univ "Al. I Cuza" s.n. (sII, b Geol) XVII:99–115

Mader D, Catalov C (1992) Comparative palaeoenvironmental modelling of Buntsandstein braided river evolution in Bulgaria and Middle Europe. Geol Balc 22:21–62

Makris J (1973) Some geophysical aspects of the evolution of the Hellenides. Bull Geol Soc Greece X:203–213

Makris J (1978) The crust and upper mantle of the Aegean region from deep seismic sounding. Tectonophysics 46:269–284

Makris J, Papoulia I, Papanikolaou D, Stavrakakis G (2001) Thinned continental crust below northern Evoikos Gulf, central Greece, detected from deep seismic soundings. Tectonophysics 341: 225–240

Mantzos IA (1991) Rb-Sr whole-rock geochronology of gneisses from Olympias, Chalkidiki. Bull Geol Soc Greece XXI:147–161

Marakis G (1970) Geochronology studies of some granites from Macedonia. Ann Geol Pays Hellén 21:121–152

Mascle J, Chaumillon E (1998) An overview of Mediterranean Ridge collisional accretionary complex as deduced from multichannel seismic data. Geo-Mar Lett 18:81–89

Mascle J, Martin L (1990) Shallow structure and recent evolution of the Aegean Sea: A synthesis based on continuous reflection profiles. Mar Geol 94:271–299

McKenzie D (1970) Plate tectonics of the Mediterranean region. Nature 226:239–243

McKenzie D (1972) Active tectonics of the Mediterranean region. Geoph JR Astron Soc 30:109–185

McKenzie D (1978) Active tectonics of the Alpine-Himalayan belt: the Aegean sea and surrounding regions. Geoph JR Astron Soc 55:217–254

Mercier J (1968) Etude géologique des zones internes des Hellenides au Macédoine centrale (Grèce). Ann Geol Pays Hellen
20:1–792

Mercier J, Vérgely P (1972) Les mélanges colores (coloured mélanges)
de la zone d'Almopias (Macedoine, Grèce). CR Somm Soc Geol
France 70–73

Migiros G, Tselepides V (1990) Der erste Nachweis von Hallstätter
Kalken in der Nord-Pindos-Decke (NW-Griechenland). N Jb
Geol Palaont H4:248–256

Mosar J, Seghedi A (1999) North Dobrogea and the Paleozoic plate
tectonics. Abstr. volume EGS Conference, Utrecht

Mountrakis D (1986) The Pelagonian zone in Greece: A polyphase
deformed fragment of the Cimmerian continent and its role in
the geotectonic evolution of the Eastern Mediterranean. J Geol
94:335–347

Mposkos E, Kostopoulos D (2001) Diamond, former coesite and
supersilicic garnet in metasedimentary, rocks from the Greek
Rhodope: a new ultrahigh-pressure metamorphic province established. Earth Plan Sci Lett 192:497–506

Neaga VI, Moroz VF (1987) Die jungpaläozoischen Rotsedimente
im Südteil des Gebietes zwischen Dnestr und Prut. Z angew
Geol 33(9):238–242

Nikishin A, Ustaomer T, Robertson AHF, Seghedi A, Ziegler PA (2001)
Role of crustal extension and basin inversion in Late Palaeozoic-
Early Tertiary Tectonic Evolution of the south margin of Eurasia
in the circum Black Sea region. EUG XI Conference, Strasbourg,
France, Abstracts Volume 6(1):316

Okay A, Monié P (1997) Early Mesozoic subduction in the Eastern
Mediterranean: evidence from Triassic eclogite in north-western Turkey. Geology 25:595–598

Pamic JJ (1984) Triassic magmatism of the Dinarides in Yugoslavia. Tectonophysics 109:273–307

Pana I, Rado G (1982) Dobrogea during the Neogene; paleogeo-
graphic connexions between the Dacic and Euxinic basins. Lucr
Ses St "Gr. Cobalcescu", Univ "Al. I. Cuza" Iasi, pp 325–335 (in
Romanian)

Papanikolaou D (1979a) Stratigraphy and structure of the Paleozoic
rocks in Greece: An Introduction. In: Sassi FP (ed) IGCP No 5,
Newsletter 1:93–102

Papanikolaou D (1979b) Unîtes tectoniques et phases de déformation
dans l'île de Samos, Mer Egee, Grèce. Bull Soc Geol France (7),
6:745–752

Papanikolaou D (1980a) Geotraverse Southern Rhodope-Crete
(preliminary results; with the contribution by N. Scarpelis). In:
Sassi FP (ed) IGCP No 5, Newsletter 2:41–48

Papanikolaou D (1980b) The metamorphic Hellenides. Proc 26th
Int Geo. Congr, Paris, Abs 1:371

Papanikolaou D (1984) The three metamorphic belts of the Hellenides: a review and a kinematic interpretation. Spec Publ Geol
Soc London 17:551–561

Papanikolaou D (1986a) Late Cretaceous Paleogeography of the
Metamorphic Hellenides. Geol Geoph Res IGME, Special issue, pp 315–328

Papanikolaou D (1986b) The Medial Tectonometamorphic Belt of
the Hellenides. 3rd Congress, Geol Soc Greece, May 1986, Bull
Geol Soc Greece 20/1:101–120

Papanikolaou D (1987) Tectonic evolution of the Cycladic blueschist
belt (Aegean sea, Greece). In: Helgeson HC (ed) Chemical transport in metasomatic processes. Reidel Publishers, Dordrecht,
pp 429–450

Papanikolaou D (1988a) Precambrian in the Hellenides (Pelagonian,
Cyclades, Peloponnesus-Crete). In: Zoubek V et al. (eds) Precambrian in younger fold belts. Wiley, New York, pp 821–840

Papanikolaou D (1988b) Precambrian in the Rhodope massif. (The
southern parts of the Rhodope massif). In: Zoubek V et al. (eds)
Precambrian in younger fold belts. Wiley, New York, pp 765–788

Papanikolaou D (1988c) Triassic in the Hellenides. Atti del 74°
Congresso Soc Geol It A:529–530

Papanikolaou D (1988d) Introduction to the Geology of Crete. IGCP
project No 276, 1st Field meeting, Crete, October, 1988, Guidebook 3–16

Papanikolaou D (1989a) Are the Medial Crystalline Massifs of the
Eastern Mediterranean drifted Gondwanan fragments? Geol
Soc Greece Spec Publ 1:63–90

Papanikolaou D (1989b) Occurrence of Arvi, Western Thessaly and
Orliakas type formations in Argolis. Bull Geol Soc Greece 24:99

Papanikolaou D (1990) Probable geodynamic interpretation of the
schist-chert formations in the Hellenides. Bull Geol Soc Greece
XXIV:135–148

Papanikolaou D (1993) Geotectonic evolution of the Aegean. 6th
Congress of the Geological Society of Greece, Athens 1992, Bull
Geol Soc Greece 28/1:33–48

Papanikolaou D (1997) The tectonostratigraphic terranes of the
Hellenides. Ann Geol Soc Hellen 37:495–514

Papanikolaou D (1999) The Triassic ophiolites of Lesvos Island within the Cimmeride orogenic event. EUG 10, Strasbourg, Abs 315

Papanikolaou D, Baud A (1982) Complexes à blocs et séries a
caractère flysch au passage Permien-Triasen Attique (Grèce
orientale). RAST Paris

Papanikolaou D, Demirtasli E (1987) Geological Correlation between the Alpine segments of the Hellenides-Balkanides and
Taurides-Pontides. Mineralia Slovaca-Monography, pp 387–396

Papanikolaou D, Panagopoulos A (1981) On the structural style of
Southern Rhodope. Geol Balc 11:13–22

Papanikolaou D, Sideris C (1979) Sur la signification des zones
"ultrapindique" et "béotienne" d'après la géologie de la région
de Karditsa: l'unité de Thessalie Occidentale. Ecl geol Helv 72/1:
251–261

Papanikolaou D, Sideris C (1983a) Le Paleozoique de l'autochthone
de Chios: une formations à blocs de type wild flysch d'âge
Permien (pre parte). CR Acad Sci Paris 297:603–606

Papanikolaou D, Sideris C (1983b) Contribution to the Paleozoic
of the Aegean Area. In: Sassi FP (ed) IGCP No 5, Newsletter 5:
138–145

Papanikolaou D, Stojanov R (1983) Geological Correlation between
the Greek and the Ygaslave part of the Pelagonian Metamorphic Belt. In: Sassi FP (ed) IGCP No 5, Newsletter 5:146–152

Papanikolaou D, Zambetakis-Lekkas A (1980) Nouvelles observations et datations de la base de la série pélagonienne (s.s.) dans
la région de Kastoria, Grèce. CR Acad Sci Paris 291:155–158

Papanikolaou D, Sassi FP, Scarpelis N (1982) Outlines of the Pre-
Alpine metamorphisms in Greece. In: Sassi FP (ed) IGCP No 5,
Newsletter 4:56–62 and Ann Géol Pays Hell 31/1:16–31

Papanikolaou D, Alexandri S, Nomikou P, Ballas D (2002) Morpho-
tectonic structure of the western part of the north Aegean basin
based on swath bathymetry. Mar Geol 190:465–492

Papazachos BC, Karakostas VG, Papazachos CB, Skordilis EM (2000)
The geometry of the Wadati-Benioff zone and lithospheric kinematics in the Hellenic Arc. Tectonophysics 319:275–300

Paraschiv D (1985) The present state of knowledge of Precambrian
and Paleozoic formations from the Moldavian Platform. Mine
Petrol si Gaze 11 (in Romanian)

Paraschiv D, Paraschiv C, Danet T, Baltes N, Danet N, Motas L (1983)
On the pre-Neogene formations in the North-Dobrogean Promontory. An Inst Geol Geofiz LIX:19–28, Bucuresti

Patrulius D, Iordan M (1974) On the presence of the Pogonophore
Sabellidites cambriensis Ian. and the algae Vendotaenia antiqua
Gnil. in the pre-Silurian deposits from the Moldavian High-
lands. D.S. LX/4 (in Romanian)

Pătrut I, Costea I, Dănet T (1983) The pre-Albian Cretaceous sedimentary in the foreland of the Romanian Carpathian Mountains. An Inst Geol Geofiz 59:119–126

Pe-Piper G (1982) Geochemistry, tectonic setting and metamorphism of the mid-Triassic volcanic rocks of Greece. Tectonophysics 85:253–272

Pe-Piper G (1998) The nature of Triassic extension-related magmatism in Greece: evidence from Nd and Pb isotope geochemistry.
Geol Mag 135:331–378

Petrova A, Vassilev E, Mikhailova L, Simeonov A, Chelebiev E (1980)
Lithostratigraphy of a part of the Upper Cretaceous in Bourgas
region. – Geol Balc 10:23–67 (in Bulgarian)

Philippson A (1898) La tectonique de l'Egéide. Ann Géogr 112–141

Philippson A (1959) Die Griechischen Landschaften, vol I–IV. Klo-
stermann, Frankfurt

Pickett E, Robertson AHF (1996) Formation of the Late Paleozoic
– Early Meosozoic Karakaya complex and related ophiolites in
NW Turkey by Paleotethyan subduction – accretion. J Geol Soc
London 153:995–1009

Radulescu F, Constantinescu P, Pompilian A, Ibadof N, Sova A (1979) Structura scoartei terestre pe profilul Galati-Oradea, determinata prin cercetari seismice. Studii tehnice si economice, D, prospectiuni geofizice, 12:69–78

Reilinger R, McClusky S, Oral M, King R, Toksoz N, Barka A, Kinik I, Lenk O, Sanli I (1997) Global Positioning System measurements of present-day crustal movements in the Arabia-Africa-Eurasia plate collision zone. J Geophys Res 102:9983–9999

Reilinger R, Toksoz N, McClusky S (2000) 1999 Izmit earthquake, Turkey was no surprise. GSA Today 10:1–6

Reischmann T (1998) Pre-Alpine origin of tectonic units from the metamorphic complex of Naxos, Greece, identified by simple zircon Pb/Pb dating. Bull Geol Soc Greece 32/3:101–111

Renz C (1940) Die Tektonik der Griechischen Gebirge. Prakt Acad Athinon, vol 8

Renz C (1955) Die vorneogene Stratigraphie der normal sedimentären Formationen Griechenlands. IGSR, Athens

Robert U, Bonneau M (1982) Les basaltes des nappes du Pinde et d'Arvi et leur signification dans l'évolution géodynamique de la Méditerranée orientale. Ann Geol Pays Hellen 31:373–408

Robertson AHF (1998) Tectonic significance of the Eratosthenes Seamount: a continental fragment in the process of collision with a subduction zone in the eastern Mediterranean (Ocean Drilling Program Leg 160). Tectonophysics 298:63–82

Robertson AHF, Degnan T (1992) Kerassia-Milia complex: evidence of a Mesozoic-Early Tertiary oceanic basin between the Apulian continental margin and the Parnassos carbonate platform in western Greece. Bull Geol Soc Greece 28/1:233–246

Robertson AHF, Dixon JE (1984) Introduction: aspects of the geological evolution of the Eastern Mediterranean. Geol Soc London Spec Publ 17:1–74

Robertson AHF, Pickett E (2000) Paleozoic-Early Tertiary Tethyan evolution of mélanges, rift and passive margin units in the Karaburun peninsula (western Turkey) and Chios Island, (Greece). Geol Soc London Spec Publ 173:43–82

Robertson AHF, Dixon JE, Brown S, Collins A, Morris A, Pickett E, Sharp I, Ustaomer T (1996) Alternative tectonic models for the Late Paleozoic – Early Tertiary development of Tethys in the Mediterranean region. Geol Soc London Sp Publ 105:239–263

Sandulescu M (1984) Geotectonics of Romania. Ed Tehn, Bucharest (in Romanian)

Savov S, Dabovski C, Budurov K (1995) New data on the stratigraphy of the Tethyan Triassic in the Strandzha Mts., SE Bulgaria. Geol Balc 25:33–42

Schermer ER, Lux D, Burchfiel BC (1989) Age and tectonic significance of metamorphic events in the Mt. Olympos region, Greece. Bull Geol Soc Greece 23/1:13–27

Schliestedt M, Altherr R, Matthews A (1987) Evolution of the Cycladic Crystalline complex: Petrology, isotope geochemistry and geochronology. In: Helgeson HG (ed) Chemical transport in metasomatic processes, Reidel Publishers, Dordrecht, pp 389–428

Seghedi A (2001) The North Dobrogea orogenic belt (Romania): a review. In: Ziegler PA, Cavazza W, Robertson AHF (eds) Peri-Tethyan rift/wrench basins and passive margins. Mém Mus National Hist Nat 186:237–257

Seghedi A, Oaie G (1995) Paleozoic evolution of North Dobrogea. IGCP Project 369, Mamaia, 1995, Field Guidebook, Central and North Dobrogea

Seghedi A, Ciulavu M, Oaie G, Radan S (1999) Hercynian deformation and very low grade metamorphism of the Paleozoic formations in North Dobrogea. Rom J Tectonics Reg Geology 77, suppl 1, 74

Seghedi A, Oaie G, Radan S, Codarcea V (2000a) Source Areas for the Late Proterozoic Turbidites from the Moesian Block. TESZ & PACE Conference, Zakopane, Poland, Abstracts volume, p 75

Seghedi A, Kasper HU, Maruntiu M (2000b) Neoproterozoic intraplate magmatism in Moesia: petrologic and geochemical data. TESZ & PACE Conference, Zakopane, Poland, Abstracts volume, p 76

Seghedi A, Oaie G, Iordan M, Vaida M (2001) Correlation of the Vendian Basins along the Southern Margin of Baltica. EUROPROBE Conference Ankara, Turkey, Abstracts, pp 70–71

Seghedi A, Spear F, Storm L (2003) The Metamorphic Evolution of the Hercynian Basement of North Dobrogea: Constraints from Petrological Studies and Monazite Dating. The Fourth Stephan Mueller Conference of the EGS, Romania, Abstracts Book, pp 82

Seidel E, Okrush M, Kreuzer H, Raschka H, Harre W (1981) Eo-Alpine metamorphism in the uppermost unit of the Cretan nappe system: petrology and geochronology. Synopsis of high-temperature metamorphics and associated ophiolites. Contrib Miner Petrol 76:351–36

Seidel E, Kreuzer H, Harre W (1982) A late Oligocene/Early Miocene high pressure belt in the external Hellenides. Geol Jb E 23:165–206

Sengör AMC (1984a) The Cimmeride orogenic system and the tectonics of Eurasia. Geol Soc Am Spec Paper 195

Sengör AMC, Satir M, Akkok R (1984a) Timing of tectonic events in the Menderes massif, Western Turkey: Implications for tectonic evolution and evidence for pan-African basement in Turkey. Tectonics, 3:693–707

Sengör AMC, Yilmaz Y, Sungurlu O (1984b) Tectonics of the Mediterranean Cimmerides: nature and evolution of the western termination of Palaeo-Tethys. Geol Soc London Sp Publ 17:77–112

Sideris C (1989) Late Paleozoic in Greece. Geol. prace, Bratislava, 88, pp 191–202

Sinclair HD, Juranov SG, Georgiev G, Byrne P, Mountney NP (1997) The Balkan thrust wedge and foreland basin of Eastern Bulgaria: structural and stratigraphic development. In: Robinson AG (ed) Regional and petroleum geology of the Black Sea and surrounding region. AAPG Memoir 68:91–114

Sinyovski D (1996) New nanofossil data on the age of Emine, Dvoinik and Obzor formations in the Eastern Balkan. In: New Data on the Geology of Bulgaria, Bull Geol Soc Conference, Sofia, pp 90–91 (in Bulgarian)

Sola M, Ozcicek B (1990) On the hydrocarbon prospectivity of North Cyrenaica. Petrol Res J 2:25–41

Soldatos I, Christofides G (1986) Rb-Sr geochronology and origin of the Elatia pluton, Central Rhodope, North Greece. Geol Balc 16:15–23

Spakman, W, Wortel, MJR, Vlarr NJ (1988) The Hellenic subduction zone: a tomographic image and its geodynamic implications. Geophys Res Lett 15:60–63

Spray JG, Roddick JC (1980) Petrology and ^{40}Ar/^{39}Ar geochronology of some Hellenic sub-ophiolite metamorphic rocks. Contrib Mineral Petrol 72:43–55

Stacy FD (1977) Physics of the Earth, 2nd ed. John Wiley, New York

Stampfli GM, Vavassis I, De Bono A, Rosselet F, Matti B, Belini M (2003) Remnants of the Paleotethys oceanic suture-zone in the western Tethyan area. In: Cassinis G, Decandia FA (eds) Regional reports and general correlation. Boll Soc Geol It, vol speciale 2, pp 1–23

Stanica D, Stanica M (1988) The investigation of the deep structure of the Moesian Platform (Romania) by means of electromagnetic induction methods. Gerlands Beiträge Geophysik Leipzig 98/2:155–163

Stanisheva-Vassileva G (1980) The Upper Cretaceous magmatism in Srednogorie Zone, Bulgaria: a classification attempt and some implications. Geol Balc 10:15–36

Stefanescu M and Working Group (1987) Geological cross sections in Romania, scale 1 : 200 000. Geol Inst Romania, Bucharest

Tari G, Dicea O, Faulkerson J, Georgiev G, Popov S, Stefanescu M, Weir G (1997) Cimmerian and Alpine stratigraphy and structural evolution of the Moesian Platform (Romania/Bulgaria). In: Robinson AG (ed) Regional and petroleum geology of the Black Sea and surrounding region. AAPG Memoir 68:63–90

Tatarim N, Rado G, Pana I, Hanganu E, Grigorescu D (1977) South Dobrogea during the Tertiary: biostratigraphy and paleogeography. St cerc geol geof geogr, Geol Acad RSR 22:27–38

Tataris A (1975) Some question regarding the "course" of the Sh2-formation and the relationships of Mt Pelion to Mt Olympus. Bull Geol Soc Greece 12:95–112

Tenchov Y (1989) Stratigraphic correlation forms of the Paleozoic in Bulgaria. Rend Soc Geol It 12:423–433

Thiebault F (1977) Etablissement du caractère ionien de la série des calcschistes et marbres (Plattenkalk) in fenêtre dean le massif du Taugete (Peloponnese, Grèce). CR Somm Soc Geol France 3:159–161

Thiebault I (1982) L'évolution géodynamique des Hellenides externes en Peloponnese méridional. Publ Soc Geol Nord 6, 574 pp

Tokay M (1981) On some Variscan events in the Amasra district of the Zonguldak coalfield (northern Anatolia). IGCP No 5, Newsletter 3:140–151

Vaida M, Seghedi A (1997) Palynological study of cores from the Borehole 1 Liman (Scythian Platform, Moldavia). N Jb Geol Palaeont Mh 7:399–408

Van Der Maar P, Jansen JB (1983) The geology of the polymetamorphic complex of Ios, Cyclades, Greece and its significance for the Cycladic Massif. Geol Rund 72:283–299

Vavassis I, De Bono A, Stampfli G, Giorgis D, Valloton A, Amelia Y (2000) U-Pb and Ar-Ar geochronological data from the Pelagonian basement in Evia (Greece): geodynamic implications for the evolution of Paleotethys. Schweiz Min Petr Mitt 80:21–43

Vérgely P (1976) Chevauchement vers l'ouest et retrocharriage vers l'est des ophiolites: deux phases tectoniques au cours du Jurassique supérieur-Crétacé dans les Héllenides internes. Bull Geol Soc France, 18:231–244

Vérgely P (1984) Tectonique des ophiolites dans les Hellenides internes (déformation, métamorphisme et phénomènes sédimentaires). Conséquences sur l'évolution des régions tethysiennes occidentales. PhD thesis, Université Paris-Sud

Visarion M, Maier O, Nedelcu-Ion C, Alexandrescu R (1979) Modelul structural al metamorfitelor de la Palazu Mare, rezultat din studiul integrat al datelor geologice, geofizice si petrografice. St Cerc Geol Geofiz 17:95–113

Visarion M, Săndulescu M, Rosca V, Stănica M, Atanasiu L (1990) La Dobrogea dans le cadre de l'avant-pays carpatique. Rev Roum Geophys 34:55–65

Volvovsky IS, Starostenko VI (eds) (1996) Geophysical parameters of the lithosphere in the southern sector of the Alpine orogen. Kiev, Naukova dumka (in Russian)

Wawrzenitz N, Krohe A (1998) Exhumation and doming of the Thassos metamorphic core complex (S. Rhodope, Greece); structural and geomorphological constraints. Tectonophysics 285:301–332

Wawrzenitz N, Mposkos E (1997) First evidence for lower Cretaceous HP/LT metamorphism in NE Greece. Eur J Mineral 9:659–664

Winchester JA, Pharaoh TC, Verniers J (2002) Accretion of Avalonia and Armorican terrane assemblage to the East European Craton. ELD Symposium, Abstracts, Stockholm, pp 58

Wortel MJR, Spakman W (2000) Subduction and slab-detachment in the Mediterranean-Carpathian region. Science 290:1910–1917

Yanev S (1992) Contribution to the elucidation of the pre-Alpine evolution of Bulgaria, based on sedimentological data from the marine Paleozoic. Geol Balc 22/2:3–31

Yanev S (1993) Gondwana Palaeozoic terranes in the Alpine collage system of the Balkans. J Himalayan Geol 4:257–270

Yanev Y, Innocenti F, Manetti P, Serri G (1998) Upper Eocene-Oligocene collision-related volcanism in Eastern Rhodopes (Bulgaria) – Western Thrace (Greece): petrogenic affinity and geodynamic significance. Acta Vulcanol 10:279–291

Yarwood GA, Aftalion M (1976) Field relation and U-Pb geochronology of a granite from the Pelagonian zone of the Hellenides (High Pieria, Greece). Bull Soc Geol France, 18:259–264

Yarwood GA, Dixon JE (1979) Lower Cretaceous and younger thrusting in the Pelagonian Rocks of the High Pieria, Greece. VI Coll Geol Aegean Reg Athens 1977, I, pp 259

Zagorchev I (1992) Neotectonics of the central parts of Balkan Peninsula: basic features and concepts. Geol Rund 81:635–654

Zagorchev I (1998) Rhodope controversies. Episodes 21:159–166

Zagorchev I (2002) Radioisotopic data and geodynamic interpretations in the eastern part of the Balkan Peninsula. Geol Balc 32:2–4

Zagorchev I, Lilov P, Moorbath S (1989) Results of Rb/Sr and K/Ar radiochronological studies on metamorphic rocks in south Bulgaria. Geol Balc 19/3:41–54

Zelazniewicz A, Seghedi A, Jachowicz M, Bobinski W, Bula Z, Cwojdzinski S (2001) U-Pb Shrimp Data Confirm the Presence of a Vendian Foreland Flysch basin next to the East European Craton. Abstracts of the joint meeting of EUROPROBE, TESZ, Timpebar, Uralides and SW-Iberia Projects, Ankara, Turkey, pp 98–101

Zelazniewicz A, Seghedi A, Bula Z, Bobinski W (2003) Fragments of a Neoproterozoic orogen derived from a Grenvillian-type crust at the Teisseyre-Tornquist margin of Baltica: U-Pb SHRIMP data from southern Poland and Romania (submitted)

Zervas S (1980) Age determination by the Rb-Sr method of some pegmatites in the area of Lagada (Macedonia, Greece). Ann Geol Pays Hellen 30/1:143–153

Transect VIII: Eastern European Craton – Crimea – Black Sea – Anatolia – Cyprus – Levant Sea – Sinai – Red Sea

Aksu AE, Ulug A, Piper DWJ, Konuk YT, Turgut S (1992) Quaternary sedimentary history of Adana, Cilicia and Iskenderun Basins: northeastern Mediterranean Sea. Mar Geol 104:55–71

Akyol Z, Arpat E, Erdogan B, Göger E, Güney Y, Saroglu F, Sentürk I, Tütüncü K, Uysal S (1974) Geological map of the Cide-Kurucasile region, scale 1:50 000. Maden Tetkik ve Arama Enstitüsü, Ankara

Ambreseys NN, Adams RD (1993) Seismicity of the Cyprus region. Terra Nova 5:8–94

Andrew T, Robertson AHF (2002) The Beysehir-Hoyran Hadim (B-H-H) Nappes, genesis and emplacement of Mesozoic marginal and oceanic units of a Northerly Neotethys in Southern Turkey. J Geol Soc London 159:529–543

Artemieva I (2003) Lithospheric structure, composition, and thermal regime of the East European Craton: implications for the subsidence of the Russian platform. Earth Plan Sci Lett 213:431–446

Baldridge WS, Eyal Ye, Bartov Y, Steinitz G, Eyal M (1991) Miocene magmatism of Sinai related to the opening of the Red Sea. Tectonophysics 197:181–201

Banks CJ, Robinson AG (1997) Mesozoic strike-slip back-arc basins of the Western Black Sea. In: Robinson AG (Eed) Regional and petroleum geology of the Black Sea and surrounding region. AAPG Memoir 68:53–62

Baroz F (1980) Volcanism and continent-island arc collision in the Pentadaktylos range, Cyprus. In: Panayiotou A (ed) Proc Int Symp, Nicosia, Cyprus, pp 73–75

Beljaevskiy NA, Volvovsky BS, Volvovsky IS, Egorkin AV, Polshkov MK, Popov EA (1977) Seismic section of the Earth's crust of Eastern Europe (Profile, Black Sea-northeastern European part of the USSR). In: The structure of the Earth's crust from seismic investigation data, pp 7–19 (in Russian)

Belousov VV, Volvovsky BS (eds) (1989) Structure and evolution of the Earth's crust and upper mantle of the Black Sea. Nauka, Moscow (in Russian)

Belousov VV, Volvovsky BS (eds) (1992) Structure and evolution of the Earth's crust of the Black Sea. Nauka, Moscow (in Russian)

Belousov VV, Volvovsky BS, Arkhipov IV, Buryanov VB, Evsyukov YD, Goncharov VP, Gordienko VV, Ismagilov DF, Kislov GK, Kogan LI, Kondyurin AV, Kozlov VN, Lebedev LI, Lokholatnikov VM, Malovitsky YP, Moskalenko VN, Neprochnov YP, Ostisty BK, Rusakov OM, Shimkus KM, Shlezinger AE, Sochelnicov VV, Sollogub VB, Solovyev VD, Starostenko VI, Starovoitov AF, Terechov AA, Volvovsky IS, Shigunov AS, Zolotarev VG (1988) Structure and evolution of the Earth's crust and upper mantle of the Black Sea. Bollettino di Geofisica Teorica e Applicata 30:109–196

Belov AA (1981) Tectonic development of the Alpine fold area in the Palaeozoic. Nauka, Moscow

Ben-Avraham Z, Garfunkel Z (1994) Structures and kinematics in the northeastern Mediterranean: a study of an irregular plate boundary. Tectonophysics 234:19–32

Ben-Avraham Z, Shoham Y, Ginzburg A (1976) Magnetic anomalies in the eastern Mediterranean and the tectonic setting of the Eratosthenes Seamount. Geophys J Roy Astr Soc 45:313–331

Ben-Avraham Z, Ginzburg A, Makris J, Eppelbaum L (2002) Crustal structure of the Levant Basin, eastern Mediterranean. Tectonophysics 346:23–43

Bogayets AT (1976) Southern boundary of the East European Platform and the structure of the late Precambrian complex in the south of the USSR. Geotektonika 6:33–44 (in Russian)

Bogayets AT, Samarsky AD, Shnyukov VI (1986) Geology of the Black Sea. In: Shpak PF (ed) Oil and gas. Geologiya shelfa USSR. Izd. Naukova Dumka, Kyiv (in Russian)

Bogdanova SV, Pashkevich IK, Gorbatschev R, Orlyuk MI (1996) Riphean rifting and major Paleoproterozoic crustal boundaries in the basement of the East European Craton, geology and geophysics. Tectonophysics 268:1–21

Bonatti E (1985) Punctiform initiation of seafloor spreading in the Red Sea during transition from a continental to an oceanic rift. Nature 316:33–37

Bulanzhe YuD, Muratov MV, Subbotin SI, Balavadze BK (eds) (1975) The Earth's crust and the evolution of the Black Sea basin. Nauka, Moscow (in Russian)

Burakovskiy VE, Gurevich BL (1970) On the East-European Platform in the north-western sector of the Black Sea and the southern part of the country between the Prut and Dnestr rivers. Doklady Academii Nauk USSR. 193(3):656–658 (in Russian)

Canca N (1994) Geological maps of the coal-bearing basins in the Western Black Sea, scale 1:100 000. Maden Tetkik ve Arama Genel Müdürlüğü

Çemen I, Göncüoglu MC, Dirik K (1999) Structural evolution of the Tuz Gölü Basin in central Anatolia, Turkey. J Geol 107:693–706

Chekunov AV (ed) (1972) Structure of the Earth's Crust and Tectonics of the South of the European Part of the USSR. Naukova dumka, Kyiv

Chekunov AV (1990) Problems of the Black Sea depression. Geophys J (Kiev) 4:471–508

Chekunov AV (ed) (1994) The lithosphere of the Central and Eastern Europe – young platforms and the Alpine fold belt. Naukova dumka, Kiev (in Russian)

Chekunov AV, Gavrish VK, Kutas RI, Ryabchun LI (1992) Dniepr-Donets paleorift. In: Ziegler PA (ed) Geodynamics of rifting, vol I: case history studies on rifts, Europe and Asia. Tectonophysics 208:257–272

Chen F, Siebel W, Satir M, Terzioglu MN, Saka K (2002) Geochronology of the Karadere basement (NW Turkey) and implications for the geological evolution of the Istanbul Zone. Int J Earth Sci 91:469–481

Cochran JR, Gaulier JM, Le Pichon X (1991) Crustal structure and the mechanism of extension in the northern Red Sea, constraints from gravity anomalies. Tectonics 10:1018–1037

Çoruh C, Costain JK, Demirbag E, Saatçiler R (1990) Seismic imaging of crustal blocks, preliminary results from the First Turkish Geotraverse. EOS, Trans Am Geophysical Union 71:557

Dean WT, Monod O, Rickards RB, Demir O, Bultynck P (2000) Lower Palaeozoic stratigraphy and palaeontology, Karadere-Zirze area, Pontus Mountains, northern Turkey. Geol Mag 137:555–582

Dercourt J, Zonenshain LP, Ricou LE, Kazmin VG, Le Pichon X, Knipper AL, Grandjacquet C, Sbortshhikov IM, Geyssant J, Lepvrier C, Perchersky DH, Boulin J, Sibuet J-C, Savostin LA, Sorokhtin O, Westphal M, Bazhrnov ML, Lauer J-P, Biju-Duval B (1986) Geological evolution of the Tethys belt from the Atlantic to the Pamirs since the Lias. Tectonophysics 123:241–315

Dercourt J, Ricou LE, Vrielynck B (eds) (1993) Atlas Tethys Palaeoenvironmental Maps. Gauthier-Villars, Paris

Dercourt J, Gaetani M, Vrielynck B, Barrier E, Biju-Duval B, Brunet MF, Cadet JP, Crasquin S, Sandulescu M (eds) (2000) Peri-Tethys Palaeogeographical Atlas

DOBREfraction '99 Working Group (2003) "DOBREfraction '99" – Velocity model of the crust and upper mantle beneath the Donbas Foldbelt (east Ukraine). Tectonophysics 371:81–110

Dolenko GN, Pavlyuk MI (1974) On the formation of the boundary of the East-European Platform on the border of the central Pre-Black Sea. Doklady Academii Nauk UkrSSR Series B, 307–310 (in Ukrainian)

Ducloz C (1972) The geology of the Bellapais-Kythrea area of the central Kyrenia Range. Cyprus Geol Surv Bull 6:75

Eaton S, Robertson AHF (1993) The Miocene Pakhna Formation, southern Cyprus and it relationship the Neogene tectonic evolution of the Eastern Mediterranean. Sediment Geol 8:273–296

El-Metwally AA, Ibrahim ME, Ali MM (1996) Pan African high grade methamorphites from the Gabal Magafa, Sinai, Egypt: mineralogical and geochemical constraints. Mansoura Univ, Faculty Science Sci J 23:17–42

Emeis K-C, Robertson AHF, Richter C et al. (eds) (1996) Proceedings of the Ocean Drilling Program, Initial Reports 160. College Station, Texas (Ocean Drilling Program)

Emery KO, Neev D (1960) Mediterranean beaches of Israel. Geol Surv Israel Bull 26:1–13

Feliks VP (1969) The development history of the border zone of the East-European Platform and Scythian Platform. Izvestiya Vysshikh Uchebnykh Zavedeniy, Geologia and Razvedka 3:18–28 (in Russian)

Finetti I, Bricch, G, Del Ben A, Pipan M, Xuan Z (1988) Geophysical study of the Black Sea. Bolletino Geofisica Teorica e Applicata 30:197–324

Fokin PA, Nikishin AM, Ziegler PA (2001) Pre-Uralian and Peri-Palaeo-Tethyan Rift systems of the East European Craton. In: Ziegler PA, Cavazza W, Robertson AHF, Crasquin-Soleau S (eds) Peri-Tethyan rift/wrench basins and passive margins. Mem Mus National Hist Nat 186:347–368

Folkman Y, Bein A (1978) Geophysical evidence for a pre-late Jurassic fossil continental margin oriented east-west under central Israel. Earth Planet Sci Lett 39:335–340

Freund R, Goldberg M, Weissbrod T, Druckman Y, Derin B (1975) The Triassic-Jurassic structure of Israel and its relation to the origin of the eastern Mediterranean. Geol Surv Israel Bull 65:1–26

Garetsky RG (1972) Tectonics of young platforms of Eurasia. Nauka, Moscow, 210 pp (in Russian)

Garfunkel Z (1998) Constrains on the origin and history of the eastern Mediterranean Basin. Tectonophysics 298:5–35

Garfunkel Z, Almagor G (1985) Geology and structure of the continental margin off northern Israel and the adjacent part of the Levantine Basin. Marine Geol 62:105–131

Garfunkel Z, Derin B (1984) Permian-early Mesozoic tectonism and continental margin formation in Israel and its implications for the history of the eastern Mediterranean. In: Dixon JE, Robertson AHF (eds) The Geologic evolution of the Eastern Mediterranean. Blackwell, Oxford, pp 187–201

Gass IG (1990) Ophiolites and oceanic lithosphere. In: Malpas J, Moores EM, Panayiotou A, Xenophontos C (eds) Ophiolites: oceanic crustal analogues. Proceedings of the Symposium "Troodos 1987", Cyprus Geological Survey Department, pp 1–12

Gass IG, MacLeod CJ, Murton BJ, Panayiotou A, Simonian KO Xenophontos C (1994) The Geology of the South Troodos Transform Fault Zone. Geological Survey Department, Cyprus, Memoir 9

Gaullier V, Mart Y, Bellaiche G, Mascle J, Vendeville BC, Zitter T, second leg PRISMED II Science Party (2000) Salt tectonics in and around the Nile deep-sea fan, insights from the PRISMED II cruise. In: Vendeville BC, Mart Y, Vigneresse JL (eds) From the Arctic to the Mediterranean: salt, shale and igneous diapirs in and around Europe. Geol Soc London Spec Publ 174:111–129

Gerasimov ME (1994) Deep structure and evolution of the southern margin of the East-European Platform according to seimostratigraphical data, and in connection with oil and gas potential. VNIGRI, Moscow (in Russian)

Gerasimov ME (1995) On the Geodynamics and oil- and gas-content of the Black Sea region. Geology of Oil and Gas 8:4–11

Glushko VV (ed) (1988) Tectonic map of Ukraine and Moldavia, 1:500 000. Kiev, Department of Geology of the USSR (in Russian)

Glushko VV, Marasanova NV (eds) (1981) Geology of oil- and gas-bearing stratified reservoirs. Izd Nauka, Moscow, USSR (in Russian)

Göncüoglu MC (1986) Geochronological data from the southern part (Nigde area) of the Central Anatolian Massif. Bull Mineral Research Exploration 105/106:83–96

Gordievich VA, Kurulo GP, Luchagin GA (1966) A new concept about the geological structure of the Pre-Black Sea depression. Oil and Gas Industry 4:5–8 (in Russian)

Görür N (1988) Timing of opening of the Black Sea basin. Tectonophysics 147:247–262

Görür N, Oktay FY, Seymen I, Sengör AMC (1984) Paleotectonic evolution of the Tuzgölü basin complex, Central Anatolia, sedimentary record of a Neo-Tethyan closure. In: JE Dixon and AHF Robertson (eds) The Geological evolution of the Eastern Mediterranean. Geol Soc London Spec Publ 17:455–466

Görür N, Monod O, Okay AI, Sengör AMC, Tüysüz O, Yigitbas, E, Sakýnç M, Akkök R (1997) Palaeogeographic and tectonic position of the Carboniferous rocks of the western Pontides (Turkey) in the frame of the Variscan belt. Bull Soc Géol France 168:197–205

‌

Güleç N (1994) Rb-Sr isotope data from the Agaçören granitoid (east of Tuz Gölü), geochronological and genetical implications. Turkish J Earth Sci 3:39–43

Gürbüz C, Evans JR (1991) A seismic refraction study of the western Tuz Gölü basin, central Turkey. Geophys J Int 106:239–251

Horowitz A (1979) The Quaternary of Israel. Academic Press, New York

Horowitz A (2001) The Jordan Rift Valley. AA Balkema Pubs, Lisse

Iacumin M, Marzoli A, El-Metwally AA, Piccirillo EM (1998) Neoproterozoic dyke swarms from southern Sinai (Egypt): geochemistry and petrogenetic aspects. J Afr Earth Sci 26:49–64

Ivannikov AV, Inozemcev YuI, Stupina LV (1999) Stratigraphy of the Mesozoic and Cainozoic sediments of the Black Sea Continental Slope. In: Geology and minerals of the Black Sea. Kyiv, pp 253–262 (in Russian)

Ivanov MK (1999) Focused hydrocarbon fluxes on deep-water continental margins. PhD dissertation, Moscow State University, Moscow, Russia (in Russian)

Kabyshev BP, Krivchenkov B, Stovba S, Ziegler PA (1998) Hydrocarbon habitat of the Dniepr-Donets Depression. Marine Petrol Geol 15:177–190

Kahle HP, Straub C, Reilinger R, McClusky S, King R, Hurst K, Veis G, Kastens K, Cross P (1998) The strain rate field in the eastern Mediterranean region, estimated by repeated GPS measurements. Tectonophysics 294:237–252

Kameneckiy OYu, Silonov FA, Pokrovskaya LV (1973) Tectonic map of the southern regions of Ukraine and Moldavia and the surrounding offshore. Trudu VNIGNI 137:125–148 (in Russian)

Kazmin VG, Natapov LM (eds) (1998) Paleogeographic Atlas of Northern Eurasia. Institute of Tectonics of Lithospheric Plates, Moscow (CD-ROM)

Kempler D, Ben-Avraham Z (1987) The tectonic evolution of the Cyprean Arc, Annales Tectonicae 1:58–71

Ketin İ (1962) Explanatory text of the geological map of Turkey at 1:500 000 scale, Sinop sheet. Maden Tetkik ve Arama Enstitüsü, Ankara

Khain VYe (1984) Regional Geotectonics. The Alpine-Mediterranean Belt. Nedra. Moscow (in Russian)

Khain VE, Leonov YG (eds) (1998) International Tectonic Map of Europe, 1:5 000 000. UNESCO-CGWM-Russian Academy of Sciences-VSEGEI St Petersburg

Koçyigit A (1991) An example of an accretionary forearc basin from northern central Anatolia and its implications for the history of subduction of Neo-Tethys in Turkey. Geol Soc Am Bull 103:22–36

Koronovsky NV, Mileyev VS (1974) About the relationships of Tauric series and Eskiorda suite in the Bodrak river valley (Mountain Crimea). Vestnik Mosk. Univ., Geologiya 1:80–87 (in Russian)

Krashenninikov VA, Udintsev GB, Mouraviov V, Hall JK (1994) Geological stucture of the Eratosthenes Seamount 1994. In: Krashenninikov VA, Hall JK (eds) Geological structure of the Northeastern Mediterranean (Cruise 5 of the Research Vessel 'Akademik Nokolaj Strakhov'). Historical Productions-Hall Ltd, Jerusalem, Israel, pp 113–130

Kröner A, Eyal M, Eyal Y (1990) Early Pan-African evolution of the basement around Elat, Israel, and the Sinai Peninsula revealed by single-zircon evaporation dating, and implications for crustal accretion rates. Geology 18:545–548

Kruglov SS, Cypko AK (1988) Tectonics of Ukraine. Nedra, Moscow (in Russian)

Le Pichon X, Francheteau J (1978) A plate tectonic model of the Red Sea-Gulf of Aden area. Tectonophysics 46:369–406

Le Pichon X, Gaulier JM (1988) The rotation of Arabia and the Levant fault system. Tectonophysics 153:271–294

Lemaire MM, Westphal M, Montigny R, Gurevitch YeL, Feinberg H, Pozzi J-P, Nazarov K (1998) Palaeomagnetism and evolution of the Scythian-Turan block from the Early Permian to the Late Triassic. Comptes Rendus de l'Académie des Sciences, Series IIA (Earth and Planetary Science) 327:441–448

Letavin AI (1980) Basement of the Young Platform of the Southern Part of the USSR. Nauka, Moscow (in Russian)

Letavin AI (1987) Geology of the Pre-Caucasus basement. In: Milanovsky, EE, Koronovsky NV (eds) Geology and mineral resources of the Great Caucasus. Nauka, Moscow pp 116–124 (in Russian)

Makris J, Ben-Avraham Z, Behle A, Ginzburg A, Giese P, Steinmetz L, Whitmarsh RB, Eleftheriou S (1983) Seismic refraction profiles between Cyprus and Israel and their interpretation. Geophys J Roy Astron Soc 75:575–591

Malpas J, Moores EM, Panayiotou A, Xenophontos C (eds) (1987) Ophiolites: oceanic crustal analogues. Proceedings of the Symposium "Troodos 1987", Cyprus Geological Survey Department, Nicosia

Malpas J, Xenophontos C, Williams D (1992) The Ayia Varvara Formation of S.W. Cyprus, a product of complex collisional tectonics. Tectonophysics 212:193–211

Mart Y (1987) Superpositional tectonic patterns along the continental margin of the southeastern Mediterranean: a review. Tectonophysics 140:213–232

Mart Y (1991) The Dead Sea Rift, from continental rift to incipient ocean. Tectonophysics 197:155–179

Mart Y, Ben-Gai Y (1982) Some depositional patterns at the continental margin of the southeastern Mediterranean Sea. AAPG Bull 66:460–470

Mart Y, Dauteuil O (2000) Analogue experiments of propagation of oblique rifts. Tectonophysics 316:121–132

Mart Y, Robertson AHF (1998) Eratosthenes Seamount, an oceanographic yardstick recording late Mesozoic-Tertiary geological history of the Eastern Mediterranean. Proceedings of the Ocean Drilling Project, Scientific Results 160:701–708

Mart Y, Ryan WBF (2002) The complex tectonic regime of Cyprus Arc: a short review. Israel J Earth Sci 51:117–134

Mart Y, Sass E (1972) Geology and origin of the manganese ore of Um Bogma, Sinai. Econ Geol 67:145–155

Mart Y, Robertson AHF, Woodside JM (1997) Cretaceous tectonic setting of Eratosthenes Seamount in the eastern Mediterranean Neotethys, initial results of ODP Leg 160. Comptes Rendus de l'Académie des Sciences, Series IIA (Earth and Planetary Science) 324:127–134

May PR (1991) The Eastern Mediterranean Mesozoic basin: evolution and oil habitat. Am Ass Petrol Geol Bull 75:1215–1232

Maystrenko Yu, Stovba S, Stephenson R, Bayer U, Menyoli E, Gajewski D, Huebscher C, Rabbel W, Saintot A, Starostenko V, Thybo H, Tolkunov A (2003) Crustal-scale pop-up structure in cratonic lithosphere: DOBRE deep seismic reflection study of the Donbas Foldbelt, Ukraine. Geology 31(8):733–736

McCallum JE, Robertson AHF (1990) Pulsed uplift of the Troodos Massif-evidence from the Plio-Pleistocene Mesaoria Basin. In Malpas J, Moores EM, Panayiotou A, Xenophontos C (eds) Ophiolites: oceanic crustal analogues. Proceedings of the Symposium "Troodos 1987", Cyprus Geological Survey Department, pp 217–230

McCallum JE, Scrutton RA, Robertson AHF, Ferrari W (1993) Seismostratigraphy and Neogene-Recent depositional history of the south central continental margin of Cyprus. Marine Petrol Geol 10:426–438

Mazarovich OA, Mileyev VS (eds) (1989) Geological Structure of the Kacha Upland of the Mountain Crimea. Stratigraphy of the Mesozoic. Moscow State University Press, Moscow (in Russian)

Milanovsky EE (1987) Geology of the USSR. Moscow, Izdatelstvo of Moscow University, vol 1 (in Russian)

Milanovsky EE (1996) Geology of Russia and adjacent areas (Northern Eurasia). Moscow University Press, Moscow (in Russian)

Mileyev VS, Baraboshkin EYu., Nikitin MYu, Rozanov SB, Shalimov IV (1996) Evidence that the Upper Jurassic deposits of the Crimean Mountains are allochthons. Transactions (Doklady) of the Russian Academy of Sciences, Earth Science Sections 342(4):121–124

Mileyev VS, Rozanov SB, Baraboshkin EYu., Shalimov IV (1997) The tectonic structure and evolution of the Mountain Crimea. In: Milanovsky EE, Mileyev VS, Nikishin AM, Sokolov BA (eds) Geological study of Crimea (Otcherki geologii Krima). Geological Faculty MSU Publishers, Moscow, pp 187–206

Monod O (1977) Récherches géologique dans les Taurus occidental au sud de Beysehir (Turquie). [Thèse de Doct. Sci.,] Univ. Paris-Sud, Orsay

Moores EM, Vine FJ (1971) The Troodos Massif, Cyprus and other ophiolites as oceanic crust: evaluations and implications. Phil Trans R Soc A268:433–466

Moustafa AR, Khalil M (1989) North Sinai structures and tectonic evolution. Middle East Res Cent Ain Shams Univ, Earth Sci Ser 3:215–231

Moustafa AR, Khalil M (1995) Rejuvenation of the Tethyan passive continental margin of northern Sinai: deformation style and age (Gebel Yelleq area). Tectonophysics 241:225–238

Muratov MV (ed) (1969a) Tectonic map of Ukraine SSR and Moldavia SSR, scale 1:1 000 000. Kiyv

Muratov MV (ed) (1969b) Geology of the USSR. Volume VIII, Crimea. Part 1, Geology, Nedra, Moscow (in Russian)

Muratov MV (1979) Tectonics and history of the development of the Alpine geosyncline region of the south of the Europeam part of the USSR and surrounding countries. In: Tectonics of the USSR. Nat Acad Sci USSR, Moscow, v. 2

Neprochnov YuR, Kosminskaya IP, Malovitsky IP (1970) Structure of the crust and upper mantle of the Black and Caspian seas. Tectonophysics 10:517–538

Neprochnov YuR, Neprochnova AE, Mirlin YG (1974) Deep structure of the Black Sea basin. In: Degens ET, Ross DA (eds) The Black Sea – geology, chemistry, and biology. Am Ass Petrol Geol Mem 20:35–49

Nikishin AM, Ziegler PA, Stephenson RA, Cloetingh S, Furne AV, Fokin PA, Ershov AV, Bolotov SN, Korotaev MV, Alekseev AS, Gorbachev VI, Shipilov EV, Lankreijer A, Shalimov IV (1996) Late Precambrian to Triassic of the East-European Craton, dynamics of basin evolution. Tectonophysics 268:23–63

Nikishin AM, Cloetingh S, Brunet M-F, Stephenson R, Bolotov SN, Ershov AV (1998) Scythian Platform and Black Sea region, Mesozoic-Cenozoic tectonics and dynamics. In: Crasquin-Soleau S, Barrier E (eds) Stratigraphy and evolution of peri-Tethyan platforms. Mem Mus National Hist Nat 177:163–176

Nikishin AM, Ziegler PA, Panov DI, Nazarevich BP, Brunet M-F, Stephenson RA, Bolotov SN, Korotaev MV, Tikhomirov P (2001) Mesozoic and Cainozoic evolution of the Scythian Platform-Black Sea-Caucasus domain. In: Ziegler PA, Cavazza W, Robertson AHF, Crasquin-Soleau S (eds) Peri-Tethyan rift/wrench basins and passive margins. Mem Mus National Hist Nat 186:295–346

Nozhkin AD, Krestin YeM (1984) Radioactive elements in lower Precambrian rocks; for example, the Kursk magnetic anomaly. Trudy Instituta Geologii i Geofiziki (Novosibirsk) 601:128

Okay A, Özgül N (1984) HP/LT metamorphims and the structure of the Alanya Massif. In: Dixon JE, Robertson AHF (eds) The Geological evolution of the Eastern Mediterranean. Geol Soc London Spec Publ 17:429–440

Okay AI, Tüysüz O (1999) Tethyan sutures of northern Turkey. In: Durand B, Jolivet L, Horváth F, Séranne M (eds) The Mediterranean basins: Tertiary extension within the Alpine orogen. Geol Soc London Spec Publ 156:475–515

Okay AI, Sengör, AMC, Görür N (1994) Kinematic history of the opening of the Black Sea and its effect on the surrounding regions. Geology 22:267–270

Okay AI, Tansel I, Tüysüz O (2001) Obduction, subduction and collision as reflected in the Upper Cretaceous-Lower Eocene sedimentary record of western Turkey. Geol Mag 138:117–142

Orszag-Sperber E, Rouchy J-M, Ellion P (1989) The sedimentary expression of regional tectonic events during the Miocene-Pliocene transition in Southern Cyprus. Geol Mag 126:291–299

Özgül N (1976) Stratigraphy and tectonic evolution of the Central Taurides. In: Tekeli O, Goncuoglu MC (eds) Proc. Internat. Symp. Geology of the Taurus Belt, Ankara, pp 77–90

Panayides I, Xenophontos C, Malpas J (eds) (2000) Proc. 3rd Internat. Conf. Geology of the Eastern Mediterranean. Geological Survey Dept., Cyprus

Panayiotou A (ed) (1979) Ophiolites. Proceedings of the International Ophiolite Symposium, Nicosia, Geological Survey Dept., Cyprus

Paulssen H, Bukchin BG, Emelianov AP, Lazarenko M, Muyzert E, Snieder R, Yanovskaya TB (1999) The NARS-DEEP Project. Tectonophysics 313:1–8

Poole A, Robertson AHF (1992) Quaternary uplift and sea-level change at an active plate boundary, Cyprus. J Geol Soc London 148:909–921

Popadyuk IV, Smirnov SY (1991) The problem of the structure of the Crimean Mountain region, traditional ideas and reality. Geotectonics 25(6):489–497

Popadyuk IV, Smirnov SY (1996) Crimea orogen, a nappe interpretation. In: Ziegler PA, Horvath F (eds) Structure and prospects of Alpine basins and forelands. Mém Mus National Hist Nat 170:513–524

Popov VS (1963) Tectonics of the Donets Basin. In: Geology of coal and oil shale Deposits of the USSR, 1. Nedra, Moscow, pp 103–151 (in Russian)

Popovich VS, Stupak LA (1973) The new data about deep structure of the Northern Pre-Black Sea. Geotectonika 2:101–106 (in Russian)

Purser BH, Bosence DWJ (eds) (1998) Sedimentation and Tectonics in Rift Basins: Red Sea-Gulf of Aden. Chapman and Hall, 663 pp

Robertson AHF (1977a) The Moni Melange, Cyprus, an olistostrome formed at a destructive plate margin: J Geol Soc London 133:447–466

Robertson AHF (1977b) Tertiary uplift history of the Troodos Massif, Cyprus. Geol Soc Am Bull 88:1763–1772

Robertson AHF (1990) Tectonic evolution of Cyprus. In: Malpas J, Moores EM, Panayiotou A, Xenophontos C (eds) Ophiolites: oceanic crustal analogues. Proceedings of the Symposium "Troodos 1987", Cyprus Geological Survey Department, pp 235–252

Robertson AHF (1994) Role of the tectonic facies concept in orogenic analysis and its application to Tethys in the Eastern Mediterranean region. Earth-Sci Rev 37:139–213

Robertson AHF (1998a) Mesozoic-Tertiary tectonic evolution of the Easternmost Mediterranean area, integration of marine and land evidence. In: Robertson AHF, Emeis KC, Richter C, Camerlenghi A (eds) Proceedings of the Ocean Drilling Program, Scientific Results 160:723–782

Robertson AHF (1998b) Tectonic significance of the Eratosthenes Seamount, a continental fragment in the process of collision with a subduction zone in the eastern Mediterranean (Ocean Drilling Program Leg 160). Tectonophysics 298:63–82

Robertson AHF (2000) Mesozoic-Tertiary tectonic-sedimentation evolution of a south Tethyan ocean basin and its margins in southern Turkey. In: Bozkurt E, Winchester JA, Piper JD (eds) Tectonics and magmatism in Turkey and the surrounding area. Geol Soc London Spec Publ 173:43–82

Robertson AHF (2002) Overview of the genesis and emplacement of Mesozoic ophiolites in the Eastern Mediterranean Tethyan region. Lithos 65:1–67

Robertson AHF, Dixon JE (1984) Introduction, aspects of the geological evolution of the Eastern Mediterranean. In: Dixon JE, Robertson AHF (eds) The geological evolution of the Eastern Mediterranean. Geol Soc London Spec Publ 17:1–74

Robertson AHF, Hudson JE (1974) Pelagic sediments in the Cretaceous and Tertiary history of the Troodos Massif, Cyprus. In: Hsü KJ, Jenkyns HC (eds) Pelagic sediments on land and under the sea. Spec Publ Int Ass Sediment 1:403–436

Robertson AHF, Woodcock NH (1980) Tectonic setting of the Troodos Massif in the east Mediterranean. Proceedings International Ophiolite Symposium, Cyprus 1979, Cyprus Geological Survey Department, pp 36–49

Robertson AHF, Woodcock NH (1986) The geological evolution of the Kyrenia Range: a critical lineament in the Eastern Mediterranean. In: Reading HG, Watterson J, White SH (eds) Major crustal lineaments and their influence on the geological history of the continental lithosphere. Phil Trans R Soc London A317:141–171

Robertson AHF, Xenophontos C (1993) Development of concepts concerning the Troodos ophiolite and adjacent units in Cyprus. In: Prichard HM, Alabaster T, Harris NB, Neary CR (eds) Magmatic processes and plate tectonics. Geol Soc London Spec Publ 70:85–120

Robertson AHF, Eaton EJ, Follows EJ, Payne AJ (1995a) Depositional processes and basin analysis of Messinian evaporites in Cyprus. Terra Nova 7:233–253

Robertson AHF, Kidd RB, Ivanov MK, Limonov AF, Woodside JM, Galindo-Zaldivar J, Nieto I (1995b) Eratosthenes Seamount, easternmost Mediterranean: evidence of active collapse and thrusting beneath Cyprus. Terra Nova 7:254–264

Robertson AHF, Emeis K-C, Richter C, Camerlenghi A (eds) (1998) Proceedings ODP, Scientific Results 160 (available on-line at *http://www-odp.tamu.edu/publications/160_SR/INTRO.HTM*)

Robinson AG, Kerusov E (1997) Stratigraphic and structural development of the Gulf of Odessa, Ukrainian Black Sea, implications for petroleum exploration. In: Robinson AG (ed) Regional and petroleum geology of the Black Sea and surrounding region. Am Ass Petrol Geol Memoir 68:369–380

Robinson PT, Malpas J (1990) The Troodos Ophiolite of Cyprus: new perspectives on its origin and emplacement. In: Malpas J, Moores EM, Panayiotou A, Xenophontos C (eds) Ophiolites: oceanic crustal analogues. Proceedings of the Symposium "Troodos 1987", Cyprus Geological Survey Department, pp 13–36

Robinson AG, Rudat JH, Banks CJ, Wiles RLF (1996) Petroleum geology of the Black Sea. Marine Petrol Geol 13:195–223

Sage L, Letouzey J (1990) Convergence of the African and Eurasian plate in the eastern Mediterranean. In: Letouzey J (ed) Petroleum and tectonics in mobile belts. Ed. Technip, Paris, pp 49–68

Saintot A, Stephenson RA, Brem A, Stovba S, Privalov V (2003) Paleostress field reconstruction and revised tectonic history of the Donbas fold and thrust belt (Ukraine and Russia). Tectonics 22(5):1059, doi:10.1029/2002TC001366

Samsonov AI, Krasnoschek AYa (1969) A new concept about the tectonic construction of Pre-Dobrogea and north-western part of the Black Sea offshore. In: Geology of the offshore and bottom of the Black and Azov Seas in the scope of UkrSSR (in Russian)

Samsonov, AV, Zhuravlev DZ, Bibikova EV (1993) Geochronology and petrogenesis of the Archaean acid volcano-plutonic suite of the Verchovtsevo greenstone belt, Ukrainian Shield. Int Geol Rev 35:1166–1181

Samsonov AV, Chernyshev IV, Nutman AP, Compston W (1996) Evolution of the Archaean Aulian Gneiss Complex, Middle Dniepr gneiss-greenstone terrane, Ukrainian Shield, SHRIMP U-Pb zircon evidence. Precambrian Research

Saunders P, Priestley K, Taymaz T (1998) Variations in the crustal structure beneath western Turkey. Geophys J Int 134:373–389

Seghedi A, Stephenson RA, Neaga V, Dimitriu R, Ioane D, Stovba S (2004) The Scythian Platform north of Dobrogea (Romania, Moldova and Ukraine). Tectonics (submitted)

Sengör AMC, Yilmaz Y (1981) Tethyan evolution of Turkey, a plate tectonic approach. Tectonophysics 75:181–241

Sengör, AMC, Yilmaz Y, Sungurlu O (1984) Tectonics of the Mediterranean Cimmerides, nature and evolution of the western termination of the Paleo-Tethys. In: Dixon JE, Robertson AHF (eds) The geological evolution of the Eastern Mediterranean. Geol Soc London Spec Publ 17:77–112

Shatsky NS (1963) Selected works. Nauka, Moscow, vol 1 (in Russian)

Shcherbak NP (1991) Stratigraphic scale of the Precambrian of the Ukrainian Shield and its correlation with the All-Union and international scale of Precambrian. Geol J 4:3–9

Shcherbak NP, Artemenko GV, Bartnitsky YeN (1989) Geochronological scale of the Ukrainian Precambrian. Naukova Dumka, Kiev (in Russian)

Shcherbak NP, Artemenko GV, Bartnitsky YeN, Shpylchak VA, Tatarinova YeI (1990) Kosivtsevo Group as a fragment of Early Archaean greenstone belts in the Ukranian Shield. Dokl Akad Nauk Ukraine 10:35–39 (in Russian)

Shcherbak NP, Artemenko GV, Bartnitsky YeN, Sergienko VN, Tatarinova YeI (1992) Age of acid metavolcanics of the Aleksandrov and Korobkovsky regions of the KMA (Kursk Magnetic Anomaly). Dokl Akad Nauk Ukraine 6:120–123 (in Russian)

Shchipansky AA (1987) Granite-gneiss domes in the Early Precambrian structure of the Kursk Magnetic Anomaly region. Geotectonika 6:39–51 (in Russian)

Shchipansky AA, Bogdanova SA (1996) The Sarmatian crustal segment, Precambrian correlation between the Voronezh Massif and the Ukrainian Shield across the Dniepr-Donets Aulacogen. Tectonophysics 268:109–126

Shnyukov EF, Shcherbakov IB, Shnyukova KE (1997) Paleoisland arc in the northern part of the Black Sea. Nat Acad Sci Ukraine, Kiev (in Russian)

Slavin, VI, Khain VE (1980) Early Cimmerian geosyncline troughs of the North of the central part of the Mediterranean Belt. Bull Moscow Univ, Ser 4, Geology 2:3–14

Smolyaninova EI, Mikhailov VO, Lyakhovsky VA (1996) Numerical modelling of regional neotectonic movements in the northern Black Sea. Tectonophysics 266:221–231

Sollogub VB (1986) Lithosphere of Ukraine, Naukova Dumka, Kiev (in Russian)

Sollogub VB, Pavlenkova NI, Chekunov AV, Khilinsky LA (1966) Deep structure of the crust along the meridional section Black Sea-Voronezh Massif. Geophysical Collection 15:16–58 (in Russian)

Spadini G, Robinson A, Cloetingh S (1996) Western versus Eastern Black Sea tectonic evolution, pre-rift lithospheric controls on basin formation. Tectonophysics 266:139–154

Stampfli GM (2000) Tethyan oceans. In: Bozkurt E, Winchester JA, Piper JD (eds) Tectonics and magmatism in Turkey and the surrounding area. Geol Soc London Spec Publ 173:1–23

Stampfli GM, Borel GD (2002) A plate tectonic model for the Palaeozoic and Mesozoic constrained by dynamic plate boundaries and restored synthetic oceanic isochrones. Earth Plan Sci Lett 169:17–33

Stampfli G, Mosar J, Faure P, Pillevuit A, Vannay J-C (2001) Permo-Mesozoic evolution of the Western Tethys realm, the Neotethys East Mediterranean basin connection. In: Ziegler P, Cavazza W, Robertson AHF, Crasquin-Soleau S (eds) Peri-Tethyan rift/wrench basins and passive margins. Mém Mus National Hist Nat 186:51–108

Stampfli GM, von Raumer JF, Borel G (2002) Paleozoic evolution of pre-Variscan terranes: from Gondwana to the Variscan collision. In: Martinez Catalán JR, Hatcher RD Jr, Arenas R, Díaz García F (eds) Variscan-Appalachian dynamics: the building of the late Paleozoic basement. Geol Soc Am Spec Pap 364:263–280

Starostenko V, Buryanov V, Makarenko I, Rusakov O, Stephenson R, Nikishin A, Georgiev G, Gerasimov M, Dimitriu R, Legostaeva O, Pchelarov V, Sava C (2004) Topography of the crust-mantle boundary beneath the Black Sea Basin. Tectonophysics (in press)

Steckler MS (1985) Uplift and extension at the Gulf of Suez, indications of induced mantle convection. Nature 317:135–139

Stein M, Goldstein SL (1996) From plume head to continental lithosphere in the Arabian-Nubian shield. Nature 383:773–778

Steinitz G (1980) Rb-Sr age determinations on basement rocks from Helez Deep 1-A well. Geol Surv Israel Rep MM/1/80

Stephenson RA, Stovba SM, Starostenko VI (2001) Pripyat-Dnieper-Donets Basin: implications for dynamics of rifting and the tectonic history of the northern Peri-Tethyan Platform. In: Ziegler P, Cavazza W, Robertson AHF, Crasquin-Soleau S (eds) Peri-Tethyan rift/wrench basins and passive margins. Mem Mus National Hist Nat 186:369–406

Stern RJ (2002) Crustal evolution in the East African Orogen: a neodymium isotopic perspective. J Afr Earth Sci 34:109–117

Stovba S (2003) Tectonics on the Odessa Shelf of the northern Black Sea. Stefan Müller Conference, Chiele Buti, Romania

Stovba SM, Stephenson RA (1999) The structural relationship of the Donbas Foldbelt with the uninverted Donets segment of the Dniepr-Donets Basin, Ukraine. Tectonophysics 313:59–83

Stovba SM, Stephenson RA (2002) Style and timing of salt tectonics in the Dniepr-Donets Basin (Ukraine): implications for triggering and driving mechanisms of salt movement in sedimentary basins. Marine Petrol Geol 19:1169–1189

Stovba S, Stephenson RA, Kivshik M (1996) Structural features and evolution of the Dniepr-Donets Basin, Ukraine, from regional seismic reflection profiles. Tectonophysics 268:127–147

Sunal G, Tüysüz O (2002) Palaeostress analysis of tertiary post-collisional structures in the Western Pontides, northern Turkey. Geol Mag 139:343–359

Tekeli O (1981) Subduction complex of pre-Jurassic age, northern Anatolia, Turkey. Geology 9:68–72

Tokarsky DY (1970) Tectonics of the Pre-Black Sea-Crimea oil and gas province based on geophysical data. PhD thesis (abstract), Moscow (in Russian)

Tugolesov DA, Gorshkov AS, Meysner LB, Solovyev VV, Khakhalev YeM (1985a) The tectonics of the Black Sea trough. Geotectonics 19:435–445

Tugolesov DA, Gorshkov AS, Meysner LB, Solovyev VV, Khakhalev YM, Akilova YV, Akentieva GP, Gabidulina TI, Klomeytseva SA, Kochneva TY, Pereturina IG, Plashihina IN (1985b) Tectonics of the Mesozoic Sediments of the Black Sea Basin. Nedra, Moscow, 215 pp (in Russian)

Tüysüz O (1990) Tectonic evolution of part of the Tethyside orogenic collage, the Kargi Massif, northern Turkey. Tectonics 5:1–33

Tüysüz O (1998) Geological map of Turkey, Sinop sheet, scale 1:500 000. Istanbul

Tüysüz O (1999) Geology of the Cretaceous sedimentary basins of the Western Pontides. Geol J 34:75–93

Tüysüz O, Dellaloglu AA, Terzioglu N (1995) A magmatic belt within the Neo-Tethyan suture zone and its role in the tectonic evolution of northern Turkey. Tectonophysics 243:173–191

Ustaömer T, Robertson AHF (1993) Late Palaeozoic-Early Mesozoic marginal basins along the active southern continental margin of Eurasia: evidence from the Central Pontides (Turkey) and adjacent regions. Geol J 28:219–238

Ustaömer T, Robertson AHF (1994) Late Palaeozoic marginal basin and subduction-accretion, the Paleotethyan Küre Complex, Central Pontides, northern Turkey. J Geol Soc London 151: 291–305

Ustaömer T, Robertson AHF (1997) Tectonic-sedimentary evolution of the north Tethyan margin in the Central Pontides of northern Turkey. In: Robinson AG (Eed) Regional and petroleum geology of the Black Sea and surrounding region. Am Ass Petrol Geol Memoir 68:255–269

Ustaömer PA, Rogers G (1999) The Bolu Massif: remnant of a pre-Early Ordovician active margin in the west Pontides, northern Turkey. Geol Mag 126:579–592

Vidal N, Alvarez-Maron J, Klaeschen D (2000) Internal configuration of the Levantine Basin from seismic reflection data (eastern Mediterranean). Earth Planet Sci Lett 180:77–89

Volvovsky BS, Starostenko VI (eds) (1996) Geophysical Parameters of the South Sector of Alpine Orogen. Naukova Dumka, Kiev (in Russian)

Wdowinski S, Zilberman E (1996) Kinematic modeling of large-scale structural asymmetry across the Dead Sea Rift. Tectonophysics 266:187–201

Weissbrod T, Sneh A (2002) Sedimentology and paleogeography of the late Precambrian-early Cambrian arkosic and conglomeratic facies in the northern margisn of the Arabo-Nubian shield. Geol Survey Israel Bull 87

Whitney DL, Teyssier C, Dilek Y, Fayon AK (2001) Metamorphism of the Central Anatolian Crystalline Complex, Turkey, influence of orogen-normal collision vs. wrench-dominated tectonics on P-T-t paths. J Met Geol 19:411–433

Whitney DL, Teyssier C, Fayon AK, Hamilton MA, Heizler M (2003) Tectonic controls on metamorphism, partial melting, and intrusion, timing and duration of regional metamorphism and magmatism in the Nigde Massif, Turkey. Tectonophysics 376:37–60

Woodside, JM, Mascle J, Zitter TAC, Limonov AF, Ergün M, Vokhonskaya A, PRISMED II Science team (2002) The Florence Rise, the western bend of the Cyprus Arc. Mar Geol 185:177–194

Yegorova TP, Kozlenko VG, Makarenko IB, Starostenko VI (1996) Tectonosphere of the Black Sea region: three-dimensional density model for the lithosphere of Europe. In: Volvovsky BS, Starostenko VI (eds) Geophysical parameters of the lithosphere of the southern Sector of the Alpine orogen. Kiev, Naukova Dumka, pp 873–892 (in Russian)

Yigitbas E, Elmas A, Yýlmaz Y (1999) Pre-Cenozoic tectono-stratigraphic components of the western Pontides and their geological evolution. Geol J 34:55–74

Yýlmaz Y, Tüysüz O, Yigitbas E, Can Genç S, Sengör AMC (1997) Geology and tectonic evolution of the Pontides. In: Robinson AG (ed) Regional and petroleum geology of the Black Sea and surrounding region. AAPG Memoir 68:183–226

Yudin VV (1994) New model of Crimean geology. Priroda 6:28–31

Yudin VV (1995) The foredeep suture of Crimea. Geol J 3–4:56–61 (in Russian)

Yudin VV (1999) Structure and geodynamics of the Southern Crimea-Northern Black Sea region. In: Shnyukov EV (ed) Geology and mineral products of the Black Sea. Kyiv, pp 61–68 (in Russian)

Zakariadze GS, Karpenko SF, Bazylev BA, Adamia ShA, Oberhänsli RE, Soloveva NA, Lyalikov AV (1998) Petrology, geochemistry, and Sm-Nd age of the pre-Late Hercynian paleooceanic complex of the Dzirula Salient, Transcaucasian Massif. Petrology 6(4):388–408 (in Russian)

Zhu L, Mitchell BJ, Akyol N (2003). The 2003 Western Anatolia Seismic Experiment: an integrated study of crust/upper mantle structure and anisotropy in western Turkey (abstract). American Geophysical Union, Fall Meeting, S21C-03

Zonenshain LP, Le Pichon X (1986) Deep basins of the Black Sea and Caspian Sea as remnants of Mesozoic back-arc basins. In: Auboin J, Le Pichon X, Monin, AS (eds) Evolution of the Tethys. Tectonophysics 123:181–211

Zonenshain LP, Kuzmin MI, Natapov LM (1990) Geology of the USSR, a plate tectonic synthesis. American Geophysical Union, Geodynamic Series 21

Index